Surface Carbohydrates of the Eukaryotic Cell

Surface Carbohydrates of the Eukaryotic Cell

G. M. W. COOK

Member of the External Scientific Staff,
Medical Research Council,
Strangeways Research Laboratory, Cambridge, England.

and

R. W. STODDART

Sometime Research Fellow of Sidney Sussex College, Cambridge.
Strangeways Research Laboratory, Cambridge, England.

1973

ACADEMIC PRESS · LONDON and NEW YORK
A Subsidiary of Harcourt Brace Jovanovich, Publishers.

ACADEMIC PRESS INC. (LONDON) LTD.
24/28 Oval Road,
London NW1

United States Edition published by
ACADEMIC PRESS INC.
111 Fifth Avenue
New York, New York 10003

Library of Congress Catalog Card Number: 73-9454
ISBN: 0-12-186850-8

PRINTED IN GREAT BRITAIN BY

J. W. ARROWSMITH LTD., BRISTOL

Preface

One of the most remarkable features of biochemistry has been its extra-ordinary rate of growth. Though giving an appearance of continuity this growth has, in fact, been the result of periods of rapid development in many different areas of the subject, at various times. Very often these bursts of growth have resulted from the convergence of ideas and data from seemingly disparate disciplines. Such a development has begun in the study of cellular surfaces, and especially with regard to their carbohydrate constituents. The steady growth of information and ideas in this field over the past decade has now accelerated to such an extent that major advances should soon be possible, and it seemed to the authors desirable to collate the basic results and concepts upon which the subject will grow. This book is an attempt to do this for the eukaryotic cell.

Naturally, in a book of this size, a degree of selectivity is inevitable. An attempt has been made to cover the basic tenets upon which the subject is founded and to draw attention to those points of growth for the future. Undoubtedly certain omissions will have been made, for which the authors take full responsibility. Nevertheless, it is hoped that this book will serve as a useful contribution to an important field of biochemistry and cell biology, by attempting to provide a comprehensive view of present knowledge of carbohydrates in cell surfaces.

It is intended that this book should serve the needs both of those actively engaged in research in the field and of those wishing to enter it. In covering animal and plant cells it is hoped that a stimulus may be given to the interaction of ideas between workers upon these two kingdoms.

Only through the active support and help of our colleagues has the production of this book been possible; especially we thank Dr. Jennifer M. Allen for her encouragement from the beginning. We also thank Mr Michael Abercrombie, F.R.S., Director of this Laboratory, for his interest and our many colleagues who have helped, advised and commented upon the manuscript, especially Dr. C. W. Lloyd. The authors are very conscious of the immense amount of help given by Mr. M. F. Applin in the preparation of illustrative material; Mrs. J. M. Wilson and Miss P. E. Walker in general secretarial assistance; Mrs. M. F. Applin, Mrs. V. Curry and Mrs. E. Roche, in typing from an almost illegible manuscript, and Mrs. B. Parsison in bibliographic help.

The authors also express their thanks to the various journals cited in the text, who have given permission for the reproduction of material. Especial thanks are also due to Dr. E. L. Benedetti, C.N.R.S., Institut de Biologie Moléculaire de la Faculté des Sciences de Paris, Dr. Audrey M. Glauert, Strangeways Research Laboratory and Dr. J. A. Kiernan, Dept. of Anatomy, University of Western Ontario, for the generous provision of original micrographs.

Finally we wish to extend our thanks to Dr. E. H. Eylar, Merck Institute for Therapeutic Research, Rahway, New Jersey, U.S.A., for his interest in this project and for his kindness in writing the foreword.

G.M.W.C. and R.W.S.
Strangeways Research Laboratory,
September, 1973   *Cambridge*

Foreword

Although it is unnecessary in these times to emphasize the importance of membranes as a research topic in biological sciences, questions arise over which components are irrelevant to structural integrity of the membrane and thus serve some special function. It is suspected by many researchers that glycoproteins of the plasma membrane may serve primarily to modulate the interaction of a cell with its external world. This book addresses itself to all facets of the glycomacromolecules in, and adjoining, animal and plant cell membranes, and provides for the researcher and student valuable analyses of past efforts in this difficult field, references, and information that should give a focus to future work. Historically, the emphasis in the early sixties on protein as well as lipid components of membranes was not misplaced; it brought new fervor and fresh concepts to the quest for membrane structure (particularly by D. Green and J. Singer) and mode of membrane bioassembly.

Membrane proteins are often insoluble and scarce, and challenge the researcher further because of their complexity and inscrutibility, especially the glycoproteins. While electrophoresis in polyacrylamide gels in sodium dodecyl sulfate has proven invaluable for qualitative scanning of membrane proteins, in order further to advance our understanding of membranes, it is desirable to isolate any membrane protein which can be extracted and solubilized whether it is biologically active or not. Study of its conformation in solution by conventional protein chemistry, and its interaction with artificial membranes, might then follow. The logic and value of this approach is illustrated by two well known cases of immunologically-active proteins, the glycoprotein of the erythrocyte, and the A1 protein of myelin. Yet few membrane proteins have been isolated, even from ample sources, and it appears we have something more to learn before we can cope intelligently with controlled dissociation of these macromolecules from their lipoidal environment.

Do specific glycoproteins of the plasma membrane, directly or subtly, subserve the cell as "chemical antennae" in a wide variety of membrane-related phenomena such as tumor cell metastasis, phagocytosis, chemotaxis, cell fusion, chalone action, hormone action, antigen recognition, and fertilization? Those workers who devoted years to mundane structural studies of glycoproteins can now anticipate some satisfaction in the possible central role glycoproteins may play in specifying the interaction of cells with their environment. The work of Moscona and co-workers suggests that certain cell-cell recognitions may be mediated by membrane glycoproteins. The influence of lectins on cell division and the aggregation of tumor cells suggests a role for membrane glycoproteins. Once the basis for these phenomena have been more clearly defined, then attention might be directed to more general questions of the role of glycoproteins in control of membrane assembly, or in polar capping. The beautiful possibilities that the heterosaccharide structure offers for diverse biological specificity within a confined unit has been recognized, in contrast to the limitations of polypeptide structure.

Possibly a wide spectrum of surface glycoproteins exist to account for a variety of biological phenomena; it is now known that membrane glycoproteins are not only "mucin" types as found on the erythrocyte, but are distinguished by mannose in the PHA receptor, and N-acetylglucosamine in the receptor for wheat germ lectin. Why then have no unusual glycoproteins been found which discriminate a transformed cell from its parents, or specify the antigen responsible for immunologic rejection of tumor cells, or account for contact inhibition or metastasis? In the absence of direct data, it is possible that glycoproteins of the plasma membrane may not explicitly subserve specific surface phenomena but act mainly to bring the glycoprotein to its external position as in the case of extracellular glycoproteins. In the latter, carbohydrate units are built up from monosaccharides along a complicated pathway from ribosome to Golgi apparatus, a process which prepares most extracellular proteins for excretion, and perhaps their eventual demise in the case of plasma glycoproteins which are adsorbed by the liver upon desialization. For most extracellular proteins, therefore, the carbohydrate serves a general rather than a specific biological role. Might the same situation apply to membrane glycoproteins? Do the carbohydrate units serve the latter simply by providing the cell with a negative surface charge and a hydrophilic ectobiology?

In order to understand the biological role of membrane glycoproteins, it is not necessarily essential to isolate the glycoproteins from appropriate states of the cell. In many cases, it may be impractical because of lack of material, and indirect techniques must be used. Structure-function relationships might be inferred from continued ingenious application of cell fusion and immunologic techniques, electron microscopy, mutants, lectins coupled to polymers, characterization of receptors, and radiolabeled monosaccharides as markers for particular glycoproteins. This book should stimulate the further application of these techniques (and others), particularly Chapter I where the isolation of plasma membranes are described. The purification of certain glycosyl transferases from appropriate membranes should help clarify remaining problems in the synthesis of glycoproteins and perhaps help unravel the secrets of membrane assembly.

Much of our future hopes in all this work reside in development of appropriate tissue culture lines.

Perhaps the field as presented by this volume will provide a spark or thread that will put researchers and students on the most fruitful tack. We might soon begin to see whether membrane glycoproteins (1) play a direct role in cell-cell interaction as proposed by a current unsupported hypothesis, or (2) whether their role is a secondary influence on other components such as adenyl cyclase or enzymes regulating levels of cyclic AMP, cyclic GMP, or prostaglandins, or (3) something else. Sometimes it is better to take one step backward and look at the total picture, as provided by this book, than take a rash step forward into the forest.

E. H. Eylar
Lincroft, N.J.

August, 1973

Contents

1

The Structure of Plasma Membranes

The molecular structure of the surface of the eukaryotic cell is very complex and though a detailed elucidation of the chemistry of the cell periphery is still far from complete, it is becoming increasingly evident that macromolecules which contain carbohydrate play a vital role in a number of important biological phenomena involving the cellular surface.

It is perhaps appropriate to consider at the outset what is included within the term "cell surface", for as Weiss (1969) has pointed out such a term approximates to a two-dimensional planar structure around the cell, in contact with its environment as distinct from the cell "peripheral zone", which he defines as a three-dimensional region including the plasma membrane and/or permeability barrier/s. In this book it is not intended to cover the chemistry of the whole plasma membrane but rather to limit attention to those macromolecules confined to the outermost regions of the cell membrane and its associated structures—in other words to molecules situated at the cellular surface. Hence in the present context the surface is understood to possess some degree of depth, not necessarily encompassing the whole width of the plasma membrane.

An understanding of the molecular nature of the cell surface is dependent upon the results of studies on the composition and structure of the plasma membrane, also described as the cell membrane or plasmalemma, and it is therefore intended to cover in this chapter some of the major concepts of the chemical organization of the plasma membrane. The subject of membrane structure and function has received extensive treatment from a number of authors and it is not intended in the space of one chapter to attempt to cover all aspects of this topic, which would occupy a complete volume, but rather to concentrate on the development of some of the main ideas in the field of this book. Biological membranes have been subjected to investigation for fifty years or more and some of the now classical models suggested for the structure of· plasma membranes, as Maddy (1966) points out, are undoubtedly among the most long-lived theories of molecular biology. In almost every case these studies have largely ignored the carbohydrate content of membranes and as such do not accurately represent the molecular nature of the cell surface. Nevertheless, these models have had an important bearing on the development of studies on the cell surface and have stimulated much experimentation.

I. Fundamental Properties of Membranes

It has been recognized for about a century that the plasma membrane acts as a permeability barrier, controlling the exchange of water and solutes between the cell and its environment.

The majority of the classical models describe the plasma membrane in terms of structures which place great emphasis on those molecular components which keep the aqueous surroundings separate from the cytoplasm. Therefore it is not surprising that such models, designed largely to describe the plasmalemma in terms of a boundary between the cell and its environment, are unlikely to take account of the molecular features required to give satisfactory explanations of all surface phenomena. The cell membrane, though a barrier, was recognized not to be just inert, isolating the cell's contents from its surroundings, but a structure that was permeable to various molecules. The permeation of many solutes was shown to be related to their lipid solubility (Overton, 1899) and this led to the importance of lipid biological membranes being established, by implication, at an early date. However, cell membranes are also permeable to other organic compounds, and hence the active transport of small molecules and ions by the plasmalemma must be taken into account when considering membrane structure. Once more it is evident that a model which gives a satisfactory explanation of transport phenomena and associated carrier molecules may well be deficient in describing all the features of the cell surface. Membranes are able to transport particulates and droplets into and out of cells by the processes of endo- and exocytosis and this fact should not be forgotten in studies on membrane structure. Here again it might be expected that membrane models which are deficient in molecular details of the cellular surface could be quite sufficient to describe such cellular processes. Though this has certainly been true for a number of years, it has been recognized, in the case of phagocytosis for example, that for complete understanding of such phenomena a knowledge of the cell surface is required and indeed considerable attention is now being brought to bear on the role of surface receptors in this endocytotic process.

Though the plasma membrane does interpose a boundary between the cell and the environment and maintains supplies of metabolites to the cell, the more peripheral regions of that membrane fulfill the equally important function of conferring a recognizable individuality upon the cell. The biochemical basis of this recognition is of prime importance in explaining many of the cellular interactions that occur in the multicellular organism. As will be illustrated in this chapter, classical membrane models provide no adequate account of surface biochemistry and properties such as recognition, a deficiency that must be remedied in any model claimed to be complete.

II. Classical Models of Membrane Structure

A. Bilayer Model

The emphasis on the role of lipids within biological membranes largely arose from the work of Overton and others, in which it was implied that this class of compound was important because of the apparent correlation between the rate of permeation of various solutes into cells and their lipid solubility. At this stage, however, no suggestion as to the arrangement of the lipid or lipids present was put forward and in this respect the work of Langmuir (1917) was extremely pertinent. Langmuir's studies on unimolecular films at air-water interfaces suggested that the lipid molecules orientate themselves in such a way that the polar ends of the molecules are in association with the air. The particular relevance of these studies became evident with the application of analytical techniques to membranes, for not only did such studies strengthen the view that lipids were important in membrane structure but, coupled with the concepts of Langmuir, they enabled deductions to be made regarding the orientation of such molecules within the plasmalemma. In 1925, Gorter and Grendel extracted lipids with acetone from a number of what they termed chromocytes, erythrocytes of different mammalian species including dog, sheep, rabbit, guinea pig, goat and man. In all cases these authors found that the area occupied by the extracted lipids, when spread as a unimolecular film on water and measured by means of a Langmuir trough, was twice the surface area of the corresponding number of erythrocytes prior to lipid extraction. Considering that all the lipid is present in the red cell stroma (synonymous with the terms ghost or post-haemolytic residue), the suggestion of Gorter and Grendel (1925) that the lipids in the membrane were arranged as a bimolecular leaflet, with the polar ends of the molecules orientated toward the aqueous environment and the non-polar ends of the lipids facing towards each other, was entirely reasonable. However, some criticism of the experimental techniques used by Gorter and Grendel have been raised. Winkler and Bungenberg de Jong (1941) and Hoffman (1962) pointed out that Gorter and Grendel (1925) underestimated the surface area of the red cells by some 50%, and therefore the ratio of lipid surface area to the surface of the erythrocyte is nearer 1 : 1 than 2 : 1. On the other hand, acetone treatment does not result in the total removal of the lipid and therefore Gorter and Grendel's original suggestion may well be correct. Indeed more recent quantitative extraction data provided by Davson (1962) supports this latter contention.

2C

A

WATER

WATER

B

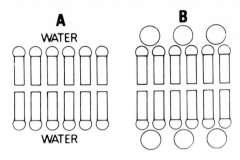

C

PROTEIN

LIPID

PROTEIN

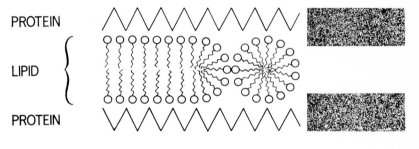

D

GLYCOPROTEIN

NHR   NHR

COOH   COOH

LIPID

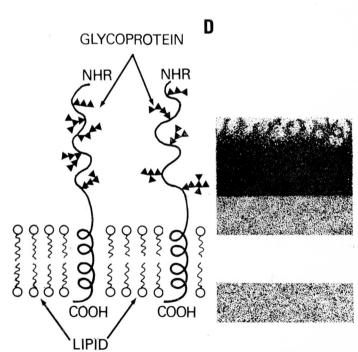

Some ten years after Gorter and Grendel (1925) made their original suggestion regarding the lipid bilayer nature of the plasma membrane Danielli and Davson (1934) made similar, though in some respects more elaborate, proposals concerning the structure of biological membranes. From surface-tension studies (Danielli and Harvey, 1934) which showed that the values obtained at cell surfaces were considerably lower than were considered possible at the time for a lipid surface, Danielli and Davson (1934) proposed a model for the cell membrane in which the lipid bilayer is sandwiched between protein. It has now been demonstrated (Haydon and Taylor, 1963) that phospholipids alone can produce low surface-tension values and the need to invoke the presence of molecules, such as proteins, to explain the observed low surface tensions at biological surfaces is, therefore, redundant.

A consequence of the cell possessing a plasma membrane of a type consisting of a continuous bilayer of lipid molecules would be the provision of an excellent permeability barrier. The suggestion that such a model represented the structure of the plasma membrane had, as was explained above, its foundation in observations on the passive entry of lipophilic solutes into the cell. To take account of active transport of ions and the facilitated diffusion of hydrophilic molecules, the modification of the bilayer model by the incorporation of pores and carriers into the structure is necessary. In the 1950s Danielli (1958) had modified his original model by the introduction of protein lined pores which traversed the width of the bilayers. The classical Danielli–Davson (1934) model which places great emphasis on lipid–lipid interactions as well as on electrostatic forces between the lipid polar head groups and the proteins covering the

Fig. 1. Models of Membrane Structure. (A) A lipid bilayer of the type proposed by Gorter and Grendel (1925). In this diagram the polar groups are represented by circular symbols and are orientated towards the aqueous phase. The rod shaped part of the diagram represents the fatty acid chains which orientate inwards towards each other. (B) Danielli–Davson (1934) model. In this model of the plasma membrane the polar head groups of the lipid are covered with hydrophilic protein (represented diagrammatically by the large circles). (C) Lucy (1968) model. This model shows globular micelles of lipid to be in dynamic equilibrium with the bimolecular leaflet structure. To the right of this diagram is illustrated the appearance of the membrane in electron micrographs of thin sections prepared by standard techniques. This structure is called the unit membrane (see Robertson, 1966). The overall thickness of this structure is 75–100 Å and the assumed relationship of the unit membrane to the lipid bilayer-structure is illustrated here. (D) A model showing in a highly diagrammatic form a glycoprotein molecule spanning the lipid bilayer. The closed triangular symbols represent glycose residues. To the right of this model is indicated the appearance of the membrane in the electron microscope following ruthenium red staining (see Chapter 2).

lipid bilayer, has in turn served as a generalized model for all biological membranes under the concept of the "unit membrane" of Robertson (see Robertson, 1966). Sjöstrand (1968) pointed out that he originally assumed that the Danielli–Davson model was generally applicable to cellular membranes and that the concept "was later forceably advocated" by Robertson who "popularized the idea" by introducing the term "unit membrane". The latter author, drawing attention to the trilamellate pattern seen in electron micrographs of thin sections of almost all membranes prepared by standard techniques, considered that the observed triple-layer pattern represented the central lipid and the two protein layers of the Danielli–Davson model. However, the unit membrane structure and its interpretation in terms of the classical plasma membrane model of Danielli–Davson has been extensively criticized. In particular Korn (1966) has put forward several arguments against the unit membrane concept, pointing out that the universality of the unit membrane theory is based upon electron microscopic evidence with its attendant difficulties of exact chemical interpretation. Although some X-ray and chemical evidence has also been advanced in support of the unit membrane concept, the former has been largely derived from data on myelin and at the time chemical evidence was obtainable from a relatively small number of preparations of varying degrees of purity. The difficulty of using myelin as a general model for biological membranes of widely differing functions is obvious, and the question of the paucity of chemical information cannot be overlooked even though considerable attention has been paid to the analysis of plasma membranes over the last three or four years. Undoubtedly, the theory has promoted much interest in the question of the reactivity of various macromolecules present in membranes with the staining reagents of electron microscopy, such as osmic acid, but it must now be considered that the universality of the unit membrane is open to serious doubt. This should, however, not be taken to imply that the classical Danielli–Davson model is being brought into question, but rather that its universality cannot be accepted with the evidence available at the present time.

It has been stressed above that although the classical bilayer model emphasizes the qualities of the cell membrane as a permeability barrier it fails to give a satisfactory explanation of the chemistry of the cell surface. Certainly the addition of protein to the surface of the structure, as suggested by Danielli and Davson (1934), was of great significance from the point of view of the biochemistry of cellular surfaces, but had little obvious effect on the interpretation of data obtained by various techniques specifically designed for investigating the cellular periphery; for example, in the case of studies by cell electrophoresis, considerable attention was paid to the suggestion that lipids were the charge determinants at the cell surface. This view held sway into the early 1960s, some years after Danielli and Davson (1934) postulated their model, but has since lost ground against new evidence for the role of the carbohydrate, sialic acid. Indeed, as will be discussed later, some authors attempted to explain the decrease in electrophoretic mobility of erythrocytes following treatment with the proteolytic enzyme trypsin in terms of the cleavage of P—N bonds in lipid structures, rather than the hydrolysis of peptide bonds in membrane proteins. This was a rather surprising situation in view of the presence of protein at the cell surface postulated in the Danielli–Davson (1934) model.

An important difficulty with the classical model of Danielli and Davson is that it presents the membrane as a symmetrical structure. The present evidence is that plasma membranes are asymmetrical structures with a significant proportion of macromolecules containing carbohydrate on the extracellular side of the membrane.

B. Subunit Model

An entirely different approach to membrane structure has been put forward by Green et al. (1967) in the form of the "subunit" model. This concept, which arises mainly from studies on mitochondrial membranes, portrays a biological membrane as being composed of cuboidal subunits. It is suggested that these cuboidal subunits have phospholipid confined to two surfaces (the other four faces do not contain lipid) and that the subunits interact hydrophobically at these surfaces to form a two-dimensional sheet of membrane. It can be seen from this short description that the subunit model for membranes is quite different from the classical bilayer concept. Stoekenius and Engelman (1969) make the point that in no case has the subunit structure for a membrane been established beyond reasonable doubt. Indeed the subunit structure has mainly been claimed for some specific membranes such as inner membrane of mitochondria, photosynthetic membranes, etc., and in their review these authors (Stoekenius and Engelman, 1969) are of the opinion that the subunit model has so far contributed little to an understanding of general membrane structure and function. More recently, however, the subunit concept has been invoked in interpreting the results of experiments on membrane biosynthesis.

It has been pointed out that the term subunit has been well defined as it applies to viruses, but there is a difficulty in defining a "subunit" in terms of membrane subunits. However, the suggestion that it is useful to distinguish between subunits of structure, subunits of function, and subunits of assembly may be of help. The authors of this suggestion (Stoekenius and Engelman, 1969) prefer to reserve the term "structural subunit" to cover a class of lipoprotein particles which are linked through identical

or equivalent binding sites, and which constitute the bulk of the membrane
material and determine its characteristic shape. A functional subunit is
considered to be comprised of all the components necessary to carry out a
particular function of the membrane. There might be different functional
subunits in any one membrane and, though these authors exclude from

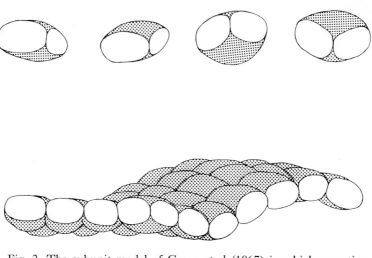

Fig. 2. The subunit model of Green *et al.* (1967) in which repeating
units (above) are shown to combine to form a membrane one unit
in thickness (below). In the diagram the lipid faces of the subunits are
indicated by stippling, while the interacting protein faces are shown as
blank areas in the diagram. Lipid depletion is postulated as reversibly
inhibiting membrane formation. At the top of the figure the single
subunits are viewed from above and below to indicate more clearly the
two faces containing lipid. Redrawn in modified form from the
scheme originally proposed by Green *et al.* (1967).

their definition components which serve to provide a general permeability
barrier, they stress that the components of a functional subunit would need
to be associated for an extended period with the membrane to come within
the scope of this term. Bearing these latter points in mind Stoekenius
and Engelman (1969) prefer to keep the general term of "subunit" for
those situations in which structural and functional subunits coincide.
A third type of subunit, namely a subunit of assembly has been defined.
Such a subunit could well be the same as a subunit of structure, though
such an assumption would have to be verified in each instance. It is
conceivable that though a membrane might be formed by an assembly
of subunits, such a mechanism should not be taken to imply that these
subunits persisted or functioned as such in the completed membrane.

Interesting though the above concept may be from the point of view of membrane structure, it is probably true to say that this model has had only minimal impact on cell surface studies. Certainly the model, in common with the bilayer theory for biological membranes, is deficient in its original form in accounting for the carbohydrates of cell membranes.

III. Isolation of Cell Membranes and Their Analysis

Before discussing more recent models of plasmalemmal structure it would seem appropriate to consider some of the recent studies performed on the isolation of plasma membranes and the chemical analysis of such material. It is investigations such as these which have, to a great extent, prompted the need to reconsider classical bilayer and subunit theories of membrane structure based, as they are, largely on the role of lipids in membrane structure. Indeed, until all the constituents of membranes have been characterized it must be emphasized that no accurate model for the plasmalemma is possible.

Up to the early 1960s very few workers had attempted to isolate plasma membranes from animal or plant cells, and analytical data on the cell membrane was consequently confined to a very few examples and then principally to erythrocytes. This paucity of analytical data has previously been alluded to when dealing with the criticisms of the unit membrane theory. In view of the ease of obtaining large quantities of red blood cells it is not surprising that this cell has been used by many investigators as a model system for studying biological membranes and indeed the fact that the human erythrocyte is basically composed of haemoglobin, encompassed by a plasma membrane, makes it an obvious candidate for membrane isolation experiments. Though the results obtained from the erythrocyte are of great value, the isolation of membranes from tissue cells is a fundamental necessity when trying to explain in molecular terms those membrane phenomena only manifested by tissue cells. Though biological membranes may, as suggested by the unit membrane theory, possess many common features of structure and chemistry, it cannot be safely assumed that all membranes are identical, or indeed to what degree they differ, until many more systems have been investigated. It is proposed, therefore, to examine the progress that has been made in the isolation of cell membranes from sources other than erythrocytes, and review in what manner the analytical investigation of such material has modified views on the structure of the plasma membrane. Initially it should, perhaps, be stated that analytical studies performed on isolated membrane fractions have done little to advance the realization that heterosaccharides are important components of cell membranes and cell surfaces, and that other techniques (see Chapter 2) have been responsible for demonstrating the importance of heterosaccharides in the cellular periphery. Apart from the study of blood group substances, the fact that analytical techniques have not until recently had a greater impact on the subject of carbohydrates in membranes is undoubtedly due to the relative insensitivity of the procedures available for the detection of the comparatively small amounts of total sugar (between 2 and 8%—Winzler, 1970) present in isolated membrane fractions. To a certain extent then, analytical studies of isolated membranes have had the important secondary role of confirming the existence of carbohydrates in cell membranes which was originally demonstrated largely by electrokinetic and microscopical techniques. Within the context of the development of ideas on the nature of biological membranes, analytical studies on isolated membrane-enriched fractions served the purpose of drawing attention to the need to consider the role of non-lipid constituents, principally proteins, in such structures.

It is essential that membrane material for chemical work should be isolated in a pure state in reasonable amounts, the preparation should not lack any membrane components present in the intact living cell and should be free of contaminants. These requirements, which represent the ideal, present a number of technical difficulties and indeed it is likely that the majority of preparations described in the literature do not completely fulfil all such conditions. Not only is it necessary to develop suitable techniques for the disruption of cells, but also procedures for the isolation of the various membrane fractions must be devised since conditions found optimal for one cell type may not hold for another.

A. Isolation of Erythrocyte Stroma

Mammalian erythrocyte stroma or post-haemolytic residue has been widely considered as a useful model for cell-membrane studies. When compared with the plasma membranes of tissue cells, erythrocyte ghosts are easy to prepare, and in addition all the human blood-group systems, with the exception of the Lewis system, appear to be an integral part of the erythrocyte stroma (Sneath and Sneath, 1959). However, the chemistry of erythrocyte stroma is complicated by the fact that the method of preparation is reflected in an altered chemical analysis. Ghosts prepared by freezing and thawing red blood cells or by the use of organic solvents as lysing agents, have an irregular haemoglobin content (Bernstein et al., 1937). Furthermore the product obtained may not only be dependent on the method of lysis, but also on the pH of the environment used in the preparation. For example, whilst the use of media of pH 9 overcomes the problem of an irregular haemoglobin content of erythrocyte stroma, this suffers from the disadvantage that the ghosts progressively diminish in volume. Dodge et al. (1963) have described a method using 20 ideal milliosmolar phosphate buffer at pH 7·4 for the preparation of haemoglobin-free ghosts of human erythrocytes, essentially all the lipid was recovered in the ghosts but non-haemoglobin nitrogen containing substances were lost. Jamieson and Groh (1971) find that extensive washing of stroma (Dodge et al., 1963) causes alterations in their isoelectric point which they suggest may be due to the loss of anionic material or to conformational changes.

B. Isolation of Plasma Membranes From Tissue Cells

Turning to tissue cells, a variety of methods have been devised since the early 1960s for the isolation of membranes and the isolation of plasma membranes has received special attention. Cells of several types have been fragmented by a number of techniques including gentle mechanical rupture (Neville, 1960; Herzenberg and Herzenberg, 1961) or ultrasonic disruption (Rajam and Jackson, 1958). Under these conditions plasma membranes may be recovered largely from a crude "nuclear" fraction obtained by low-speed centrifugation of the homogenate. In contrast, by the use of nitrogen cavitation (Kamat and Wallach, 1965), rupture in the French Pressure Cell (Cook et al., 1965) or the use of more vigorous mechanical breakage in some tissue homogenizers (Graham et al., 1968), the plasma membrane may be reduced to small vesicular fragments which sediment with microsomal material in the ultracentrifuge. Plasma membranes may then be isolated from nuclear or microsomal fractions by means of centrifugation in media of high density such as sucrose or Ficoll solutions.

In the technique described by Neville (1960) rat liver cells were subjected to gentle homogenization in a Dounce homogenizer in a hypotonic medium, containing bicarbonate which broke the nuclear envelope. A combination of washing and sucrose density gradient techniques then yielded a plasma membrane fraction consisting of large sheets of material. Emmelot et al. (1964) subsequently modified the procedure of Neville (1960), introducing additional sucrose density layers in order to achieve a more effective separation between the membranes and the few mitochondria persisting in the pregradient stage. This modification was also useful for providing purer preparations of plasma membranes from primary rat hepatomata as opposed to transplanted rat hepatomata. The preparations consisted predominantly of large sheets of membranes of various adjacent cells still attached by means of desmosomes, terminal bars and various junctional complexes. It is likely that the presence of junctional complexes gives a degree of stability to the system, and makes it possible to isolate the plasma membrane in sheet form. Portions of the plasmalemma not so stabilized might be expected to be reduced to vesicles which would sediment with microsomal material on differential centrifugation. In this respect the work of Graham et al. (1968) is of particular

2C*

interest; these authors made a comparative study of the isolation of plasma membrane fragments of rat liver using (a) the modification by Emmelot *et al.* (1964) of Neville's method (Neville, 1960) and (b) the technique of Kamat and Wallach (1965) as devised for the isolation of plasma membranes from Ehrlich ascites cells, with the exception that a Potter–Elvehjem homogenizer was substituted for nitrogen cavitation. The latter method yielded liver plasma membranes in the form of closed vesicles, as opposed to the planar sheets obtained when using the former procedure. Graham *et al.* (1968) point out that the Neville (1960) technique does not work for isolated parenchymal cells where the junctional complexes have been broken, although a vesicular preparation of plasma membranes may still be obtained. Further, they make the important point that such a preparation may be more useful for binding and permeability studies than open sheets of planar material, especially if the vesicles are sealed with the original outer surface to the outside. It should be noted, however, that material prepared by Neville's procedure (Neville, 1960), with the junctional complexes *in situ*, has certain advantages in allowing one to decide which is the outer surface of the plasma membrane in isolated fractions, and this has been of prime importance in assigning heterosaccharides to the exterior of the cell membrane in electron microscopical preparations (see Chapter 2). A disadvantage is that the method may be rather selective for certain areas of the plasma membrane and Wallach (1967) has estimated that in such methods, at the most, 14% of the entire surface membrane is isolated.

The danger that the conditions prevailing in the isolation procedure may influence the analytical results of the final isolated material has already been noted. In the case of rat liver Takeuchi and Terayama (1965) were particularly concerned that the treatment of cells with hypotonic media (as in the Neville (1960) procedure and Emmelot *et al.* (1964) modification) might cause drastic alteration of the cellular structure, and they therefore tried to separate cell membranes under milder conditions using an isotonic sucrose medium containing calcium chloride. The use of isoosmotic conditions have also been investigated by Nigam *et al.* (1971). It is interesting to note that Takeuchi and Terayama (1965) also succeeded in isolating plasma membranes as sheets of unit membranes still held in contact, presumably by means of the desmosome-like structures observed in their preparations, as in the original intercellular adhesion. A direct comparison of the chemistry of these various preparations is not possible as the authors have reported their analytical results in differing ways, either expressing the values obtained for the various components as a percentage of the total preparation or as mg or μmol membrane bound phosphorus or as μmol per mg protein. In the work of Emmelot *et al.* (1964) some 15–25% of the membrane cholesterol was esterified whilst Takeuchi and Terayama (1965) found cholesteryl ester to be only about 5–10% of the total cholesterol. Both authors give values for hexosamines and sialic acids though the Japanese authors report a higher ratio of hexosamine to sialic acid (Takeuchi and Terayama, 1965). A possible explanation of this might be that components leak out of the isolated membranes. Emmelot *et al.* (1964) have shown that some 26% of the membrane isolated protein becomes solubilized in saline at room temperature, though in their experiments they report that all the hexosamine and 95% of the sialic acid was recovered in the saline-insoluble fraction.

A completely different approach to the problem of isolating plasma membrane fragments has been developed by Kamat and Wallach (1965). They used nitrogen cavitation to disrupt the plasma membranes of Ehrlich ascites carcinoma cells into small vesicles which were then recovered in the microsomal pellet after differential ultracentrifugation. The microsomal fraction, the composition of which may vary according to tissue origin and method of homogenization, consisted of vesicular fragments of plasma membranes and endoplasmic reticulum as well as free ribosomal material. It was then washed in hypotonic buffers to produce a transient leakiness of the vesicles, with the release of intravesicular, soluble proteins. Presumably this soluble protein represents cytoplasmic material trapped on homogenization and not membrane-derived protein, but this assumption needs rigorous investigation, especially when this technique is applied to material derived from cells other than the Ehrlich ascites carcinoma. After the washing procedure, the vesicles derived from the endoplasmic reticulum, were preferentially shrunk and aggregated by Mg^{2+} ions at pH 8·6, and then separated from the plasma membrane vesicles by means of Ficoll density gradient techniques. In Ficoll gradients a volume decrease, as experienced by endoplasmic reticulum vesicles, leads to an increase in density[1]. Hence, assuming that the effect of the divalent cations is preferentially directed towards fragments derived from structures other than the cell membrane, it can be seen how a purification of plasma membrane fragments may be achieved. The method of Kamat and Wallach (1965) has been criticized by Bosmann *et al.* (1968) who point out that the method "results in extensive fragmentation of the membranes yielding micellar artifacts that may be unsuitable for structural studies". This point should certainly be borne in mind, though Kamat and Wallach (1965) used a series of enzyme and immunological markers to monitor their separation procedure and were able to demonstrate a selective enrichment in their surface membrane fraction of Na^+

---

[1] A fractionation system where membrane fragments bearing a particular receptor are separated by selectively increasing the density of the membrane vesicle by coupling a particle of very high density to the receptor has been described by Wallach *et al.* (1972). Termed "affinity-density-perturbation" by Wallach *et al.* (1972) these authors have demonstrated the method by using a density perturbant (concanavalin A bound by glutaraldehyde to coliphage K29) with fragments of pig lymphocyte membranes.

Cells disrupted quantitatively by intracytoplasmic cavitation of nitrogen gas.

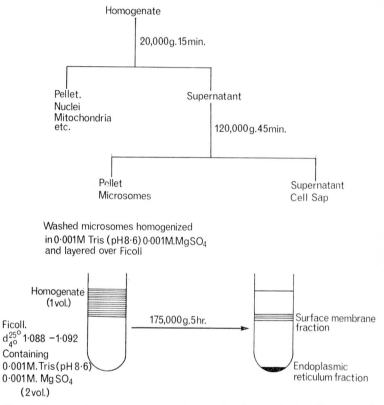

Fig. 3. Preparation of plasma membranes by the method of Kamat and Wallach (1965). In their original paper Kamat and Wallach (1965) prepared a nuclear/mitochondrial pellet by centrifuging homogenates at 12,500 rev/min (rotor 9RA, Lourdes Betafuge) for 15 min. Microsomes were sedimented at 40,000 rev/min (rotor No. 40 Spinco L-2 preparative ultracentrifuge) for 45 min. The values quoted in the above figure are g max values which the authors have used to obtain comparable fractions using different equipment (Measuring and Scientific Equipment Ltd: 3 × 23 ml swing cut rotor in a Super Speed 50 machine). In the final fractionation step the authors have used an SW39 Spinco rotor as described by Kamat and Wallach.

and K$^+$ activated ATPase as well as surface antigens. Their method has also been applied to a variety of other cell types including thyroid cell membranes (Stanbury et al., 1969), BHK$_{21}$ cells (Gahmberg and Simons, 1970), pig and calf lymphocytes (Ferber et al., 1972) and, in the authors' laboratory, murine lymphocytes (Warley and Cook, in preparation).

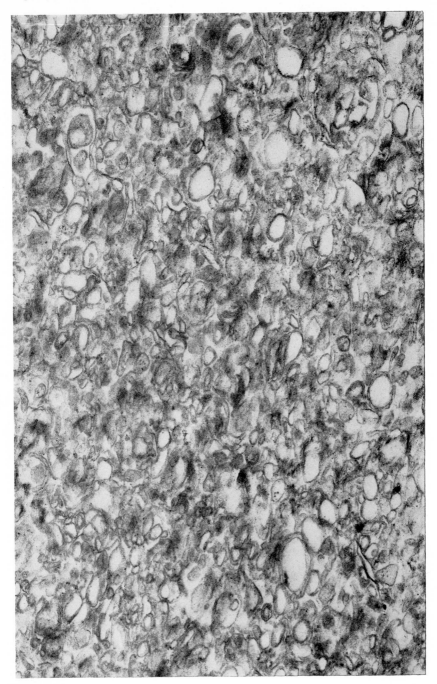

Fig. 4. Electron micrograph of plasma membrane fraction prepared from murine lymphocytes by the method of Kamat and Wallach (1965). The preparation consists of vesicular membranous structures as opposed to sheets of material (see Fig. 6a, b). (Warley and Cook in preparation ×40,000.)

An entirely different approach to the isolation of surface membranes was devised by Warren *et al.* (1966). In the methods described earlier it was noted that the presence of certain features, such as cell junctions, had the possible effect of strengthening the structure yielding a preparation consisting largely of sheets of membrane, whilst in the absence of any supporting structure the plasmalemma was reduced to smooth surfaced

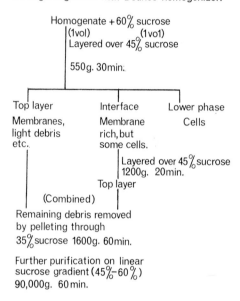

Plasma membrane strengthened using various reagents (FMA etc.). Surface membrane removed as large fragments with Dounce homogenizer.

Homogenate +60% sucrose
(1vol)   (1vol)
Layered over 45% sucrose

550g. 30min.

Top layer   Interface   Lower phase
Membranes,   Membrane   Cells
light debris   rich, but
etc.   some cells.

Layered over 45% sucrose
1200g. 20min.
Top layer

(Combined)
Remaining debris removed
by pelleting through
35% sucrose 1600g. 60min.

Further purification on linear
sucrose gradient (45%-60%)
90,000g. 60 min.

Fig. 5. Preparation of plasma membranes by the technique of Warren *et al.* (1966).

vesicles. Realizing the importance of strengthening the surface membrane to enable it to withstand the rigors of the separation and isolation procedures, Warren *et al.* (1966) approached the problem of plasma membrane isolation by stabilizing the cell surface membrane by a number of chemical treatments, followed by removal of the plasmalemma intact, or as large fragments, in a Dounce homogenizer. The membrane-enriched fraction can then be separated and purified upon density gradients. In their procedures a variety of chemical treatments including the sulphydryl blocking agents, fluorescein mercuric acetate (FMA) and 5,5'-dithiobis-(2-nitrobenzoic acid), as well as zinc chloride and tris buffer all in hypotonic solution were employed. The use of a hypotonic medium causes the plasma membrane to rise from the underlying cytoplasm and enables

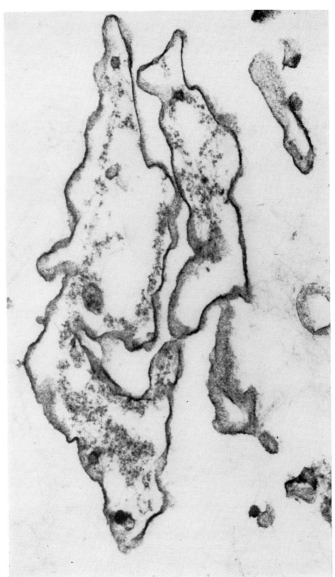

Fig. 6a. Electron micrograph of surface membranes isolated from murine lymphocytes by the FMA method of Warren *et al.* (1966). The preparation consists of sheets of membranous material as opposed to vesicular fragments (see Fig. 4.), the inner surface of the membrane being distinguished from the outer surface by attached cytoplasmic material. From the Doctoral Dissertation (University of Cambridge, 1970) of O. S. Kochhar by permission of the author. ×40,000.

the isolation of free membrane to be attained more easily on homogenization. Unfortunately, according to the authors (Warren *et al.*, 1966) the
mechanism by which the various agents employed bring about stabilization
of the surface membranes is not known, and it is only assumed that the
sulphydryl blocking agents act by reacting with sulphydryl groups in the

Fig. 6b. Electron micrograph of surface membranes isolated from murine lymphocytes by
the FMA method of Warren *et al.* (1966). In
this preparation the membrane has rolled up in
a manner described by Warren and his colleagues
and appears as a scroll in cross-section. From the
Doctoral Dissertation (University of Cambridge,
1970) of O. S. Kochhar by permission of the
author. ×40,000.

surface structure. However, the authors do suggest that the fluorescein
mercuric acetate, which has a very broad range of effective use, might
work by the virtue of its binding to sulphydryl groups while the organic
moiety is soluble in the lipid phase of the membranes. It should also be noted
that some of the stabilizing agents used may be injurious to surface
enzymes. Nevertheless, as Warren and his colleagues point out, much
useful chemical analysis can be carried out on such material and some

realistic metabolic studies can be performed. The reagent used may not cause irreversible damage to all surface enzymes and in the case of some of the stabilizing agents, for example Zn^{2+}, phosphohydrolyase activity could be recovered in good yield from chick embryo fibroblast membranes with EDTA (Perdue and Sneider, 1970).

This study (Perdue and Sneider, 1970) is of particular interest as a comparison was made, between membranes of chick embryo fibroblast cells, prepared by the Zn^{2+} method of Warren et al. (1966) where very pure preparations of plasma membrane were isolated as large membrane sheets, and material prepared in absence of any chemical treatment. In the latter case almost the same distribution of membrane material was obtained upon purification as that obtained from the treated cells, though by contrast smaller membrane vesicles and fragments were produced.

The zinc ion method of Warren et al. (1966) has also been examined by Brunette and Till (1971) again using L-cells. However, this work differed from that of the original method in that instead of using difference of density or sedimentation characteristics to achieve a separation of the surface membrane from the other organelles, methods which Brunette and Till (1971) point out were not always satisfactory, these authors used a dextran-polyethylene glycol aqueous two-phase system which has the ability of separating particles on the basis of differences in their surface properties. This system has the advantage that once the homogenate has been prepared the isolation of surface membranes may be achieved very quickly. In these authors' hands (Brunette and Till, 1971) separations were achieved in less than two hours following the preparation of an homogenate.

In general the techniques of Warren et al. (1966) have been applied to suspensions of individual cells rather than solid tissues. Indeed, one of the needs for developing chemical strengthening techniques of the surface membranes is that stabilizing surface connections are only likely to be present in quantity in solid tissue and absent in the case of cell suspensions. However, Warren et al. (1966) express the belief that surface membranes may be isolated from solid tissues if the tissues are dispersed into individual cells before adding the stabilizing agents. Where cells were grown on glass, they were removed from the glass by means of a rubber policeman. The cells were then subjected to membrane-stabilization by zinc ion or FMA treatments and the surface membranes isolated in the normal manner. In addition, fluorescein mercuric acetate solution could be spread over the monolayer of cells for a short period of time before their removal by the rubber policeman. Homogenization and isolation techniques could then proceed. An alternative rapid and simple technique for isolating plasma membranes from monolayer and multilayer tissue culture cells has been described by Barland and Schroeder (1970) who used a modification of the zinc ion technique of Warren *et al.* (1966) followed by treatment with FMA. In their treatment with zinc ion, dimethyl sulphoxide was added, presumably as an aid to penetration. Following this consecutive treatment with zinc and FMA the culture was shaken on a rotating platform, whereupon an increasing number of cell ghosts were seen in the suspending medium when examined by phase-contrast microscopy. Undoubtedly this technique is quick and simple to perform and may be of especial interest because it appears that only that portion of the surface membrane in contact with the culture medium is removed, since membranes are found adhering to the substratum. Such a technique might be useful for investigating possible chemical differences between the upper surfaces of the cell and that part of the membrane in contact with the culture vessel. However, the use of dimethyl sulphoxide and its effect on biological membranes should be given particular attention in view of the possible disruptive effect of this agent on such a structure. The authors of this procedure also make the point that it is probable that very little biological activity remains in membranes isolated with zinc chloride and fluorescein mercuric acetate, though of course useful analytical data may be forthcoming from such material.

It will be clear from the above descriptions that, in the absence of special strengthening structures or artificially induced stability by the use of various chemical reagents, on homogenization the cell membrane is reduced to small vesicular fragments. McCollester (1970) noting that in the latter case the cytoplasmic membranes also "break down to form particles or vesicles which are often indistinguishable from the particles derived from the disrupted surface membrane", presumably had in mind morphological identity as Kamat and Wallach (1965) have shown differences in enzyme content between the various vesicular populations. These latter fragments could only be identified by exacting centrifugal separation conditions, described earlier. In McCollester's (1970) method an attempt has been made to design a method in which large fragments of easily identifiable cell surface membrane are obtained but without resorting to the use of harsh chemical procedures. In his method (McCollester, 1970), which he largely applied to Meth A ascites tumour cells carried in Balb/c mice, as well as some description of its application to leukaemic cells of AKR/J mice, McCollester extracted suspensions of the cells with alkaline borate/ethylenediamine tetra-acetate (EDTA) mixtures. With this treatment phase contrast microscopy shows that the cell swells, and bursts with the expulsion of the nucleus and intracellular particles. Sometimes the nucleus remains attached to the ghost. Cells that do not undergo this swelling and bursting phenomenon, are, to judge from the lack of Brownian movement of their interacellular particles, probably dead. At the completion of this extraction stage the addition of more borate precipitates a great deal of cellular material, mostly DNA; the extruded nuclei coalesce as do those ghosts which still retain their attached nuclei; but membranes free of nuclei and cytoplasmic contents are unaffected. The ghosts are then harvested, following a series of centrifugation and washing procedures, with a loss of cell particles which are retained in the supernatants. In the method, care has to be exercised with regard to the conditions under which the stroma are handled as severe shearing soon reduces the ghosts to vesicular particles. Certainly this technique enables one to prepare from Meth A cells plasma membranes which from the published electron micrography and chemical data appear to be present in a considerable degree of purity. Unfortunately neither enzyme activity nor recovery data of various chemical classes was provided by McCollester (1970). However, his assertion, based on the ease with which all the steps can be monitored microscopically, that his method permits the recovery of surface membrane is undoubtedly fully justified. It is regrettable in view of the very laudable objective (McCollester, 1970) of producing a mild method for obtaining large fragments of surface membrane which is applicable to diverse cell types, that the author did not give any chemical or electron microscopic data of his preparation of leukaemic cell membranes, though a phase contrast picture showing the cells before extraction and the resulting ghosts were given. The exact mechanism underlying this extraction technique is not entirely resolved, though the essential role of borate was clearly demonstrated by McCollester (1970). Apparently it is suggested that in order to maintain the cytoskeleton (endoplasmic reticulum) FAD is important since the removal of FAD by apo-D-amino acid oxidase results in cytoskeletal breakdown. From a survey of inhibitors of FAD-dependent enzymes it was found that EDTA and borate effectively break down the cytoskeleton. McCollester (1970) observes, however, that borate not only promotes extractability of the cells, but is also probably important in preventing intracellular structures from adhering to the separated ghosts. It was suggested by McCollester (1970) that the borate combines with the carbohydrate residues on the surface of both the isolated surface membrane and the intracellular particles, thereby enhancing the negative charge on each and favouring mutual repulsion as opposed to adhesion.

Recently the technique of McCollester has been re-examined (A. Warley, personal communication) with regard to its application to leukaemic tissue derived from AKR mice. It was found that though the cells responded to borate/EDTA in exactly the manner described by McCollester, the membrane fraction subsequently isolated by this author's technique was contaminated with mitochondria and other organelles with only a limited degree of purification as assessed by measurement of marker activities (5'-nucleotidase). This fraction was, however, readily purified

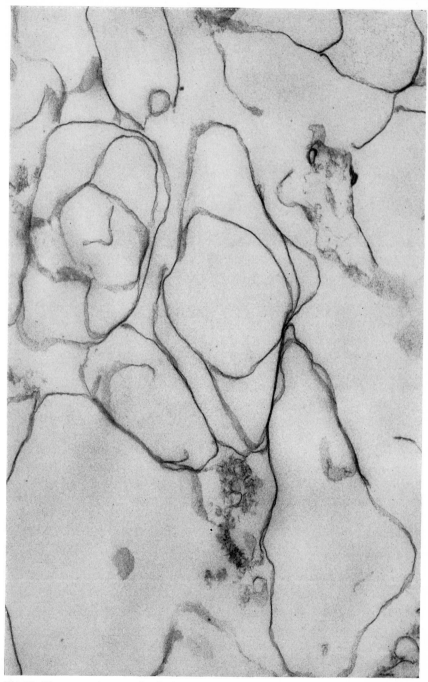

Fig. 7. Electron micrograph of plasma membranes prepared from acute lympho-blastic leukaemic cells of AKR mice by a modification of the method of McCollester (1970). From Warley and Cook, in preparation. ($\times$ 40,000.)

(A. Warley) by passing the material through a column of glass beads (Superbrite No. 150, 3M Company), which had previously been soaked in 0·1 N HCl and washed in distilled water ahead of the contaminating organelles and cytoplasmic debris. The fractions containing eluted ghosts as assessed by phase contrast microscopy were pooled and membranes collected by centrifugation. Electron microscopic examination shows that the preparations consist of large sheets of membranes with an absence of contaminating structures and the appropriate enzyme markers also indicate that a considerable degree of purification has been achieved. The glass bead technique was originally described by Warren et al. (1966) for purifying L-cell surface membranes stabilized with the zinc ion method, and the amalgamation of this method with the McCollester (1970) fractionation has been particularly advantageous in preparing good yields of membranes in a high state of purity in a minimal time (three hours). In addition to leukaemic lymphoblasts Warley and Cook (in preparation) have demonstrated that the above modification of the McCollester (1970) technique may be used to prepare highly purified preparations of murine lymphocyte plasma membranes. This is particularly interesting, for as pointed out by Allan and Crumpton (1970) relatively little attention has been paid to the isolation of well-characterized plasma membranes of lymphocytes, a particularly surprising situation in view of the importance of this cell in many important immunological phenomena. These latter authors (Allan and Crumpton, 1970) had previously described for the first time a method for isolating plasma membranes from lymphocytes. Using pig mesenteric lymph nodes, they prepared a mince which consisted of largely intact nuclei together with large granules and sizeable fragments of plasma membranes and endoplasmic reticulum. The plasma membranes were purified by a combination of differential centrifugation and sucrose density gradient procedures. Allan and Crumpton (1970) obtained 30 mg (dry weight) of plasma membranes from 30 g (wet weight) of lymph nodes. This fraction, which was vesicular by nature, was shown to have a molar ratio of cholesterol to phospholipid of 1·01 which is characteristic of plasma membrane fractions with a high degree of purity (Coleman and Finean, 1966). In addition, the membrane was shown to possess the appropriate (5′-nucleotidase) marker activities and to have a high capacity to adsorb lymphoagglutinins from anti-lymphocyte serum. Ferber et al. (1972) also described the preparation of plasma membranes from pig mesenteric, as well as from calf mediastinal, lymph nodes. These latter authors extended the nitrogen cavitation technique of Kamat and Wallach (1965), to the homogenization of lymphocytes. They found (Ferber et al., 1972) that the conditions for equilibration with nitrogen are critical and that it was necessary to lower the gas pressure used at the time of equilibration to prevent nuclear disintegration. These authors (Ferber et al.,

1972) consider that the equilibration conditions are particularly important because of the small cytoplasmic space available in the lymphocyte; in this respect it is interesting that in a comparison of the homogenization of murine normal and leukaemic lymphoid cells by nitrogen cavitation (Kamat and Wallach, 1965) performed some years previously (see Cook, 1970) the method resulted, in contrast to the malignant lymphoblasts, in considerable nuclear damage in the case of normal lymphoid cells. The normal lymphoid cells do, as pointed out by Ferber *et al.* (1972) possess a reduced ratio of cytoplasmic space to nucleus when compared with the leukaemic cells. Apart from this, Ferber *et al.* (1972) also observed that abrupt suspension of lymphocytes in sucrose solutions caused irreversible cell aggregation and it was necessary to add sucrose to lymphocytes suspended in an isotonic salt solution. Allan and Crumpton (1970) also found sucrose undesirable for the disruption of lymphocytes because of the problem of extensive breakage and aggregating of nuclei; in the case of their method for plasma membrane isolation sucrose was not used during the initial stages of the fractionation procedure in order to avoid the problems associated with sucrose. In the case of McCollester's (1970) method as applied by Warley and Cook (in preparation) to lymphocytes there is the considerable advantage that the use of sucrose is completely avoided.

The methods available for the isolation of well characterized plasma membrane fractions are quite numerous with the result that a considerable amount of analytical and biosynthetic studies on this organelle are now possible.

C. Analytical Examination of Isolated Plasma Membrane Fractions

1. CHARACTERIZATION OF PLASMA MEMBRANE FRACTIONS

Within the last few years a large body of analytical data has been accumulated from biochemical studies on plasma or surface membrane fractions isolated in the ways described above. It is apparent from the foregoing section that the method of isolation of the membrane may well influence the form that the final preparation takes and, presumably, may also have an effect on the composition of the separated fraction. In addition, analytical data have been widely used to determine whether isolated fragments of membrane are truly derived from the plasma membrane. This is especially important in those cases where there is an absence of any morphological criterion of purity and origin, for example where the membrane has been reduced to vesicles as opposed to an almost intact ghost or envelope, or those exceptional cases where some identifiable surface structure is present, as exemplified by the liver, where it is possible to see remains of bile canaliculi, and by the amoeba with its characteristic surface coat.

a. *Enzyme Markers*

A considerable number of workers have turned their attention to an examination of the amounts of alkali metal-activated adenosine triphosphatase present in their preparations. It is now widely held (see Glynn, 1968) that the active transport of Na^+ and K^+ ions across animal plasma membranes is coupled with the adenosine triphosphatase activity of these membranes; the hydrolysis of adenosine triphosphate generates the energy for the pump. The accurate estimation of the ouabain sensitive, Na^+K^+ activated Mg^{2+} dependent ATPase may be difficult, especially if the activity is present in very small amounts. Indeed Gahmberg and Simons (1970) who reported on Na^+K^+ activated ATPase activity present in preparations of BHK_{21} plasma membranes, comment that the activity measured by them in this cell is low, especially when compared (see Gahmberg, 1971) to other tissues such as brain, muscle and testis. Considerable variation in the activity found in different experiments performed on the same tissue, as experienced by Stanbury *et al.* (1969) with thyroid cell membranes, may also complicate matters, though in the example quoted the authors generally found a higher concentration of the activity in their light membrane (plasma membrane) fraction compared with the pellet of material from the endoplasmic reticulum. These authors also reported that the assay for 5′-nucleotidase gave a better discrimination between their two membrane fractions. This latter activity has been used by several authors as a marker enzyme for plasma membranes and Bosmann *et al.* (1968) have shown that this enzyme is found almost exclusively in the plasma membrane fraction of the HeLa cells. Conversely, Gahmberg and Simons (1970) (using BHK_{21} cells) found that the activity was not very much concentrated during the isolation procedure and suggest that 5′-nucleotidase may not be specific for plasma membranes in these particular cells. It is possible, as Gahmberg and Simons (1970) point out, that the activity may be selectively lost during isolation. Another problem is that the measurement of 5′-nucleotidase may be complicated by non-specific phosphatase activity. Stanbury *et al.* (1969) acknowledge this difficulty with their work on thyroid membranes but are careful to draw attention to the ratio of specific activity of 5′-nucleotidase between their light membrane and pellet fractions, which are always greater than for phosphatase activity at pH 7·8 or alkaline phosphatase measured at pH 10, suggesting that they are measuring a specific 5′-AMPase. As an alternative to making a comparison with measured non-specific phosphatase activity, it should be possible to estimate the 5′-nucleotidase using adenosine deaminase and by diverting the action of the non-specific phosphatase with excess β-glycerophosphate. Certainly the use of enzyme markers for characterizing membrane fractions, though widely employed, should be approached with due caution as regards interpretation of the results. For a detailed description of enzymes and various membrane systems the review of Reid (1967) should be consulted. Ideally, the identification of a fraction as being derived from the plasmalemma should be established by reference to criteria in addition to those purely dependent upon enzymatic activity.

b. *Immunological and Chemical Markers*

In addition to enzyme markers, use has been made of immunological and chemical markers. Kamat and Wallach (1965), for instance, have used surface antigens to monitor the separation and purification of plasma membrane fragments from Ehrlich ascites carcinoma cells. The H2 antigen has also been used to characterize plasma membrane fractions: Evans and Bruning (1970) have demonstrated that mouse liver plasma membrane may be fractionated into two sub-fractions, the lighter of which contains more H2 antigen than the heavier sub-fraction. However, both these sub-fractions contained more antigens than in material derived from endoplasmic reticulum. Earlier Hertzenberg and Hertzenberg (1961) had used histocompatability antigens as markers of plasma membrane in the liver instead of using morphological criteria. A possible difficulty associated with the use of the histocompatability antigen is that they have been found in various other cell fractions (Kandutsch, 1960), though naturally it can be argued that the presence of histocompatability antigens in nuclear and mitochondrial fractions, for example, may indicate contamination of these fractions by cell membrane. Indeed, the recent general view (see Winzler, 1970) is that the histocompatability antigens are glycoproteins which are primarily located on the cellular surface.

A means of resolving the problem of the location of a membrane marker to only one particular cellular site, such as the plasma membrane, would be to label the surface of the intact cell with a chemical which is easily detectable in small quantities. Such a reagent would need to react readily with the cell under physiological conditions to form a covalent linkage without permeating the membrane or causing membrane lysis. The early work of Maddy (1964) on the design of a true chemical marker of the surface membrane, 4-acetamido-4'-isothiocyanostilbene-2,2'-disulphonic acid (SITS) is particularly interesting. Maddy (1964) points out that SITS possesses all the three requirements of a good marker of plasma membrane. The reagent possesses reactivity through its isothiocyanate group reacting with accessible amino, histidyl and guanido groups of the membrane proteins to form a stable product; it is impermeable by virtue of its sulphonic acid groups; and is easily detected by means of the fluorescent properties conferred on the molecule by its stilbene nucleus. This reagent, which is now available commercially, has been used successfully by Maddy (1964) with ox erythrocytes. The compound reacted with the outer surface of such cells from which fluorescent ghosts could be prepared. It is rather surprising that this reagent has not found wider use in the literature though Marinetti and Gray (1967) have indicated that this reagent may be used successfully to label the plasma membranes of liver cells. In addition to SITS Berg (1969) and Carraway et al. (1971) have investigated the use of diazotized ^{35}S sulphanilic acid, a protein reagent which penetrates slowly the membranes of intact erythrocytes, as a surface marker.

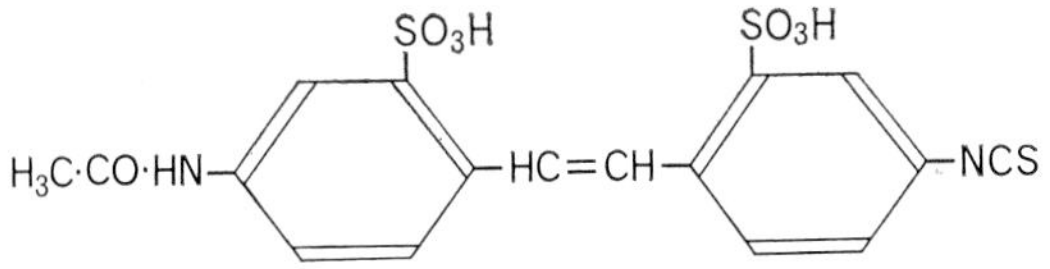

Fig. 8. 4-acetamido, 4′-isothiocyanostilbene-2,2′- disulphonic acid (SITS).

The use of an alternative reagent formylmethionyl sulphone methyl phosphate has recently been described by Bretscher (1971a). This acylating agent certainly appears to be unable to pass through the cell membrane of human erythrocytes (Bretscher, 1971a), readily reacts with surface glycoprotein and, in contrast to SITS, with lipid. The reagent has been used

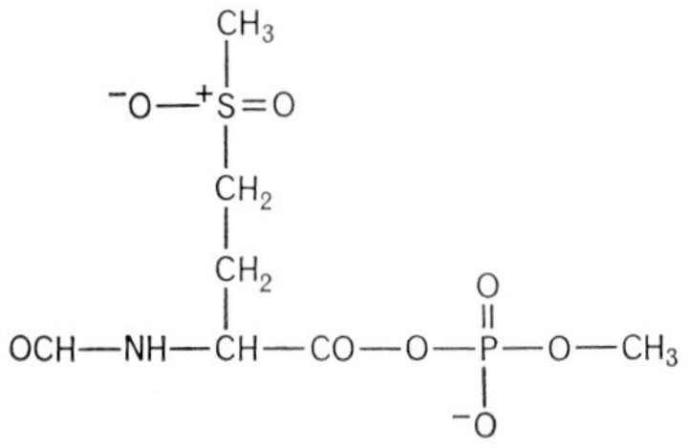

Fig. 9. Formylmethionyl sulphone methyl phosphate.

as a radioactive, ^{35}S labelled preparation of very high specific activity (about 10 Ci/mmol) and may thus be easily located. An alternative added chemical marker system has been the iodination of the plasma membrane using a lactoperoxidase technique (Phillips and Morrison, 1970). These particular methods will be discussed in detail elsewhere.

2. CHARACTERIZATION OF OTHER MEMBRANE SYSTEMS

In the above section particular attention was paid to the question of characterizing those membrane fragments present in cellular homogenates which are derived from the plasma membrane. Such homogenates will also contain pieces of other membrane systems such as fragments derived from the Golgi apparatus, rough endoplasmic reticulum and perhaps the nuclear envelope. Here, as in the case of the plasma membrane, enzymatic markers have been widely used. NADH-diaphorase (Kamat and Wallach, 1965) and glucose-6-phosphatase (Morré et al., 1969) have, for instance, been used as markers for endoplasmic reticulum. By contrast, membranes from the Golgi apparatus are identified by the lack of those very enzymatic activities considered to be specific for endoplasmic reticulum and plasma membrane. The characterization of fractions derived from the Golgi apparatus have been described in detail elsewhere (Cook, 1972).

3. COMPOSITION OF ISOLATED PLASMA MEMBRANE FRACTIONS

a. General Remarks

The classical membrane models have emphasized their lipid, and to some extent their protein, nature and a number of studies on the gross composition of erythrocyte ghosts, which are to a large extent responsible for these models of the plasma membrane, have been recorded. With the isolation of a number of surface membrane preparations, by means of the techniques described above, it has become possible to extend such analytical studies to a much wider range of cells.

The composition of erythrocyte ghosts depends to a certain extent upon the method of preparation used. Basically, human stroma consists of protein and lipid with the former present in greater amounts. Depending on the method of preparation, up to 40% of lipid may be present and similar amounts are found in the stroma of other species. The lipids of human erythrocyte ghosts are largely phospholipid though neutral lipids, mostly cholesterol in unesterified form, are present. Also present in erythrocyte stroma is a glycolipid fraction, these materials which have been examined chemically by various workers (Klenk and Lauenstein, 1952; Yamakawa et al., 1956) are particularly interesting as they have been found to possess blood-group activity (Hakomori and Jeanloz, 1961).

In the case of rat liver plasma membranes a gross composition of 41–54% protein, 26–32% phospholipid and 4–7% cholesterol has been reported (Takeuchi and Terayama, 1965), compared with 60% protein and 40% lipid for mouse liver plasma membranes (Hertzenberg and Hertzenberg, 1961). While in the case of the plasma membranes of the HeLa cells 61% protein and 50% lipid has been found (Bosmann et al., 1968). These figures are quite representative examples of the gross composition of animal cell plasma membranes. It should, however, be pointed out that Emmelot and Benedetti (1967) who report a figure of 65% protein for rat liver plasma membranes showed that some 20–25% of this protein is soluble in physiological saline, this soluble material however, may be derived to a considerable extent, if not exclusively, from the cytoplasm and intercellular fluid.

Turning to carbohydrate content it seems that a general figure of between 1 and 8 % carbohydrate may be expected. Apart from erythrocytes, where a sufficient quantity of material is available for chemical analysis, very few detailed analyses of the carbohydrate present in other membrane preparations have been made. It should be possible for more information to be obtained now that various techniques for preparing and characterizing plasma membranes are available. The problem of the sensitivity of the analytical method may, however, still be an important factor and in this regard the isotope dilution method for assaying sugars of mammalian cells described by Shen and Ginsburg (1967) is potentially valuable.

The lipid protein ratio of rat liver cells (Weinstein et al., 1969) and HeLa plasma membranes are approximately the same. In the case of the HeLa cell there is a high molar ratio of cholesterol to phospholipid which Bosmann et al. (1968) suggest is a good criterion of the purity of their preparation, though Weinstein et al. (1969) suggest that this result may indicate that the HeLa cell surface membranes may belong to a different class of surface membrane from those of the L-cell and rat liver. The HeLa cell, they point out, appears to be more closely related to erythrocyte membranes and myelin. What is particularly interesting is that Weinstein et al. (1969) suggest that surface membranes possessing high cholesterol-phospholipid ratios may also have high glycolipid contents, and that it may be possible to classify surface membranes according to their glycolipid content.

b. *Proteins*

However, at the present time considerable attention is being focused on the proteins of membranes and the subject has recently been reviewed by Guidotti (1972). Proteins which are the predominant components of membranes are receiving particular emphasis in modern models of cell membrane structure. Indeed Singer and Nicolson (1972) have recently suggested when refuting the idea that it is not possible for useful generalizations to be made about the gross structure of membranes "that an analogy exists between the problems of the structure of membranes and the structure of proteins". They further point out that each kind of protein is structurally unique but that generalizations about protein structure have been very useful for reaching an understanding of the properties and functions of proteins. However, apart from such analogies, it is important that these major components of biological membranes be considered in their

Table I   Isolation of Plasma Membranes of Animal Cells

Tissue	Method	Chemical analysis			Enzyme activity			Comments	Reference
		Protein (%)	Lipid (%)	Carbohydrate (%)	5'-Nucleotidase (EC 3.1.3.5)	Na^+K^+ ATPase	Mg^{2+} ATPase (EC 3.6.1.4)		
					μmol substrate hydrolysed/mg protein/h.				
Methods utilizing stabilizing surface structures									
Rat liver	Cells homogenized and membrane isolated from "nuclear" fraction by flotation on sucrose solutions	ND	ND	ND	ND	—	—	Identification from morphological criteria	Neville (1960)
Rat liver	Modification of above technique	55[a]	ND	1·2[a][b]	32·2 ± 6·8	11·6 ± 1·6	44·2 ± 6·4	Identification based on morphological criteria and immunological techniques	Emmelot et al. (1964)
Rat liver	Modifications of above method. Isotonic medium used in place of hypotonic bicarbonate	47·5	34·5	1·1[b]	ND	ND	ND	Phase contrast and electron microscopy performed	Takeuchi and Terayama (1965)
Vesicular preparation									
Ehrlich ascites carcinoma cells	Nitrogen cavitation. Vesicular membrane fragments isolated from "microsomal" fraction by Ficoll density gradient techniques	ND	ND	ND	—	7·95	—	Immunologic and enzyme markers used to monitor separation procedures	Kamat and Wallach (1965)

Method involving chemically induced stabilization

Method								Reference
Mouse fibroblasts	Chemical stabilization of plasma membrane. Followed by homogenization and isolation of sheets of membranes by sucrose density gradient techniques	ND	ND	ND	ND	0.7 – 3.0(c)	1.1 – 3.0(c)	Identification based on morphological criteria	Warren et al. (1966) Warren et al. (1967)
Other methods									
HeLa cells	Cells ruptured in a Dounce homogenizer. Membrane prepared from 4,000 g supernatant fluid by discontinuous sucrose gradient techniques	60	40	1.5(b)	35.2	35.8	—	Morphological and enzymatic characterization	Bosmann et al. (1968)
Pig mesenteric lymph node cells	Tissue minced and membrane fragments isolated by differential centrifugation and gradient techniques	57 ±1	42 ±1	—	10.1±1.5	—	—	Electron microscopy and immunological techniques employed in characterization	Allan and Crumpton (1970)
Murine Meth A ascites tumour cells	Ghosts produced by extracting cells with EDTA/borate. Isolated by differential centrifugation techniques	42·5	40	4·5(d)	ND	ND	ND	Method monitored by phase contrast and electron microscopy	McCollester (1970)

(a) Details abstracted from original paper and data quoted by Winzler (1970).
(b) Sialic acid and hexosamine only.
(c) Range varies depending on method of chemical stabilization used.
(d) Includes CHCl$_3$ insoluble and soluble carbohydrate (as glucose).
ND = Not determined.

own right. A major question, from the point of view of both membrane structure and function, is whether membrane proteins fulfil primarily an inert structural, or a catalytic role. It is certainly well established that enzymatic proteins are an important constituent of membranes, though it is not entirely clear what percentage of the protein is enzymatic and to what extent proteins are involved in a structural capacity. The concept of "structural protein" in membranes has been widely discussed and owes much of its foundation to Green et al. (1961), whose subunit theory of plasma membrane structure (Green, et al., 1967) has already been discussed earlier in this chapter. The basis of the theory is that lipoprotein subunits, in the form of parallelpipeds, interact at the faces which do not contain lipid to form a membrane one subunit in thickness. When lipid is absent the subunits can interact at all six faces leading to the formation of amorphous aggregates. According to this concept, membrane-forming protein must be capable of polymerization into three-dimensional aggregates, of interacting asymmetrically with phospholipid and, when so coated, of binding hydrophobically with other such units to form a two-dimensional membranous sheet. Originally an enzymatically inactive, apparently homogenous protein of 20,000–30,000 molecular weight was isolated by Green et al. (1961) from beef heart mitochondria and apparently accounted for up to 50–75 % of the total particulate protein of this organelle. Following on from this work, "structural protein" similar in properties to the above-mentioned material has been isolated from a wide variety of membrane systems supporting the concept of related structural proteins common to various types of membranes. However, in 1968 Green and his group (Green et al., 1968) demonstrated that standard preparations of structural protein were heterogenous as analysed by acrylamide gel electrophoresis. Indeed, with the means of obtaining well defined plasma membrane fractions from a variety of cell types, it is now possible to make a comparative study of plasma membrane proteins, bearing in mind that there is the possibility that the results obtained may be artefact of the solubilization procedure (see Cook, 1971). It has been pointed out by Winzler (1970) that a number of careful studies on solubilized membrane proteins have now been carried out and an exceedingly heterogenous group of some twenty different proteins is seen. Winzler (1970) has stated that the concept of a single major structural protein in any cell type seems very doubtful at the present. However, he points out that there are strong indications in isolated membrane proteins of homology: for example lipophilic amino acid content solubility in detergents and ability to bind phospholipids. Although it appears unlikely that there is a single major structural protein, even for a single cell type, homology of structure and properties would suggest that several of the proteins are able to form associations with plasma membrane lipids.

Using isolation procedures already described, various studies on the characterization of membrane proteins using principally disc electrophoretic techniques have been reported. Working with the surface membrane of rat liver, for example, Neville (1968) has isolated, by preparative disc gel electrophoresis of alkali extracts of plasma membrane, an organ specific single protein with a minimal weight of 70,000 that accounts for about 10% of the total membrane protein. Neville (1968) believes that his protein, which is localized at the cell surface, may be important in recognition phenomena, with each cell type having a characteristic protein at the cell surface for cellular recognition purposes. Such a protein Neville (1968) refers to as an "eigen protein". More recently, Neville and Glossman (1971) have reported on a comparative disc gel electrophoretic study of the proteins of well defined plasma membrane fractions from rat liver, kidney and erythrocytes, compared with liver and kidney mitochondria together with liver smooth endoplasmic reticulum. Although each membrane displayed its own unique protein subunit pattern in sodium dodecyl sulphate (SDS) gel electrophoresis, they report on the presence of a common band of molecular weight 48,000 in the plasma membranes of kidney and liver and to a less prominent extent in the erythrocyte stroma. In the liver and kidney this subunit represents 3–10% of the total membrane protein and Neville and Glossman (1971) suggest that the role of this component is primarily structural. Further, these authors point out that though their results support the idea of a common structural protein for plasma membranes, the failure to find this subunit in other membrane systems (mitochondria and smooth endoplasmic reticulum) makes it seem unlikely that there exists a structural protein common to all membrane systems. Certainly, as this work demonstrates, there is no reason to suppose that protein cannot have a structural role in membranes, though not necessarily along the lines originally proposed for "structural protein". Indeed an earlier and detailed study by Kiehn and Holland (1970) on membrane proteins of various mammalian cells, in particular of murine tissues, is of interest here. Kiehn and Holland (1970), while not ruling out the possibility that non-catalytic proteins could play a significant role in membrane structure, did, in agreement with the work discussed above, show that no single protein or small group of proteins make up the bulk of membrane structure in cells.

c. Carbohydrates

A growing awareness of the role of carbohydrate in membrane structure is reflected in the increased attention given to the analysis for carbohydrates in preparations of this organelle. It is not intended to catalogue the various results which are now being reported in the literature (some are listed in the accompanying tables), but rather to concentrate on a few representative

Table II　Carbohydrates in Plasma Membranes of Animal Cells

Gross analysis:

Source	Neutral sugar (%)	Hexosamine (%)	Sialic acid (%)	Reference
Rat liver		1	0·1	Takeuchi and Terayama (1965)
HeLa cells		1·1	0·3	Bosmann et al. (1968)
Pig lymphocytes	3·4		0·6	Allan and Crumpton (1970)
Meth A tumour cells	4·5			McCollester (1970)

Some more detailed analysis of constituent sugars of plasma membranes:

Source	Fucose	Mannose	Galactose [nmol/mg protein]	Hexosamine	Sialic acid	Reference
Mouse fibroblasts (L-cells)						
Zn^{2+} method[a]	7·6±7·1	23·7±6·6	59·4±8·2	53·5±12·6	15·0±2·2	Glick et al. (1970)
FMA method[a]	4·7±3·6	40·0±12·3	83·6±23·3	35·7±11·7	24·3±1·2	
Rat ascites hepatoma (AH 7974F)						
	33·6	57·2	97·8	52·2 (D-GalNH$_2$) 59·2 (D-GluNH$_2$)	28·8	Shimizu and
Tris method[a]	16·1[b]	57·2[b]	43·2[b]	19·9 (D-GalNH$_2$)[b] 52·2 (D-GluNH$_2$)[b]	20·7[b]	Funakoshi (1970)

[a] Methods after Warren et al. (1966).

[b] Figures given in lower rank are for analysis performed on lipid-free membranes, as opposed to intact membranes (upper rank). The analysis for glucose is 56·4 nmol/mg protein (intact membranes) and 11·4 nmol/mg protein (lipid-free membranes).

papers which have been specifically concerned with the carbohydrate composition of plasma membranes, and which may serve to indicate how such studies can give a wider insight into the role of carbohydrates in cell membranes.

The special case of the carbohydrate composition and structure of the erythrocyte will be detailed in Chapter 3. One of the first examples of a plasma membrane to be isolated from a non-erythroid cell was the liver cell plasma membrane and in the earlier papers on the isolation of this membrane (Emmelot *et al.*, 1964; Takeuchi and Terayama, 1965) some data for the sialic acid and hexosamine content were given. More recently (Emmelot and Bos, 1972), the sialic acid containing materials present in preparations of the plasma membranes of rat and mouse liver and hepatomas have been further described. In the case of rat liver and hepatoma membranes, Emmelot and Bos (1972) found that the sialic acid content of the preparation depended on the homogenization medium used; generally membranes prepared with media containing citric acid possessed reduced amounts of sialic acid compared with material prepared using a bicarbonate-containing medium. The latter medium suffered from the disadvantage that with some tumours it caused nuclear disruption and the resulting nucleohistone gel makes membrane isolation difficult. Although citric acid causes a reduction in sialic acid content of normal liver (from 33 nmol/mg to 19 nmol/mg protein) and hepatoma (hepatoma−484, 45 nmol/mg reduced to 34 mmol/mg protein) plasma membranes, the relative decrease being less marked in the case of the hepatomas, the sialic acid to protein content was greater in the case of the hepatomas. The difference, using bicarbonate medium, in sialic acid to protein content in the case of mouse hepatomas and normal plasma membranes was less marked and was of a similar order to that found in normal rat liver plasma membrane. The sensitivity to neuraminidase of the sialic acid-containing material in these various plasma membrane preparations within the species was similar, with the exception of plasma membranes of mouse hepatoma 147042, which appears to have a decreased sensitivity towards the enzyme when compared to other murine membranes. Following neuraminidase treatment and a post enzyme wash with bicarbonate, further material containing sialic acid could be solubilized by trypsin treatment. In these studies it is surprising that a proportion of the sialic acid remained resistant to solubilization by sequential treatments with neuraminidase (which releases the free acid) and trypsin (sialic acid released in bound form), especially as the authors demonstrate that some 95 % of the sialic acid and hexosamine in the case of rat liver membranes is present as glycoprotein rather than glycolipid. Emmelot and Bos (1972) measure the resistance to neuraminidase under conditions in which all the sialic acid of the human erythrocyte is sensitive to this enzyme. It would not be unreasonable to assume

3C

that the conditions of enzyme treatment found to be optimal for one membrane type may not necessarily be optimal for another cell.

The finding by Emmelot and Bos (1972) that the gel profiles of sialoglycopeptides released by trypsin from rat liver and hepatoma plasma membranes were quite different is of special note. Indeed analytical data, expressed in terms of the content of a particular sugar (in the example quoted above, sialic acid) may only serve to indicate the presence of this class of molecule in the membrane. The identification of differences between "normal" and "malignant" from larger fragments, such as glycopeptides, serves to heighten the usefulness of analytical data performed on isolated membrane preparations (see Chapter 6). Chemical differences such as may be evident in the glycopeptides obtained from purified membranes may be of significance when one attempts an explanation as to the changes in the cell surface in respect to the neoplastic character of a cell.

The analytical data described by Emmelot and Bos (1972) was confined to sialic acid and hexosamine though they refer to unpublished experiments where, using thin layer chromatography, glucose, galactose and mannose have been demonstrated to be present in their various membrane preparations. In this respect the findings of Shimizu and Funakoshi (1970) on the carbohydrate composition of plasma membranes of the AH 7974F rat ascites hepatoma, prepared by the technique of Warren et al. (1966) is interesting. These authors (Shimizu and Funakoshi, 1970) showed that the greater part of the glucose and about half the fucose, galactose and galactosamine were present as glycolipid, while the same sugars together with mannose and glucosamine were also present as glycoprotein. Though these authors contrast their results with those of others (such as the results described above) and note that "the carbohydrate composition of glycoproteins in plasma membranes seems to differ from one cell to another", they point out the usefulness of structural studies on glycopeptides. It cannot be stressed too greatly that studies detailing carbohydrate composition of isolated plasma membranes, while important in demonstrating the presence of these molecules in this organelle, may not alone readily aid one in arriving at the functional significance of these materials in cellular interaction phenomenon and that the use of larger fragments may be more profitable.

Turning to studies on plasma membranes prepared from cells other than liver, Glick et al. (1970) have published a very detailed account of the carbohydrate composition of plasma membranes (Warren et al., 1966) from mouse L-cell fibroblasts. They find that the average amounts of sialic acid, hexosamine, galactose and fucose were considerably less in membranes prepared using fluorescein mercuric acetate (FMA) than in surface membranes prepared by their Zn^{2+} procedure. However, membranes prepared by the latter procedure contain more of the cell protein than those prepared with the fluorescein mercuric acetate technique. Hence when the results are expressed per mg of protein the FMA membranes contain, with the exceptions of hexosamine and fucose, a higher concentration of carbohydrate. The authors raise the possibility that there may be a loss of glycoproteins in the FMA treated membranes. They also suggest that the presence of mannose at approximately the same levels in both membrane preparations may indicate that this sugar is part of a "background structure", while some of the other glycoproteins are more loosely associated with the surface membrane. Certainly these results serve to illustrate how the method of isolation adopted may well influence the analytical data; reference has already been made to this problem in the case of the human erythrocyte and liver membranes, and the data of Glick et al. (1970, again amplifies this problem.

In the studies reported by Glick et al. (1970) they found that the sialic acid content of the L-cell varies considerably and this can apparently be linked to rate of growth of the culture. Variations within the whole cell were also reflected in the individual membrane preparations (made by Zn^{2+} method) suggestive of a fluctuation in the amount of membrane glycosubstances. In common with some other authors (see above) Glick et al. (1970) found that about one quarter to one third of the sialic acid of the membranes was not susceptible to neuraminidase and, in contrast to the results with liver membranes, could be correlated with the presence of 21–24% of the membrane sialic acids being present as glycolipid (Weinstein et al., 1970). Since haematoside and disialoganglioside in the extracted state are susceptible to neuraminidase it is possible that the glycolipids are in some way shielded from the enzyme.

Besides the problem associated with the fluctuation in analytical values obtained with different membrane preparations, the variation in the amounts of glycosubstances of the intact cell and the isolated membranes show how interpretation of biological behaviour based on such chemical analysis must only be made with great caution. In those cases where tissue culture cells are used for the preparation of plasma membranes for analytical studies rigidly controlled conditions of culture must be maintained.

IV. Modern Models of Membrane Structure

It will be apparent from the preceding sections that analytical evidence demonstrates that biological membranes consist not only of lipid and protein, but also of significant quantities of carbohydrate.

With the increased awareness of the importance of protein in membranes several workers have turned their attention to a consideration of the organization of proteins in relation to lipids in membranes. These ideas have resulted in substantially similar modifications to the Danielli–Davson (1934) models being postulated by both Wallach and Zahler (1966) and

Lenard and Singer (1966). Using the techniques of optical rotary dis-
persion and circular dichroism to investigate the protein conformations in
various membrane preparations, including human erythrocyte stroma and
fragments of plasma membranes of Ehrlich ascites carcinoma cells, these
workers derived models which take account of the evidence that a portion
of the membrane protein is present in helical conformation as opposed
to random coil. Wallach and Zahler (1966) suggested that the hydrophilic
portions of membrane polypeptide are located on both surfaces of the
membrane and are joined by hydrophobic rods which cross the apolar
core of the membrane. The material at the surface is considered to be in
the form of an irregular coil while that portion of the protein which penetrates the width of the membrane is predominantly in the form of helical
rods. These authors point out that the hydrophobic units could be single
or they could form aggregates. In the latter case it is envisaged that the
formation of a "microtubule" is a possibility. Such a microtubule, which
might have a polar interior, could be of importance in membrane transport
phenomena. Lenard and Singer (1966) also suggest similar ideas of membrane structure and in one of their models consider the situation where the

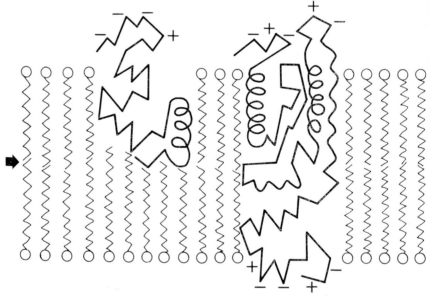

Fig. 10. The fluid mosaic model of the structure of cell membranes. This schematic
cross-sectional view of a biological membrane, redrawn from the scheme of
Singer and Nicolson (1972) depicts a discontinuous bilayer of lipid in which are
embedded the integral globular protein molecules. The extent to which the
integral proteins are embedded will depend on the size and structure of the
molecules. The arrow on the left of the diagram indicates the expected plane of
cleavage in freeze-etching experiments.

helical polypeptide chains traverse the membrane. They were careful however, to point out there was no evidence bearing on this idea; such is no longer the case (see below). These modifications to the Danielli–Davson (1934) model have been dealt with here because it is from these models that the fluid mosaic model of Singer and Nicolson (1972) has evolved. This structure is particularly interesting because the model consisting of a mosaic of alternating globular proteins and phospholipid bilayer introduces the concept of a fluid or dynamic nature of membranes. The dynamic aspects of membranes are so often overlooked, and the classical models only tend to emphasize a structure of a static nature which is inconsistent with so many of the biological properties of membranes. Nevertheless, it is perhaps understandable (see Cook, 1971) that the classical models embodied a static nature for membranes when it is considered that so many of the physical techniques that have been used to examine membrane structure involved chemical fixation procedures, prior to the making of the observations. As well as drawing on recent experimental evidence, Singer and Nicolson (1972) present several theoretical considerations to support their contentions and point out that the classical model of a plasma membrane (consisting of a lipid bilayer sandwiched between two monolayers of protein) is thermodynamically unstable. In the classical model, the membrane protein is shown to have non-polar amino acid residues exposed to water and the ionic and polar groups of the lipid are sequestered from contact with water by the layer of protein. However, in such a model neither hydrophobic nor hydrophilic interactions are maximized (Singer and Nicolson, 1972).

In addition these authors are careful to distinguish between what they term "peripheral" and "integral" proteins of membranes. The former type of proteins are categorized as those which require only mild treatments, such as increase in ionic strength or use of chelating agents, to dissociate the intact molecule from the rest of the membrane. These "peripheral" proteins, examples of which are cytochrome C of mitochondrial membranes and spectrin of erythrocyte membranes, are held only to the membrane by weak non-covalent interactions and, as opposed to the "integral" proteins, are not strongly associated with the lipids of the membrane. The "integral" proteins, which constitute the major protein within the membrane, require much more drastic treatments to dissociate them from the membrane than the "peripheral" proteins. The separation of membrane proteins into two classes is believed to be particularly important when considering membrane structure, for as Singer and Nicolson (1972) point out the "peripheral" proteins may not be directly relevant to the structural question. This is not to say that "integral" protein may be correlated with the "structural" protein concept of Green et al. (1961) where membranes are considered to possess a predominant protein which fulfills a purely structural role. Indeed, more recent evidence indicates that structural protein preparations are heterogeneous when examined by acrylamide gel electrophoresis (for review see Cook, 1971) which rules against the "structural" protein concept. In view of the preponderance of "integral" proteins in membranes, the appreciable amount of helical polypeptide shown to be associated with membrane proteins are assumed to apply directly to "integral" protein (Singer and Nicolson, 1972). Further, as most soluble globular proteins exhibit less helical character in their native structure, it is suggested that the "integral" proteins of the cell membrane are largely globular in shape and are not spread out as monolayers. In sum, the fluid mosaic model considers these globular "integral" proteins to be present in a matrix of lipid, the latter being in the form of a bilayer. The idea of lipids being in a bilayer, Singer and Nicolson (1972) point out, is in accord with various physical data, such as X-ray diffraction, which favours such an orientation. However, from such data it is not possible to deduce whether such a bilayer is continuous or whether it could be interpreted in the way they suggest, or, indeed, whether the lipid is all in a bilayer form. Some of the lipids may be in the form of globular micelles (Lucy, 1968), and it is suggested that transitions between a micellar arrangement and the bimolecular leaflet are important in both physiological and pathological phenomena. Singer and Nicolson (1972) consider it very unlikely that globular protein could be present on the outer surface of the bilayer because the thickness of the membrane would be so much greater than is perceived experimentally. It is suggested that some small proportion of the lipid is strongly bound to protein while the major portion is in the form of a bilayer and not strongly interacting with protein. In support of the suggestion that there is a close association of some lipid with membrane protein, Singer and Nicolson (1972) draw attention to the fact that many membrane enzymes require specific phospholipids for the expression of their activities. In addition, purified preparations of phospholipase C have a profound effect on the lipids of membranes without having any effect on the conformation of the proteins. Glaser et al. (1970) studied the action of purified phospholipase C preparations on human erythrocyte membranes both by chemical means and by the techniques of circular dichroism and proton magnetic resonance. They found that while the lipids are profoundly affected by this treatment, circular dichroism indicates that the conformation of the membrane protein is unaffected. This leads to the conclusion that lipid and protein are able to change their structure independently of one another. Further evidence, not quoted by Singer and Nicolson but which supports their ideas of membrane structure, has come from the studies of Ottolenghi and Bowman (1970) who studied the action of phospholipase C on rat kidney mitochondria and human erythrocytes by means of interference and electron microscopy.

They found discrete areas of digestion which did not coalesce, and this, they suggest, is compatable with a membrane model in which protein does not cover the entire phospholipid surface.

The globular protein units are considered as being amphipathic, with their highly polar region being associated with the aqueous phase, while the non-polar region of the molecule is embedded in the hydrophobic region of the membrane. A protein of sufficient size and structure could traverse the width of the membrane. In contrast to the integral proteins, it is unlikely that peripheral proteins will be amphipathic.

A study by Randall *et al.* (1972) throws light upon the distinction between "peripheral" and "integral" proteins and their relation to the surface carbohydrates of the erythrocyte. Various fractions of protein and lipo-protein were isolated from human erythrocyte ghosts by sequential washing, and their ability to interact with benzyl alcohol was followed both by nuclear magnetic resonance and by fluorescence-quenching. The binding of benzyl alcohol to the intact ghost is of a characteristic form and it is believed that the sites on the membrane protein that interact with the alcohol are intimately connected with lipid-protein interactions. These authors (Randall *et al.*, 1972) found that a great deal of protein could be removed by relatively mild washing and that this showed no binding of benzyl alcohol. A residual fraction retained all the binding sites and also all the sialic acid originally present in the ghosts. None of the fractions was a single protein, but each was simpler than the whole ghost, indicating considerable fractionation.

The proposals of Singer and Nicolson (1972) have been dealt with at some length because, although they do not go into the finer details of membrane structure, their proposals on the gross arrangement of lipids and proteins, including glycoprotein, within a cell membrane represent an attempt to evolve a general model which lays emphasis on the dynamic aspects of biological membranes. What other additional evidence is there in favour of such a model? Evidence for proteins' being present in the cell membrane, independent of the major portion of the lipid, has already been discussed, but the possibility that protein is embedded within the lipid of the plasmalemma is equally relevant to a fluid model of membrane structure. Freeze-etching in particular has provided detailed information on the interior of membranes and the results of studies on red cells with this technique are particularly pertinent to the problem of whether protein is present within the membrane. In the freeze-etching technique a frozen specimen is fractured with a microtome knife, the surface can then be shadowed with metal and the replica examined in the electron microscope. With biological membranes the fracture plane is along the interior of the membrane splitting the membrane approximately in two halves. The true outer surface of the membrane may be exposed by a process of deep etching in which some of the ice is sublimed away from the specimen before shadowing. It is appropriate to mention here that the contention that the fracture plane is within the membrane was not without some degree of controversy; Moor and Mühlethaler (1963) suggested that the outer and inner surfaces of the membrane are revealed, while Branton (1966) argued that the cleavage takes place along the interior of the membrane. This problem was resolved by a number of elegant experiments. For example, Pinto da Silva and Branton (1970) covalently linked ferritin to both sides of the membrane of erythrocyte ghosts and then demonstrated that ferritin was never observed on the fracture faces. Ferritin could, however, be demonstrated on the surface of the membrane following deep etching. In a similar study, in which the human erythrocyte ghost was labelled with fibrous actin, Tillack and Marchesi (1970) confirmed that cleavage takes place within the membrane.

An examination of the fractures exposed by the cleavage shows them to consist of numerous globular structures (4,000 μ^2 or 500,000 per erythrocyte: see Marchesi et al., 1971) of uniform size distributed evenly over the surface of the cell and separated by a smooth matrix. This picture would certainly add weight to the fluid mosaic model assuming that the globular units could be demonstrated to be protein. A number of experiments have now been performed which do indeed indicate that these structures are associated with protein, and what is particularly interesting is that there is evidence that glycoproteins are associated with these intramembranous particles. When the isolated erythrocyte membranes are treated with the proteolytic enzymes trypsin and pronase, treatments which are known to degrade the surface glycoproteins of the human red cell (Cook et al., 1960; Cook and Eylar, 1965; see Chapter 2), these globular particles are seen to clump into irregular patterns, although the mechanism of this clumping is not entirely clear. Marchesi et al. (1971) suggest that the particles are normally anchored to specific sites in the membrane, perhaps to other proteins which are affected by the proteolytic enzymes. These latter authors have explored the relationship of these particles to glycoproteins by a series of labelling and mapping techniques with or without prior treatment of the membrane with trypsin. Labelling of the glycoproteins was achieved using either a conjugate of ferritin and phytohaemagglutinin, a lectin which is known to react with glycoproteins of the membrane (see Chapters 2 and 3), or purified influenza viruses which react with the terminal sialic acid residues of the surface glycoproteins (see Chapter 2). Marchesi et al. (1971) showed that both these labels were distributed evenly over the cell surface in the same general pattern as adopted by the intramembrane globular particles, while after treatment with trypsin, a treatment which only removes some of the phytohaemagglutinin binding sites and half the membrane sialic acid, the clustering of the intra-

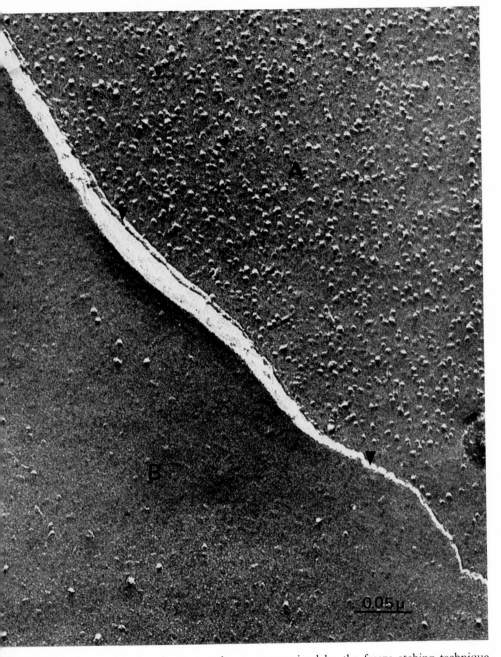

Fig. 11. The appearance of a plasma membrane as examined by the freeze-etching technique illustrated in this figure of a replica of fibroblast like variant line from Balb/c mouse plasmocytoma MOPC 173. Large fracture surfaces of the plasma membranes of two adjoining cells are exposed. The arrow heads are pointing to the intercellular space. The outwardly direct fracture face A of the inner leaflet of one plasma membrane is characterized by numerous globular particles. The inwardly directed fracture face B of the outer leaflet of the adjoining plasma membrane shows very few particles. (By courtesy of Dr. E. L. Benedetti.)

membrane particles is also reflected in the distribution of the mapping particles. That this is not purely coincidence is suggested from studies with the virus labelling technique where imprints caused at the membrane binding site are found to correspond with the regions containing the clumped particles. On the basis of this evidence it is tempting to speculate that while the membrane glycoproteins are associated with the globular particles visualized by the freeze-etching technique, it is unlikely on size considerations that the hydrophobic portion of the red cell glycoproteins are responsible for all the material of the globular particles. The latter point is readily acknowledged by Marchesi and his colleagues (Marchesi *et al.*, 1971), however, they stress that the exact size of the particles is difficult to determine accurately in view of the layer of platinum deposited at the replica stage.

In addition to the freeze-etching technique, there is a growing body of chemical data which would suggest that membrane glycoprotein is not only embedded in the matrix of the membrane but naturally spans the width of the membrane. The concept of membrane glycoprotein being embedded in the membrane is not entirely new, for in 1964 Morawiecki drew attention to the presence of hydrophilic and hydrophobic regions of the MN blood group glycoproteins. It was suggested that the hydrophilic portion is oriented in the extra-cellular space, while the hydrophobic region is embedded in the lipid. Bretscher (1971a) has developed formyl methionyl sulphone methyl phosphate as a chemical reagent for the labelling of membrane surfaces (see above). By use of this reagent with intact human erythrocytes, Bretscher (1971b) was able to demonstrate that two proteins are exposed on the outer surface of the human red cell surface, of 105,000 and 90,000 molecular weight, according to studies by way of SDS gel electrophoresis. The latter protein carried most of the cell surface sialic acid and carbohydrate. As will be described in the following chapter, a decade earlier to this work the technique of cell electrophoresis had established the presence of sialic acid-containing carbohydrate moieties, linked to protein, at the surface of the erythrocyte. In an accompanying paper to that of Bretscher (1971a), Bender *et al.* (1971) came to some very interesting and similar conclusions as to the accessibility of protein in the intact human erythrocyte membrane with respect to this cell's surface. These latter authors made use of the wide specificity protease from *Streptomyces griseus*, pronase, to investigate surface availability of membrane proteins. It is known from the work of Cook and Eylar (1965) that this enzyme is capable of removing all the sialic acid present in this cell in the form of glycopeptides. Bender *et al.* (1971) demonstrated, by means of SDS gel electrophoresis, that approximately one-sixth of the protein of the human red cell moves with a molecular weight of about 125,000, and that at least one of the components in this region contains sialic acid. These authors further showed that nearly all the protein in this region of the gel is sensitive to pronase, which considerably reduces the size of the native protein and leaves residues of 50,000 to 100,000 molecular weight behind in the membrane. These fragments are protected from the action of the enzyme in the intact cell. However, in the case of the erythrocyte ghost all the protein is degradable which it was suggested indicated that the fragments were protected by other membrane components such as lipid, or they could go all the way through the membrane, though at the time no evidence was available on the latter point. Of considerable importance in this type of work is the criticism that in the process of cell lysis to produce the red cell ghost some reorientation or unfolding of membrane material takes place; to date no entirely satisfactory answer to this problem has been produced. In addition to the results of Bender *et al.* (1971), work performed by Phillips and Morrison (1970) which shows that proteins of the red cell membrane in the 90,000 molecular weight class are exposed to the cell surface and can be catalytically iodinated by the non-penetrating enzyme lacto-peroxidase is of particular interest. These authors (Phillips and Morrison, 1971) showed that three membrane proteins of molecular weights 90,000, 95,000 and 105,000 were sensitive to pronase digestion and the fragments formed which remain with the membrane, where they are inert to further enzymatic hydrolysis, are of molecular weights 65,000. Certainly all these results not only indicate that at least two proteins, including the sialo-glycoprotein, of the human erythrocyte are present at the surface of this cell, but would also indicate that portions of the membrane, insensitive to proteolytic degradation, are buried within the membrane structure. Admittedly, there is some variation amongst the different authors in the reported molecular weights of the various components. Wallach (1972) points out, for example, that the molecular weight estimates of Bender *et al.* (1971) are probably high as these authors used SDS-polyacrylamide gel electrophoresis without disulphide reduction, while Bretscher (1971b) draws attention to the anomalous mobility of glyco-proteins in SDS gels. However, a general consensus emerges from these reports that parts of the red cell membrane proteins are exposed and that others are buried within the membrane structure. To what extent the proteins of the human erythrocyte membrane is buried within the membrane has been examined by Bretscher (1971a), who also made use of the fact that pronase digestion of the intact erythrocyte removed much of the surface carbohydrate of this cell (Cook and Eylar, 1965). Bretscher (1971a, b) found that after the pronase digestion of intact cells the poly-peptide material of molecular weight 105,000 is replaced by a component of 70,000. In addition, pronase action prevented labelling of the glyco-protein with formylmethionyl sulphone methylphosphate. In the intact cell label is apparently only attached to the "glyco" part of the glycoprotein since the radioactivity in finger prints of such material is found as a "smear around the origin". The protein of 105,000 molecular weight is labelled to a greater extent when present in the erythrocyte ghosts. Bretscher (1971b) suggests this is caused by his reagent's labelling, in addition, those parts of the protein which are present on the inside surface at the membrane. In addition, the polypeptide of the glycoprotein is only labelled when ghosts are used. Indeed, comparison of the finger prints of material from the glycoprotein regions of gels of labelled ghosts of untreated and labelled ghosts of pronase digested erythrocytes, revealed that two peptides were deleted in the latter case. Bretscher's interpretation of this result is that these two peptides are on the interior side of the membrane, which it is not possible to label in the intact cell. That they can be removed from the outer surface with pronase leads him to suggest that the peptides are attached to a protein, of which part is on the outer surface of the cell. Certainly, finger prints of purified glycoprotein derived from stroma labelled on "both sides" shows these two peptides to be the principal labelled components apart from a "smear around the origin". The view that glycoprotein spans the width of the membrane is certainly a challenging idea and adds support to the dynamic fluid mosaic model of membrane structure. Nevertheless, the question of reorientation within the membrane when ghosts are prepared cannot be overlooked. It is possible that the two peptides, considered by Bretscher (1971b) to be derived from the interior surface, may be made more available to the reagent by the lytic process without necessarily being exposed on the cytoplasmic side of the cell membrane.

With regard to components present on the cytoplasmic sides of plasma membranes, the studies of Wallach and his colleagues (Steck et al., 1970) on "right-side-out" and "inside-out" vesicular preparations of human erythrocyte membranes are of particular importance. In 1970 Steck et al. published a method in which erythrocyte ghosts were encouraged to undergo spontaneous endocytosis, or budding of the plasma membrane into the interior of the ghost, with the accumulation of a number of "inside-out" vesicles in the parent ghost. These inverted vesicles could be liberated by gentle homogenization. In the case of ghosts kept in the presence of divalent cations the membranes were found to be stabilized against spontaneous endocytosis. On homogenization such preparations were found to vesiculate largely by a process of exocytosis, that is to say by a process whereby the membrane buds off in the extracellular surroundings. Such vesicles are present as "right-side-out" vesicles having retained the orientation of the parent membrane. The two types of vesicles could be separated from each other by means of equilibrium dextran density-gradient centrifugation. Vesicles formed in the absence of divalent ions, "inside-out" vesicles, were present in the upper low-density zone of the gradient, presumably as a result of the internalizing of the excess negative charge which causes the formation of a vesicle of lower density than the parent membrane. These two preparations of vesicles were characterized chemically by treating the populations with the enzymes neuraminidase and trypsin. It was reasoned that in the case of "right-side-out" preparations the sialoglycoprotein would be accessible to the action of these enzymes in the same way as the surface of the normal intact cell (Cook et al., 1960; Cook et al., 1961; see also Chapter 2). In contrast, the sialic acid residues localized on the external surface of the cell (Eylar et al., 1962; Cook and Eylar, 1965) would be expected to be shielded from the action of these enzymes, in the case of inside-out preparations. This reasoning was verified by the actual analytical results which showed that only some 15 % and 12 % of the sialic acid was released in the free and bound form by neuraminidase and trypsin respectively, from the inside-out vesicles though their specific sialic acid content was almost identical to the vesicles from the denser regions of the gradient. These latter membrane fragments were considerably more sensitive to hydrolysis by both enzymes, though it should be pointed out that the quantities of sialic acid released are not in exact agreement with the amounts released by the two enzymes for intact cells. The percentage of sialic acid released from that available in these ghosts by neuraminidase is 79 %, while in the intact cell this enzyme will release nearly 100 % of the sialic acid present in the membrane. In the case of the proteolytic enzyme trypsin Steck et al. (1970) found that 73 % of the available sialic acid was released whilst it has been found in the intact cell that this enzyme is only capable of releasing some 50 % of the available sialic acid. These results might indicate, as Steck et al. (1970) point out, that, apart from the problem of cross contamination of the preparations, there may be small amounts of sialic acid in the "inner aspect" of the erythrocyte membrane or partial permeability of the vesicles to the enzymes. It would seem as though the second alternative is the more likely explanation, especially as morphological data from freeze-etching experiments indicates a greater degree of purity than does the analytical data. This again draws attention to the point made earlier, that when preparing ghosts the possibility that membrane constituents undergo some degree of reorientation cannot be entirely dismissed. Nevertheless, the availability of these two populations of vesicular membranes does give a unique opportunity to examine the cytoplasmic side of a cell membrane. Steck et al. (1971) found that whole ghosts and normally orientated right-side-out vesicles were more susceptible to digestion with proteolytic enzymes, as monitored by the production of protons, than the inside-out vesicles. One protein component was found to be resistant to proteolytic degradation, while all the other major proteins were found to be digested at the outer surface, but only one was susceptible from the inside-out configuration. With regard to the sialoglycoproteins, these authors report (Fairbanks *et al.*, 1971) three sialoglycoproteins in their gels, which they found were susceptible to proteolytic attack from either side. These authors therefore concluded that the membrane of the red cell has a highly asymmetric arrangement of orientated proteins, some of which appear to traverse the width of the membrane. This result would seem to be in agreement with the ideas of other authors, though for reasons explained above the extrapolation of data obtained from stroma to that on intact membranes must be made with due caution.

Certainly these studies show the importance of proteins in membrane structure and add weight to the contention that their conformation may well be contrary to that indicated in the classical models of plasma membrane structure. The extention of the techniques described for investigating protein distribution within erythrocyte stroma to the surface membranes of tissue cells (see Poduslo *et al.*, 1972) will be of particular importance in ascertaining how far the ideas obtained from red blood cells may be extended to other cell types.

V. Relation of Ideas on Membrane Structure to the Understanding of the Cell Surface

The purpose of studying the molecular structure of cell membranes must surely be to interpret biological phenomena in molecular terms. A considerable number of biological events take place at the cell surface so a knowledge of the chemistry of the cell membrane, and consequently its periphery, is particularly desirable when attempting an explanation of several fundamental cellular events.

Classical models of membrane structure, as it has been observed, completely ignore the presence of carbohydrates and as such are quite unsuitable models for explaining a number of important cell surface phenomena. On the other hand, the fluid mosaic model does take account of the presence of glycosubstances in the plasmalemma. Chemical evidence in favour of this model has been described in the preceding section. In their original article Singer and Nicolson (1972) provide additional biological evidence which they suggest substantiates the notion that the membrane proteins are in a fluid state in the intact membrane. Of the additional evidence, those pieces of work which relate directly to plasma membranes will be dealt with in this section, not only because they provide further evidence in favour of the fluid mosaic model, but also because such evidence draws attention to the need for a model which considers cell membranes in dynamic terms, rather than the static quality of the classical models. Predictions of the mechanisms underlying many biological surface phenomena are certainly more easily understood if one can regard the plasmalemma as a dynamic entity.

Undoubtedly the experiments of Frye and Edidin (1970) on the rapid intermixing of surface antigens following the formation of mouse–human heterokaryons are of prime importance in drawing attention to the fluid aspects of plasma membrane structure. In a series of very elegant experiments Frye and Edidin (1970) used the indirect fluorescent antibody technique to follow the distributi⋯ ᴼᶠ surface antigens in heterokaryons formed between a thymidine–kinase negative (TK-) subline of the mouse L-cell, cllD, and VA-2 cells, with Sendai virus as the fusing agent. The VA-2 cells are a line of human origin being an 8-azaguanine-resistant sub-clone obtained from W-18-VA-2 cells, an SV40 virus transformed human line which has been free of infective virus for a number of years. The distribution in the heterokaryons of mouse, H-2, histocompatibility antigens and human antigens was followed by treatment with a mixture of fluorescein-labelled goat anti-mouse IgG and tetramethylrhodamine-labelled goat anti-rabbit IgG (anti-Fc). Using this combination of marker reagents mouse H-2 antigens were visualized in their studies by a green fluorescence and the human surface antigens as a red fluorescence. Frye and Edidin (1970) found the degree of fusion was not great in their system and several fields had to be examined; the mouse and human antigens were largely confined to separate halves of the heterokaryon. However, after 40 minutes at 37°, the antigens on the cell surface were essentially mixed to give the appearance of completely double-staining (mosaic) cells. The mosaic appeared to form through the stages[2] $M\frac{1}{2}–H\frac{1}{2} \rightarrow M\frac{1}{2}–H1 \rightarrow M1–H1$, the formation of $M\frac{1}{2}–H1$ class of cells was caused by the apparently faster rate of mixing of the human antigens which Frye and Edidin (1970) attributed to a concentration effect. Since the human marker was represented by all the antigens it was therefore present in higher amounts per unit area on the original membrane. Singer and Nicolson (1972), however, point out that the histocompatibility antigens occur as large aggregates and might be expected to diffuse more slowly than a complex mixture of largely unaggregated human antigens. Frye and Edidin (1970) put forward four possible mechanisms to explain their observation: that there is a very rapid matabolic turnover of the antigens; integration of material into the membrane from a cytoplasmic precursor pool; movement or "diffusion" of antigen in the plane of the membrane; and lastly movement of existing antigen from one membrane site into the cytoplasm and its subsequent emergence at a new position in the membrane. On the basis of experiments performed at lower temperatures, as well as in the presence of various inhibitors, Frye

---

[2] The same notation as used by Frye and Edidin (1970) to describe the double-staining category of cells is followed here. $M\frac{1}{2}–H\frac{1}{2}$, a heterokaryon showing unmixed partial ring reactions for each fluorochrome $M\frac{1}{2}–H1$, a complete ring reaction for human markers but only a partial reaction for mouse H2 antigens. $M1–H1$ refers to a mosaic in which complete ring reactions for both fluorochromes is found. The $M1–H\frac{1}{2}$ situation, being the reverse of $M\frac{1}{2}–H1$, was only rarely observed.

and Edidin (1970) favoured a membrane structure in which the surface components are rigidly held in place, but which is sufficiently fluid to allow free diffusion of surface antigens. The explanation of rapid metabolic turnover or synthesis of new antigen molecules to explain the distribution of antigens in the newly formed heterokaryons, can be discounted as various metabolic inhibitors including puromycin, which has been shown (Kraemer, 1966, 1967) to prevent the appearance of surface components in the cell membrane, as well as the glutamine antagonist, 6-diazo-oxo-norleucine, have no effect on the mosaic formation. The latter inhibitor is of special interest as it has been shown to be an effective inhibitor of the synthesis of amino sugar components of cell surfaces (Oppenheimer et al., 1969). One treatment which does effectively inhibit the intermixing of the surface antigen with the formation of mosaics, and which is consistent with the fluid mosaic concept of cell membrane structure, was lowered temperature. The results of this treatment are in accord with the suggestion that the antigens are diffusing in a viscous solvent such as lipid. The curve showing per cent of mosaic at 40 minutes against temperature, the authors point out, is similar to those given for the penetration of glycerol into liquid crystal "liposomes" against temperature. Both systems show a distinct melting temperature above which the process may occur. Bearing in mind that the experiments of Frye and Edidin (1970) apply only to those components accessible at the cell surface, and does not necessarily give an insight into whether intermixing occurs on the cytoplasmic side of the cell membrane, their results strongly support the ideas of the fluid mosaic structure for cell membranes.

Assuming that the fluid mosaic model of the structure of plasma membranes is a more accurate portrayal of the molecular arrangement of the components of the cell membrane than the description given in the classical models, it is pertinent to enquire how such a structure enables one to make predictions on the likely mechanism of various cell surface phenomena. For example, does this model improve understanding of such phenomena as pinocytosis and phagocytosis or surface changes associated with malignant transformation?

The process of endocytosis may well be a case in point where assumption of the fluid mosaic model of cell membranes allows one to predict the possible functional mechanism. Phagocytosis may be conveniently divided into two main stages: attachment of the particle to the surface of the phagocyte; and invagination of the plasma membrane with the subsequent digestion of the adhering particle (Rabinovitch, 1967, 1968). Certain immunoglobulins will enhance the rate and extent of the cellular response to an adherent particle. The nature of the receptor sites at the cell surface with which such particles react to form the initial attachment is, however, not known, though the role of membrane glycoproteins has been investigated (Allen and Cook, 1970). Evidence has been obtained (Allen *et al.*, 1971) which indicates that even if sugars are not directly involved in the receptor site, then it is situated very near to glycoprotein. Of particular interest from the point of view of membrane structure, and the way in which prediction can be made in their relevance to surface phenomena, are the suggestions of Allison and his colleagues (Allison *et al.*, 1971) that the normal stimulus to endocytosis is attachment to receptors on the macrophage surface of immunoglobulin which has undergone a con-figurational change after contact with antigen. They suggest that this attachment might generate the formation of ion-conducting channels, presumably by a clustering of membrane proteins which traverse the plasmalemma, which allow the entry of calcium ions into the cell, and that these ions actuate the contraction of actomyosin-like microfilaments in the vicinity. It is interesting that treatment of the phagocyte with the lectin, concanavalin A, inhibits phagocytosis (Allen *et al.*, 1971; Berlin, 1972) which indicates that by interacting with membrane receptors this lectin may be able to restrict the mobility of membrane components. Conversely, when lectins were adsorbed to the sheep erythrocytes they were phagocytosed by *Acanthamoeba* spp. at a greater rate than in the case of the control experiments (Rabinovitch and De Stefano, 1971). However, in this particular example evidence was advanced which indicated that the lectins did not bind to the phagocytes. That immunoglobulins bound at the cell surface can cause movement of components within the membrane is well documented in the case of lymphocytes. Taylor *et al.* (1971) have demonstrated that the addition of antibody, reactive with surface immunoglobulin molecules, induces these molecules to gather over one pole of the cell; the process has been termed "cap" formation. Subsequently, the immunoglobulins are cleared from the cell surface by pinocytosis into the Golgi region. The effect does not occur if the experi-ment is performed at 0° or if univalent Fab fragments are used in place of bivalent antibody. Singer and Nicolson (1972), in a note added in proof to their original paper, suggest that the results of Taylor *et al.* (1971) can be explained by the bivalent antibody causing a clustering of membrane immunoglobulin molecules which then triggers off pinocytosis. Such a mechanism would imply that the immunoglobulin molecules are free to diffuse within the plane of the membrane, a suggestion in keeping with the tenets of the fluid mosaic model for plasma membrane. Of special interest in view of the inhibiting effect of concanavalin A on phagocytosis, and the suggestion that reorientation of membrane protein may act as a trigger to the process, is the demonstration by Yahara and Edelman (1972) that "cap" formation can be inhibited by this lectin—it is suggested by preventing the free diffusion of immunoglobulin receptors in the membrane.

Considerably more work will have to be performed before a complete understanding of the wide range of cell surface phenomena can be attempted in molecular terms. In this respect those models of membrane structure which allow for the presence of all the components known from chemical evidence to be present are of considerable importance, especially if such models underline the dynamic qualities of this organelle.

References

ALLAN, D. and CRUMPTON, M. J. (1970). *Biochem. J.* **120**, 133–143.

ALLEN, J. M. and COOK, G. M. W. (1970). *Expl. Cell Res.* **59**, 105–116.

ALLEN, J. M., COOK, G. M. W. and POOLE, A. R. (1971). *Expl. Cell Res.* **68**, 466–471.

ALLISON, A. C., DAVIES, P. and DE PETRIS, S. (1971). *Nature New Biology* **232**, 153–155.

BARLAND, P. and SCHROEDER, E. A. (1970). *J. Cell Biol.* **45**, 662–628.

BENDER, W. W., GARAN, H. and BERG, H. C. (1971). *J. molec. Biol.* **58**, 783–797.

BERG, H. C. (1969). *Biochim. biophys. Acta* **183**, 65–78.

BERLIN, R. D. (1972). *Nature New Biology* **235**, 44–45.

BERNSTEIN, S. S., JONES, R. L., ERICKSON, B. N., WILLIAMS, H. H., AVRIN, I. and MACY, I. G. (1937). *J. biol. Chem.* **122**, 507–528.

BOSMANN, H. B., HAGOPIAN, A. and EYLAR, E. H. (1968). *Archs. Biochem. Biophys.* **128**, 51–69.

BRANTON, D. (1966). *Proc. natn. Acad. Sci. U.S.A.* **55**, 1048–1056.

BRETSCHER, M. S. (1971a). *J. molec. Biol.* **58**, 775–781.

BRETSCHER, M. S. (1971b). *Nature New Biology* **231**, 229–232.

BRUNETTE, D. M. and TILL, J. E. (1971). *J. Membrane Biol.* **5**, 215–224.

CARRAWAY, K. L., KOBYLKA, D. and TRIPLETT, R. B. (1971). *Biochim. biophys. Acta* **241**, 934–940.

COLEMAN, R. and FINEAN, J. B. (1966). *Biochim. biophys. Acta* **125**, 197–206.

COOK, G. M. W. (1970). *In* "Molecular Biology" (A. Haidemenakis, ed.), pp. 179–207, Gordon and Breach, New York, London and Paris.

COOK, G. M. W. (1971). *A. Rev. Pl. Physiol.* **22**, 97–120.

COOK, G. M. W. (1972). *In* "Lysomes in Biology and Pathology" (J. T. Dingle, ed.), Vol. 3, pp. 237–277, North-Holland Publishing Company, Amsterdam.

COOK, G. M. W. and EYLAR, E. H. (1965). *Biochim. biophys. Acta* **101**, 57–66.

COOK, G. M. W., HEARD, D. H. and SEAMAN, G. V. F. (1960). *Nature, Lond.* **188**, 1011–1012.

COOK, G. M. W., HEARD, D. H. and SEAMAN, G. V. F. (1961). *Nature, Lond.* **191**, 47–47.

COOK, G. M. W., EYLAR, E. H. and LAICO, M. T. (1965). *Proc. natn. Acad. Sci. U.S.A.* **54**, 247–252.

DANIELLI, J. F. (1958). *In* "Surface Phenomena in Chemistry and Biology" (J. F. Danielli, K. G. A. Pankhurst and A. C. Riddiford, eds), pp. 246–265, Pergamon Press Inc., New York.

DANIELLI, J. F. and DAVSON, H. (1934). *J. cell. comp. Physiol.* **5**, 495–508.

DANIELLI, J. F. and HARVEY, E. N. (1934). *J. cell. comp. Physiol.* **5**, 483–494.
DAVSON, H. (1962). *Circulation* **26**, 1022–1037.
DODGE, J. T., MITCHELL, C. and HANAHAN, D. T. (1963). *Archs. Biochem. Biophys.* **100**, 119–130.
EMMELOT, P. and BENEDETTI, E. L. (1967). *In* "Protides of the Biological Fluids" (H. Peeters, ed.), Vol **15**, pp. 315–326, Elsevier, Amsterdam.
EMMELOT, P. and BOS, C. J. (1972). *J. Membrane Biol.* **9**, 83–104.
EMMELOT, P., BOS, C. J., BENEDETTI, E. L. and RÜMKE, PH. (1964). *Biochim. biophys. Acta* **90**, 126–145.
EVANS, W. H. and BRUNING, J. W. (1970). *Immunology* **19**, 735–741.
EYLAR, E. H., MADOFF, M. A., BRODY, O. V. and ONCLEY, J. L. (1962). *J. biol. Chem.* **237**, 1992–2000.
FAIRBANKS, G., STECK, T. L. and WALLACH, D. F. H. (1971). *Biochemistry, N.Y.* **10**, 2606–2617.
FERBER, E., RESCH, K., WALLACH, D. F. H. and IMM, W. (1972). *Biochim. biophys. Acta* **266**, 494–504.
FRYE, L. D. and EDIDIN, M. (1970). *J. Cell Sci.* **7**, 319–335.
GAHMBERG, C. G. (1971). Academic Dissertation, University of Helsinki, 1971.
GAHMBERG, C. G. and SIMONS, K. (1970). *Acta path. microbiol. scand.*, Section B, **78**, 176–182.
GLASER, M., SIMPKINS, H., SINGER, S. J., SHEETZ, M. and CAHN, S. I. (1970). *Proc. natn. Acad. Sci. U.S.A.* **65**, 721–728.
GLICK, M. C., COMSTOCK, C. and WARREN, L. (1970). *Biochim. biophys. Acta* **219**, 290–300.
GLYN, I. (1968). *Br. med. Bull.* **24**, 165–169.
GORTER, E. and GRENDEL, F. (1925). *J. exp. Med.* **41**, 439–443.
GRAHAM, J. M., HIGGINS, J. A. and GREEN, C. (1968). *Biochim. biophys. Acta* **150**, 303–305.
GREEN, D. E., TISDALE, H. D., CRIDDLE, R. S. and BOCK, R. M. (1961). *Biochem. biophys. Res. Commun.* **5**, 81–84.
GREEN, D. E., ALLMAN, D. W., BACHMANN, E., BAUM, H., KOPACZYK, K., KORMAN, E. F., LIPTON, S., MacLENNAN, D. H., McCONNELL, D. G., PERDUE, J. F., RIESKE, J. S. and TZAGOLOFF, A. (1967). *Archs. Biochem. Biophys.* **119**, 312–335.
GREEN, D. E., HOARD, N. F., LENAZ, G. and SILMAN, H. I. (1968). *Proc. natn. Acad. Sci. U.S.A.* **60**, 227–284.
GUIDOTTI, G. (1972). *A. Rev. Biochem.* **41**, 731–752.
HAKOMORI, S. and JEANLOZ, R. W. (1961). *J. biol. Chem.* **236**, 2827–2834.
HAYDON, D. A. and TAYLOR, J. M. (1963). *J. theor. Biol.* **4**, 281–296.
HERZENBERG, L. A. and HERZENBERG, L. A. (1961). *Proc. natn. Acad. Sci. U.S.A.* **47**, 762–767.
HOFFMAN, J. F. (1962). *Circulation* **26**, 1073.
JAMIESON, G. A. and GROH, N. (1971). *Analyt. Biochem.* **43**, 259–268.
KAMAT, V. B. and WALLACH, D. F. H. (1965). *Science N.Y.* **148**, 1343–1345.
KANDUTSCH, A. A. (1960). *Cancer Res.* **20**, 264–268.
KIEHN, E. D. and HOLLAND, J. J. (1970). *Biochemistry N.Y.* **9**, 1716–1728.
KLENK, E. and LAUENSTEIN, K. (1952). *Z. physiol. Chem.* **291**, 249–258.

KORN, E. D. (1966). *Science N.Y.* **153**, 1491–1498.

KRAEMER, P. M. (1966). *J. cell. Physiol.* **68**, 85–90.

KRAEMER, P. M. (1967). *J. cell. Physiol.* **69**, 199–208.

LANGMUIR, I. (1917). *J. Am. chem. Soc.* **37**, 1848–1906.

LENARD, J. and SINGER, S. J. (1966). *Proc. natn. Acad. Sci. U.S.A.* **56**, 1828–1835.

LUCY, J. A. (1968). *Br. med. Bull.* **24**, 127–129.

McCOLLESTER, D. L. (1970). *Cancer Res.* **30**, 2832–2840.

MADDY, A. H. (1964). *Biochim. biophys. Acta* **88**, 390–399.

MADDY, A. H. (1966). *Int. Rev. Cytol.* **20**, 1–65.

MARCHESI, V. T., TILLACK, T. W. and SCOTT, R. E. (1971). *In* "Glycoproteins of Blood Cells and Plasma" (G. A. Jamieson and T. J. Greenwalt, eds), pp. 94–105, Lippincott, Philadelphia.

MARINETTI, G. V. and GRAY, G. M. (1967). *Biochim. biophys. Acta* **135**, 580–590.

MORAWIECKI, A. (1964). *Biochim. biophys. Acta* **83**, 339–347.

MOOR, H. and MÜHLETHALER, K. (1963). *J. Cell Biol.* **17**, 609–628.

MORRÉ, D. J., MERLIN, L. M. and KEENAN, T. W. (1969). *Biochem. biophys. Res. Commun.* **37**, 813–819.

NEVILLE, D. M. JNR. (1960). *J. biophys. biochem. Cytol.* **8**, 413–422.

NEVILLE, D. M. JNR. (1968). *Biochim. biophys. Acta* **154**, 540–552.

NEVILLE, D. M. JNR. and GLOSSMANN, H. (1971). *J. biol. Chem.* **246**, 6335–6338.

NIGAM, V. N., MORAIS, R. and KARASAKI, S. (1971). *Biochim. biophys. Acta* **249**, 34–40.

OPPENHEIMER, S. B., EDIDIN, M., ORR, C. W. and ROSEMAN, S. (1969). *Proc. natn. Acad. Sci. U.S.A.* **63**, 1395–1402.

OTTOLENGHI, A. C. and BOWMAN, M. H. (1970). *J. Membrane Biol.* **2**, 180–191.

OVERTON, E. (1899). *Vierteljahresschr. Naturforsch. Ges. Zürich* **44**, 88–135.

PERDUE, J. F. and SNEIDER, J. (1970). *Biochim. biophys. Acta* **196**, 125–140.

PHILLIPS, D. R. and MORRISON, M. (1970). *Biochem. biophys. Res. Commun.* **40**, 284–289.

PHILLIPS, D. R. and MORRISON, M. (1971). *Biochem. biophys. Res. Commun.* **45**, 1103–1108.

PINTO DA SILVA, P. and BRANTON, D. (1970). *J. Cell Biol.* **45**, 598–605.

PODUSLO, J. F., GREENBERG, C. S. and GLICK, M. C. (1972). *Biochemistry, N.Y.* **11**, 2616–2621.

RABINOVITCH, M. (1967). *Proc. Soc. exp. Biol. Med.* **124**, 396–399.

RABINOVITCH, M. (1968). *Proc. Soc. exp. Biol. Med.* **127**, 351–355.

RABINOVITCH, M. and DE STEFANO, M. J. (1971). *Nature, Lond.* **234**, 414–415.

RAJAM, P. C. and JACKSON, A. L. (1958). *Nature, Lond.* **181**, 1670.

RANDALL, R. F., STODDART, R. W., METCALFE, S. M. and METCALFE, J. C. (1972). *Biochim. biophys. Acta* **255**, 888–899.

REID, E. (1967). *In* "Enzyme Cytology" (D. B. Roodyn, ed.), pp. 321–406, Academic Press, London.

ROBERTSON, J. D. (1966). *In* "Principles of Biomolecular Organisation", a Ciba Found. Symp. (G. E. W. Wolstenholme and M. O'Connor, eds), pp. 357–417, J. and A. Churchill, London.

SHEN, L. and GINSBURG, V. (1967). *Archs. Biochem. Biophys.* **122**, 474–480.

SHIMIZU, S. and FUNAKOSHI, I. (1970). *Biochim. biophys. Acta* **203**, 167–169.

SINGER, S. J. and NICOLSON, G. L. (1972). *Science N.Y.* **175**, 720–731.
SJÖSTRAND, F. S. (1968). *In* "Regulatory Functions of Biological Membranes",
B.B.A. Library (J. Jäarnefelt, ed.), Vol. **11**, pp. 1–20, Elsevier, Amsterdam.
SNEATH, J. S. and SNEATH, P. H. A. (1959). *Br. med. Bull.* **15**, 154–157.
STANBURY, J. B., WICKEN, J. V. and LAFFERTY, M. A. (1969). *J. Membrane Biol.*
1, 459–467.
STECK, T. L., WEINSTEIN, R. S., STRAUS, J. H. and WALLACH, D. F. H. (1970).
Science N.Y. **168**, 255–257.
STECK, T. L., FAIRBANKS, G. and WALLACH, D. F. H. (1971). *Biochemistry, N.Y.*
10, 2617–2624.
STOECKENIUS, W. and ENGELMAN, D. M. (1969). *J. Cell Biol.* **42**, 613–646.
TAKEUCHI, M. and TERAYAMA, H. (1965). *Expl. Cell Res.* **40**, 32–44.
TAYLOR, R. B., DUFFUS, W. P. H., RAFF, M. C. and DE PETRIS, S. (1971). *Nature
New Biology* **233**, 225–229.
TILLACK, T. W. and MARCHESI, V. T. (1970). *J. Cell Biol.* **45**, 649–653.
WALLACH, D. F. H. (1967). *In* "The Specificity of Cell Surfaces" (B. D. Davis
and L. Warren, eds), pp. 129–163, Prentice-Hall Inc., New Jersey.
WALLACH, D. F. H. (1972). *Biochim. biophys. Acta* **265**, 61–83.
WALLACH, D. F. H. and ZAHLER, P. H. (1966). *Proc. natn. Acad. Sci. U.S.A.*
56, 1552–1559.
WALLACH, D. F. H., KRANZ, B., FERBER, E. and FISCHER, H. (1972). *FEBS
Letters* **21**, 29–33.
WARREN, L., GLICK, M. C. and NASS, M. K. (1966). *J. cell. Physiol.* **68**, 269–287.
WARREN, L., GLICK, M. C. and NASS, M. K. (1967). *In* "The Specificity of
Cell Surfaces" (B. D. Davis and L. Warren, eds), pp. 109–127, Prentice-
Hall, New Jersey.
WEINSTEIN, D. B., MARCH, J. B., GLICK, M. C. and WARREN, L. (1969). *J. biol.
Chem.* **244**, 4103–4111.
WEINSTEIN, D. B., MARCH, J. B., GLICK, M. C. and WARREN, L. (1970). *J. biol.
Chem.* **245**, 3928–3937.
WEISS, L. (1969). *Int. Rev. Cytol.* **26**, 63–105.
WINKLER, K. C. and BUNGENBERG DE JONG, H. G. (1941). *Archs. néerl. Physiol.*
25, 431–466.
WINZLER, R. J. (1970). *Int. Rev. Cytol.* **29**, 77–125.
YAHARA, I. and EDELMAN, G. M. (1972). *Proc. natn. Acad. Sci. U.S.A.* **69**,
608–612.
YAMAKAWA, T., MATSUMOTO, M., SUZUKI, S. and IIDA, T. (1956). *J. Biochem.,
Tokyo* **43**, 41–52.

2
Carbohydrates of Plasma Membranes and Surfaces of Animal Cells: Evidence

In the preceding chapter it was shown that, with the evolution of ideas about the structure of membranes, there is a growing realization that materials containing carbohydrates are important components in cell membranes. As yet, little is known of the structural details of membrane glycosubstances (though see Chapter 3), but there is an increasing awareness that heterosaccharides are probably involved in a number of biologically important surface properties of the plasmalemma. This is in marked contrast to the situation a decade ago. Even in 1968 it was still necessary to stress (Cook, 1968) "that models of membranes, especially of plasma membranes, that do not take into account the presence of carbohydrate-complexes must be regarded as incomplete." It is those techniques which have drawn attention to the presence of oligosaccharides at the surfaces of animal cells, that will be considered in this chapter.

Three major fields of investigation have provided evidence for the presence of carbohydrates at the cell periphery: immunological studies, electrophoresis of intact cells and microscopical investigations. It is surprising that the classical models of membrane structure ignore the carbohydrates, although the chemical nature of the blood-group substances has long indicated the need to consider heterosaccharides in membrane biochemistry. It seems strange, on reflection, that so little correlation was drawn between the results of immunochemistry and membrane biophysics; possibly this was partly caused by the tendency for investigations on immunologically active molecules to be performed on material derived from sources other than intact cells (for example, ovarian cyst fluids).

The need to consider the rôle of glycosubstances in membranes, and especially the cell surface, has received great impetus from electrokinetic studies performed on viable animal cells, and the conclusion that glycosubstances, especially those containing sialic acid, are important surface species has been considerably strengthened by subsequent microscopical observations.

I. Electrokinetic Studies

The electrophoretic mobility of particles, including biological cells, is considered to reflect the zeta potential: that is, the potential at the hydrodynamic surface of shear of the particle. This mobility is a complete measurement derived from the true surface potential (caused by charged groups at the free surface of the particle and within it, but accessible by diffusion) and possibly adsorbed ions. In the case of biological cells suspended in a saline solution at physiological pH, the application of an electric field moves the cells towards the positive electrode. Indeed it has been found that the surface of all mammalian cells so far examined carries a net negative surface charge under these conditions.

As defined above, the potential responsible for this movement is located at the hydrodynamic surface of shear, and the ionizable species that cause it must be very near that plane of shear in order to exchange protons with the environment rapidly enough. Therefore electrokinetic measurements performed on biological cells relate to the outermost regions of the cell membrane and, in the absence of ion adsorption, the measurements can be regarded as reflecting properties caused by ionogenic groupings in the periphery of the membrane. Seaman and Cook (1965) have demonstrated that aldehyde-stabilized human erythrocytes, the anionogenic groups of which have been esterified with diazomethane, possess zero electrophoretic mobility over a pH range where saponification is unlikely to take place. This result may indicate that ion adsorption plays no important rôle in the electrokinetic properties of biological cells, although it should be remembered that the adsorptive properties of the native surface and those of the aldehyde modified cell may well be different. Previously, Heard and Seaman (1960) had shown that the charge densities of erythrocytes were the same in the presence of CNS^-, F^-, Cl^- and I^- ions and had therefore concluded that this cell's surface is not predominantly non-ionogenic, and also that it lacked any appreciable positive ionogenic character.

In electrokinetic studies the erythrocyte has been widely used as a model system, in keeping with other areas of membrane research. The choice of cell has been a source of criticism amongst cell biologists in view of the greater complexity of structure and function of tissue cells. Nevertheless, the erythrocyte has proved to be an extremely valuable model and many of the procedures and techniques now commonly used in the electrokinetic investigation of relatively limited amounts of tissue cells have only been made possible after rigorous development of the methods using considerably larger amounts of erythrocytes. In addition, molecules isolated from erythrocytes or tissue cells present the biochemist with similar problems with regard to the elucidation of their chemical structure, although manipulation is facilitated with erythrocytes in view of the availability of the material (see Chapter 3).

The electrokinetic properties of cells, especially red blood cells, have been studied for several years. Early work using the moving boundary technique has been described, though the microscopic method, where migration of individual cells in a known potential gradient is observed, rapidly became more popular. The latter technique is usually referred to as "cell-electrophoresis", although in some early papers the term "micro-electrophoresis" is used.

Thus, all mammalian cells so far studied by cell-electrophoresis possess a net negative surface charge, and, in the case of the human erythrocyte, it has been realized that this charge arises from strongly acidic groups. With the early emphasis placed on the rôle of phospholipids in membranes, it is not surprising that until the end of the 1950s many workers were agreed in attributing the electrokinetic behaviour of the erythrocyte's surface to a membrane component that contained ionizable phosphate groups, probably associated with a phospholipid system (Bangham et al., 1958; Engstrom and Finean, 1958). Earlier, Furchgott and Ponder (1941) had suggested that the surface of the erythrocyte was largely composed of lipid, and the acidic groups could be caused by cephalin molecules. Winkler and Bungenberg de Jong (1940) had also been led to suggest a phospholipid character for the outer layer of the erythrocyte. These studies had largely relied on physical data, such as the derivation of dissociation constants for surface ionogenic groups and charge reversal spectra, for the identification of surface charge as originating from membrane phospholipid. No doubt the classical ideas of membrane structure, discussed in the preceding chapter, added great assurance to such an interpretation, and might explain why other evidence which had been accumulating at about the same time (and which suggested that the electrokinetic properties of red blood cells might be associated with mucosubstances[1]) was largely ignored in contemporary biophysical studies. At this time knowledge of the chemistry of glycoproteins was very sparse, and in consequence the rôle of these macromolecules and the glycolipids was overlooked in the interpretation of the physical properties of the cellular surface.

As far back as the early 1940s, Hirst (1942a, b), working with chicken erythrocytes, found that these cells adsorbed influenza virus at 4° causing the cells to agglutinate. Following a short period at 37° the virus particles were eluted spontaneously, causing the agglutinates to fall apart, and though the virus retained its activity the erythrocytes were rendered inagglutinable. Hirst interpreted this result in terms of an enzyme-substrate interaction. A receptor substance was construed to be present at the surface of the erythrocyte, which attracted and bound the ions, and the virus was supposed to

---

[1] Mucosubstance (mucoprotein and mucolipid), though widely used in the late 1950s, is synonymous with the currently accepted terms of glycoprotein and glycolipid. Mucopolysaccharide has been largely supplanted by the term glycosaminoglycan.

possess an enzyme which then acted upon the receptor substances. McCrea (1947) also showed the presence of an enzyme from filtrates of *Clostridium welchii*, which could render erythrocytes inagglutinable by influenza virus. In 1947 Burnet and Stone found similar "receptor-destroying" activity in a preparation of *Vibrio cholerae*, and they termed this activity "receptor-destroying enzyme" (RDE). Following from this work, Hirst (1948) suggested, from adsorption and elution experiments with viruses, that the receptor sites on the surface of the erythrocyte might be mucoprotein. Especially important from the point of view of electrokinetic studies was the observation by Hanig (1948) that the elution of virus from the red cell surface was accompanied by a decrease in the electrophoretic mobility of the erythrocyte. In 1952 Stone and Ada examined more fully the relationship between changes in the electrophoretic mobility of erythrocytes and their agglutination after treatment with viral enzyme.

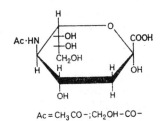

$$Ac = CH_3CO-;CH_2OH-CO-$$

Fig. 1. The sialic acids. In this figure, two of the more commonly occurring sialic acids, *N*-acetylneuraminic acid (Ac=CH$_3$CO) and *N*-glycolylneuraminic acid (Ac=CH$_2$OH—CO) are shown. The molecule can, in addition, be acylated at other positions (e.g. C4, C7) in *N-O* diacetylneuraminic acid and *N*-acetyl *O*-acetylneuraminic acids. In glycoproteins the sialic acid residue is linked α-ketosidically at C2 usually to galactose or *N*-acetyl hexosamine; this bond is sensitive to neuraminidase (EC 3.2.1.18). The carboxyl group (pKa 2·6 for *N*-acetylneuraminic acid) of the sialic acids are responsible for a significant portion of the net negative charge of animal cells.

Anderson (1948) demonstrated that an inhibitor of virus haemagglutination, shown to be a mucoid, lost its inhibitory capacity after RDE treatment. Ada and French (1959) prepared a highly purified specimen of the enzyme and Gottschalk (1957) showed this enzyme to be an α-glycosidase, neuraminidase. Further, Klenk and his colleagues (Klenk and Lempfrid, 1957; Klenk and Uhlenbruck, 1958) found that RDE liberates acylated neuraminic acids from erythrocyte stroma of various mammalian species, leading Klenk (1958) to suggest that the negative charge on the erythrocyte might be ascribed to an acylated neuraminic acid. Even at this stage this view was largely hypothetical, since no direct measurements had been made in support of the suggestion and the belief then current among workers in the field of cell electrophoresis was that the negative charge of the erythrocyte arose from phosphate groups of phospholipids. Bateman *et al.* (1956) had suggested that the decrease in electrophoretic mobility of red blood cells after treatment with the PR8 strain of influenza virus was caused by the appearance of cationic groups, and this further complicated the picture. Curtain (1953) had shown that a urinary muco-protein, which inhibited viral haemagglutination following treatment with RDE, possessed cationic groups of pK 11, and this gave weight to the ideas of Bateman *et al.* (1956). Certainly, even in 1959, the suggested rôle of glycoproteins in cellular electrokinetic phenomena met with little accept-ance and the postulated rôle of phospholipid was still firmly entrenched.

Earlier, Pickles (1946) had found that erythrocytes sensitized with incomplete anti-Rh antibody were rendered agglutinable after treatment with filtrates of *Vibrio cholerae*. Attempting to define the properties of the enzyme present in the *Vibrio cholerae* filtrate, Morton and Pickles (1947) showed that trypsin also could be used to demonstrate the presence of incomplete antibodies. Previous work by Ballentine and Parpart (1940) had shown that treatment of erythrocytes with trypsin for twenty-four hours did not cause any appreciable haemolysis, and they concluded that the permeability of the red cell had not been altered. Indeed, the very early work of Northrop (1926) showed that living cells were impermeable to a solution of trypsin, but injection of the enzymes into the cell caused rapid disintegration of the cell. The use of trypsin as a tool for investigating cell surfaces is now widely established, though in the case of animal cells Ponder's observation (Ponder, 1951) on the action of this enzyme on the electrophoretic mobility of the human erythrocyte must be one of the earliest examples. Ponder (1951) found that the electrophoretic mobility of the human erythrocyte was reduced after treatment with the enzyme; the reduction corresponded to a decrease of about 30% in the net negative surface charge density and was accompanied by an increase in volume of the cell. As an explanation of this, Ponder (1951) suggested that either trypsin was incorporated into relatively small parts of the cellular surface, or that a general "loosening up process" occurred. Following on from these studies, Pondman and Mastenbroek (1954) investigated the action of trypsin on the human erythrocyte and confirmed the results above. They went so far as to suggest that trypsin was capable of cleaving P—N bonds with a resultant loss of phospholipid from the erythrocyte membrane. This is another example of the influence that the classical membrane model may have had on the interpretation of data.

In 1960 Seaman and Heard re-examined the action of trypsin on the electrokinetic properties of the human erythrocyte, and although they confirmed all the electrokinetic data of previous workers, they were unable to find any significant release of lipid-bound phosphate from the erythrocytes after trypsin treatment. They did find that large amounts of inorganic phosphate were released from the cells into the suspending fluid by leakage of intracellular phosphate, though this inorganic phosphorus was not increased by the action of trypsin. Seaman and Heard (1960) suggested that the observed decrease in electrophoretic mobility of the cells after trypsin treatment could be brought about in two ways. Firstly, by fission of peptide bonds without loss of ionogenic material from the membrane but with an concomitant reorganization of the outermost regions of the plasma membrane, analogous to the general "structural loosening up" of Ponder (1951). Alternatively, peptide bond fission accompanied by a loss of negative ionogenic groupings from the cell periphery could take place. These authors (Seaman and Heard, 1960) suggested that in the second case "this fission may lead either to a direct loss of carboxyl groupings associated with amino acid elements in a polypeptide chain, or to loss of groups such as phosphate or sulphate from a protein-phospholipid or protein-carbohydrate complex". However, they concluded that "reduction in red cell charge on treatment with trypsin is probably not due to the removal of phosphate groups from the interface, nor to the physical adsorption of trypsin" and further that "microcrenation, reorientation, or swelling of the membrane is possible— the charge could also be produced by bond fission with a net loss of carboxyl groups from the electrophoretic plane of shear." Although compatible with some of the later classical models of plasma membrane structure which allow for the presence of peripheral protein, the suggestion of Seaman and Heard (1960) that carboxyl groups were lost from the red cell was of particular importance, because they cast doubt on the then widely accepted view that it was phosphate groups that determined the charge on the erythrocyte surface.

In continuation of this work Cook et al. (1960) examined the tryptic degradation products of the intact, saline washed, human erythrocyte, by use of paper chromatography. They showed that the decrease in electrophoretic mobility of the red blood cells was accompanied by a release of a sialomucopeptide from the membrane. This sialomucopeptide contained terminally-linked sialic acid residues. Numerous control experiments were performed to show that the glycopeptide was a constituent of the membrane and not associated with lytic products of the enzyme itself. This result provided an explanation of the specific action of trypsin on the intact erythrocytes, as opposed to stroma. When coupled with the electrophoretic data it had the added importance of localizing, for the first time, a sialoglycopeptide residue at the periphery of the cell.

Subsequent work by Cook et al. (1961) on the action of crystalline neuraminidase from *Vibrio cholerae* on the electrophoretic mobility of human erythrocytes, and by Eylar et al. (1962), using neuraminidase purified from *Vibrio cholerae* and *Diplococcus pneumoniae*, on erythrocytes from a number of different animals, demonstrated the importance of sialic acids at the surface of the erythrocyte.

In the human erythrocyte, the release of sialic acid by neuraminidase always causes a decrease of about 80% in their electrophoretic mobility when measured in 0·145 M sodium chloride. If enzyme-treated cells are allowed to react with lower aldehydes as specific reagents for potential cationogenic groups (Heard and Seaman, 1961), they show only a very small increase in electrophoretic mobility. This indicates that the initial reduction in charge is almost certainly caused by the removal of anionic groups, and not by the generation of a large population of cationic groups. This is important in view of the earlier proposals of Bateman *et al.* (1956), discussed above. Care was taken in both the investigations above to demonstrate chemically that free sialic acids are released into the incubation medium following neuraminidase treatment of the cells. If it is assumed that the carboxyl of each sialic acid residue, when bound in the erythrocyte membrane, corresponds to one unit of negative charge, the amounts liberated are greater than is necessary to explain the decrease in

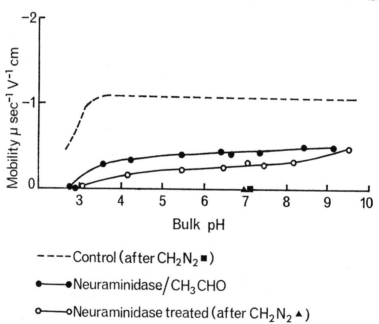

Fig. 2. pH-mobility relationships for human erythrocytes. This figure is drawn from the combined data of Cook *et al.* (1961) and Seaman and Cook (1965). The electrophoretic mobility determinations were performed in media of ionic strength 0·145 g ions/litre.

surface charge. A number of explanations have been advanced to account for the discrepancy between the amounts of sialic acid released and that which would be expected on theoretical grounds. It is possible that a number of the molecules are released from sites which are buried deep in the membrane and therefore unable to influence the electrophoretic mobility of the cell, because they cannot exchange with hydrogen ions at the hydrodynamic plane of shear, at a significant rate. Alternatively, the calculation of the surface charge-density by means of the Gouy-Chapman equation may yield low charge-densities for an ion-penetrable system (Haydon, 1961) such as the periphery of the red blood cell.

This work demonstrated conclusively that sialic acids are an important ionogenic species at the surface of the erythrocyte. Indeed, from observations upon a number of erythrocytes of different species Eylar et al. (1962) showed that there was a roughly linear relationship between sialic acid

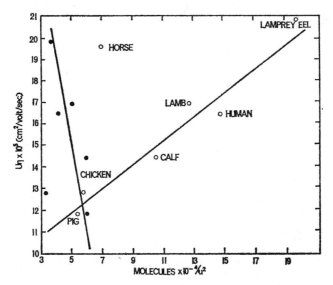

Fig. 3. The relationship between the electrophoretic mobility and the number of molecules of sialic acid (O), or phosphatidyl-serine (●) per μ^2 of surface among erythrocytes of various species. The approximately linear relationship between the sialic acid surface density and mobility illustrates the consistent position of sialic acid carboxyl groups in the general composition of the red cell surface. Lipid (phosphatidylserine) however shows a scatter of points indicative of a relationship opposite to that anticipated if phosphate groups are largely the electrokinetic charge determinant species at the cell surface of the erythrocyte. From Eylar et al. (1962) by permission of J. biol. Chem.

surface charge densities, measured as the number of sialic acid molecules per μ^2, and mobility; while lipid, as phosphatidylserine, showed a scatter of points which indicated a relationship opposite to that anticipated if phospholipids are the main determinants of surface charge. The one exception amongst the seven different species studied by these authors was the horse, which has a conspicuously high mobility which is not accounted for by the sialic acid content of its erythrocytes, though this cell has a very high sialoglycolipid content relative to other species.

Although erythrocytes were the first cells to be shown, by the electrokinetic technique, to possess sialic acid residues at their peripheries and consequently were the first example of an animal cell where surface carbohydrate was clearly demonstrated by this method, the same general pattern became quickly established for tissue cells. Sialic acids were shown to be present as an important ionogenic species at the surface of the Ehrlich ascites carcinoma cell (Wallach and Eylar, 1961; Cook et al., 1962) as well as the normal and malignant rat liver cells, cells from a human bronchial carcinoma, HeLa cells, cells from rat myeloma and L-strain mouse fibroblasts (Fuhrmann et al., 1962; Simon-Reuss et al., 1964). Forrester et al. (1962) examined normal and polyoma transformed hamster kidney fibroblasts and once more illustrated the importance of sialic acids as an ionogenic species at the cell surface. Since the early 1960s, the number of publications on cell electrophoresis has increased enormously and the use of neuraminidase treatment could now almost be considered a routine procedure in any electrokinetic characterization of the cell surface. The electrophoretic technique has provided overwhelming evidence for the presence of sialosubstances, whether glycoprotein, or in some cases glycolipids, at cell surfaces, and hence the presence of sugar-containing molecules has been clearly established as an important and general feature of the surface of the animal cell. Certain points of caution on the use of neuraminidase and the interpretation of its action upon the electrokinetic properties of cells sould be recognized. Of prime importance is the purity of the enzyme preparation used. This will of course apply not only to the use of neuraminidase in electrokinetic studies, but also to any study, be it physicochemical or biochemical, aimed at elucidating the molecular nature of cell surfaces by means of selective enzyme treatment. In some of the original electrokinetic studies (Cook et al., 1961) crystalline neuraminidase (Ada and French, 1959) of high purity was used. Crystalline neuraminidase is normally prepared by adsorption and elution from washed and packed human erythrocytes. Hence it is unlikely that the electrokinetic changes observed after treating human erythrocytes with solutions of crystalline neuraminidase are caused by the irreversible adsorption of the enzyme or its impurities onto the cellular surface. In addition, haemagglutination studies (Buzzell and Hanig, 1958) have shown that, after elution of this enzyme from red blood cells, no further adsorption of the enzyme onto these cells is possible. A number of different preparations of neuraminidase are available commercially. In some published electro-kinetic work little chemical data is given upon the particular preparations used, and these may vary considerably in effectiveness in any system under study. Further, Kraemer (1968) has drawn attention to the danger of using commercially available neuraminidase from *Clostridium perfringens* for studies on membrane-bound sialic acid, without further purification, since the preparation may be contaminated with cytotoxic, haemolytic and phospholipase activities. A further example of this problem is illustrated in the studies of Emmelot and Bos (1966) who found differing results in the action of neuraminidase preparations on the K^+-activated-p-nitrophenyl-phosphatase of liver plasma membranes. One preparation abolished this activity, though this result could not be confirmed when another prepara-tion, free of contaminating activities, was used. It should not be assumed that a decrease in electrophoretic mobility, following treatment with neuraminidase, is necessarily a true measure of the contribution of sialosyl residues to the electrokinetic properties of the cells. Likewise an apparent absence of change in the electrophoretic mobility of a treated cell is no indication that sialic acid residues are not present at the cell-surface, since sialic acids are known to be linked in different positions on the aglycone sugar which are of varying lability to enzymatic glycosidic cleavage and steric factors could also make them resistant to neuraminidases. Cook *et al.* (1963) showed that sialic acids may be enzymatically released from the solid, as opposed to the ascitic form of sarcoma 37, without a reduction in surface charge. Similar results have been obtained by Wallach and Esandi (1964) using a solid sarcoma MC_1M_{SS}. These investigations demonstrated, as in the case of sarcoma 37, that the release of sialic acids from this tumour did not produce any change in the cells' electrophoretic mobility. To be sus-ceptible to the action of neuraminidase the sialic acids must be bound –ketosidically; the cleavage of this linkage cannot directly generate any new ionogenic groups. Indeed, the use of lower aldehydes as specific blocking agents for cationic groups after neuraminidase treatment has been used, though in only a very few papers, to confirm that the reduction in electro-phoretic mobility is the result of the removal of an ionogenic group, rather than the generation of a significant number of cationic groups. However, the removal of the charged sialic acid residues may induce conformational changes in the cell membrane proteins, with the result that other charged groups may then become effective at the electrokinetic plane of shear. Such a mechanism may not be operative in all cases. Cook and Jacobson (1968) found that when leukaemic cells, derived from AKR mice suffering from acute lymphoblastic leukaemia, or normal lymphoid tissue from healthy animals, were treated consecutively with neuraminidase and aldehyde the same electrokinetic surface was produced irrespective of the sequence of these treatments. Such a result would suggest that neuraminidase treatment of these particular cells is not accompanied by significant conformational changes in the peripheral membrane material.

Bearing in mind these considerations, studies using cell electrophoresis, especially when coupled with suitably controlled neuraminidase treatments, have undoubtedly been of major importance in demonstrating that carbohydrate complexes are an important element of the cellular surface.

II. Microscopical Observations

The presence of macromolecules that are rich in carbohydrates at the surface of a wide range of cell types is also apparent from histochemical evidence.

A. Methods Involving the Use of Chemical Treatments or Cationic Colloids

Studies using both the light and electron microscope have added greatly to the evidence demonstrating that heterosaccharide materials are important constituents of cell surfaces. In these investigations chemical treatments considered specific for sugars or sugar acids have principally been used. The periodic acid reactions belong to the former class and interaction with certain basic colloids to the latter. The use of such procedures will be dealt with here, and attention will be paid to some of the crucial developments in this field which have indicated the important part played by sugars in the chemistry of the cell surface and the extent to which such materials are present in a wide range of plasma membranes of animal origin. An attempt will be made to indicate the main lines of evidence which have clearly demonstrated the importance of heterosaccharides in cellular surfaces (especially when the more recent use of lectins as specific saccharide stains for microscopy is taken into account). For additional technical references attention is drawn to the recent reviews of Martinez-Palomo (1970), Parsons and Subjeck (1972) and Rambourg (1971).

The Gasics (Gasic and Gasic, 1962a and b) were among the first to provide microscopical evidence for the presence of carbohydrate-containing materials at mammalian cell surfaces, and their investigations did much to confirm earlier electrokinetic studies which showed that glycosubstances, especially those containing sialic acids, are an important feature of the cellular surface. In their studies the Gasics made use of the Hale stain for identifying carbohydrate-containing surface mucins of ascites tumour cells. The reaction depends upon the affinity of acidic groupings for colloidal iron, which can then be visualized as Prussian blue by interaction with potassium ferrocyanide, and was early used for detection of mucopolysaccharides. Initially, they used the light microscope and subsequently they described a modified Hale procedure for use in the electron micro-scope. These investigations demonstrated that the murine ascitic mammary carcinoma, TA3, had a Hale stain-positive coating which was clearly visible at the extreme cellular periphery in the light microscope. Further-more, this coating was removed or partially destroyed by pancreatin, trypsin, pepsin and neuraminidase (described in the original paper as receptor destroying enzyme); while papain (both crude and crystalline preparations) and testicular hyaluronidase were ineffective in removing the material from the surface of these cells. The Hale-staining material could be a sialoglycoprotein, though it is perhaps surprising that such material would be relatively insensitive to papain digestion. In the same work, Gasic and Gasic (1962b) used the periodic acid-Schiff reaction (PAS) in which dialdehydes formed by the periodate oxidation of carbon bonds bearing vicinal hydroxyl groups, in such molecules as sugars, are allowed to react with leuco-fuchsin to produce a purple staining reaction. This was unaffected by neuraminidase treatment. However, tryptic proteo-lysis eliminated the periodic acid-Schiff reaction. Again these observations are compatible with the suggestion that glycoproteins are present at the cell periphery, especially when the later work by Gasic and Gasic (1963) is taken into account, which indicates that galactosyl and N-acetylgalacto-saminyl residues give rise to PAS staining. The importance of galactosyl and N-acetylgalactosaminyl residues in this latter work was demonstrated by use of a mixture of glycosidases from *Clostridium perfringens* together with the addition of various sugars in excess, to abolish the action of the individual glycosidases by product inhibition.

In addition to these histochemical observations, Gasic and Gasic (1962b) were able to demonstrate that the TA3 cells could regenerate their surface "sialomucin" within 1–2 hours following treatment with neuraminidase, an observation which is discussed in great detail elsewhere in this book (Chapter 5). As well as these studies with the light microscope, Gasic and Berwick (1963) showed that, with suitable modification and precautions, the Hale stain could be used in conjunction with electron microscopy to demonstrate sialic acid-containing mucin at the surface of TA3 cells. These authors pointed out that the usefulness of the Hale stain depends on the formation of ferric ferrocyanide at the site of "sialomucin". The ferric ferrocyanide is seen as a blue colouration in the optical microscope and they contended that on theoretical grounds it should not be necessary to use the complete Hale-stain when using the electron microscope, since the particles of bound colloidal iron could be detected directly. However, they found that such particles were extremely small and required very high magnification, and were easily mistaken for bodies of similar size of "non-specific" nature. They used the complete stain in the electron microscope, where one gets definite crystal formations which are larger than those of

5C

colloidal iron and quite distinct from non-crystalline structures. Again the specificity of the reaction between the reagent and sialic acid-containing material was suggested by the ability of neuraminidase treatment to abolish the staining at the cellular surface, and this phenomenon was clearly illustrated with TA3 cells by Gasic and Berwick (1963) in their paper. When MCIM sarcoma cells and the undifferentiated mammary adeno-carcinoma 15091A were examined (Gasic and Berwick, 1963) by the same technique, they were found to have almost no "sialomucin" or, at least, a greatly decreased amount of it on their cellular peripheries. It is notable however, that Wallach and Esandi (1964) were able to demonstrate measureable quantities of non-lipid sialic acid in MCIM tumour cells, as well as a reduced electrophoretic mobility of the ascitic form of the cells following treatment with neuraminidase. Therefore, this lack of Hale-staining material at the surface of a cell should not be taken as evidence for the absence of sialic acid-rich material at the cellular periphery.

An interesting application of the Hale stain is provided by the work of Defendi and Gasic (1963) who showed that embryonic hamster cells transformed *in vitro* by polyoma virus possessed a thick coat of "sialomucin". The reaction at the cellular periphery varied in the cell lines examined, and Defendi and Gasic (1963) suggested that the increase in "sialomucin" at the surface of the transformed cells is one of the main factors responsible for the loss of contact inhibition. Colloidal iron has also been used as a stain of isolated plasma membrane. Benedetti and Emmelot (1967) have examined, by the electron microscope, the distribution of colloidal iron hydroxide granules upon plasma membrane fragments, isolated from normal rat liver and transplanted rat hepatoma 484. These authors were able to show, from experiments using neuraminidase treated material, that the colloidal iron particles reacted with the membrane-bound sialic acid. Further, by taking membrane junctions as markers of the intra- and extracellular sites on the isolated membrane, Benedetti and Emmelot (1967) were able to demonstrate, by its asymmetric staining, that neuraminidase-sensitive sialic acid is confined to the extracellular side of the plasma membrane. This is a particularly important piece of work, not only for showing by electron microscopy the presence of carbohydrate containing material in isolated membranes, but also because of the demonstration that this material is confined to the extracellular side of the plasma membrane which, in the intact cell, is the cellular surface. Benedetti and Emmelot (1967) suggest that the staining with colloidal iron is the result of the inter-action of the positively charged iron hydroxide complex with the negatively charged carboxyl groups of the sialic acids. They were able to show that incubation of the membranes in physiological saline with the strongly basic polyamino acid, polylysine, prevented the subsequent staining reaction. The colloidal iron staining solution used had a pH of 1·7, and it is interesting

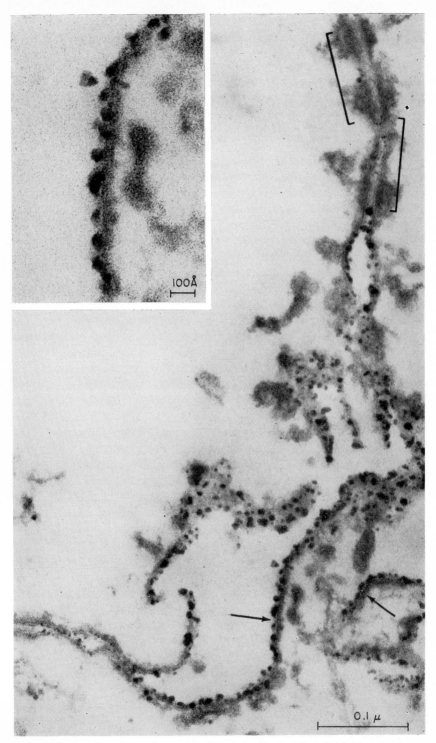

Fig. 4. Isolated rat-liver plasma membranes stained with colloidal iron hydroxide. The electron-dense granules are restricted to the outer leaflet of the membranes (insert). Junctional complexes (brackets) are not stained. From Benedetti and Emmelot (1967) with permission of *J. Cell Sci*.

that the sialosyl residues of the liver membrane still show appreciable ionisation of carboxyl groups at this low pH value. The work of Marikovsky and Danon (1969) is particularly useful when explaining the chemistry of the binding of colloidal iron hydroxide particles. These workers demonstrated that young human and rabbit erythrocytes, separated from old cells by differential flotation, have a raised electrophoretic mobility compared with old cells, indicating that young cells possess an increased negative charge density. The young cells, on treatment with positively charged colloid, showed a dense and regular deposition of particles at the surface when examined in this section in the electron microscope. This labelling was markedly decreased, as well as being irregular in location, on the surface in the case of the older erythrocytes, and this decreased labelling correlates with the electrokinetic properties of the cell. Treatment of young cells with neuraminidase partially reduced the binding of positively charged colloid, and in electrophoretic behaviour, enzyme treated cells behaved like old erythrocytes. On doubling the concentration of neuraminidase, the electrophoretic mobility of the cells was reduced by 70 and 80% for rabbit and human cells respectively, and the labelling with positively charged colloid was abolished.

In addition to colloidal iron, other "cationic dyes" have been used to examine the surfaces of animal cells in the electron microscope. Among them are colloidal thorium (Revel, 1964), ruthenium red (Luft, 1964) and alcian blue. The last stain has been used by Behnke (1968) in the fixative to give greatly improved preservation of surface staining material.

Ruthenium red, which is sometimes described as "ammoniated ruthenium oxychloride", and is usually prepared with chloride as the counter ion, has the proposed (Fletcher *et al.*, 1961) structural formula:

$$(NH_3)_5 \, Ru—O—Ru \, (NH_3)_4—O—Ru \, (NH_3)_5{}^{6+}$$

The average oxidation number of the Ru atoms is $+10/3$ and the cation of the complex may be easily oxidised to ruthenium brown in which the average oxidation number of the Ru atoms is $+11/3$. The compound may be repeatedly cycled between the red and brown forms. A mechanism by which ruthenium red (when used in conjunction with osmium tetroxide[2]) is thought to act as a label for electron microscopic visualization of tissue polysaccharide has been put forward (Luft, 1971a). It is thought that this complex first binds to the acidic mucopolysaccharides and related substances of the tissue, and is then oxidized by the osmium tetroxide of the

---

[2] The use of ruthenium red with osmium tetroxide is particularly important when this dye is used in electron microscopy. Osmium tetroxide contributes "an essential amplification of the reaction to produce sufficient density to be visible easily in the electron microscope", Luft (1971a). The ruthenium red may be conveniently added to the fixatives used for processing the tissue, being incorporated into the initial glutaraldehyde and subsequent osmium tetroxide fixation stages. Though ruthenium red may be omitted from the glutaraldehyde fixative, the use of the ruthenium red–osmium tetroxide mixture is critical.

fixative into ruthenium brown which in turn oxidizes the saccharide material to which it is bound with subsequent reduction back to ruthenium red, coupled with the reduction of osmium tetroxide into lower insoluble products; such as an inorganic polymer similar to that postulated for ruthenium dioxide. The ruthenium red may thus be regarded as a catalyst in which one molecule of the complex is able to reduce several molecules of osmium tetroxide. These insoluble products are considered to be responsible for the increase in contrast seen in the electron microscope. Alternatively, the reaction may not be catalytic in the sense that the ruthenium red transfers many electrons from the substrates to osmium tetroxide, but may follow a "self propagating" mechanism whereby the oxidation of the polysaccharide generates a new carboxyl group which binds a new ruthenium red molecule, and so on. Certainly the mechanism of its action is complex.

Ruthenium red has been used alone as a stain for optical microscopy since the end of the last century, especially for the study of acid plant polysaccharides (Mangin, 1893), and has attracted much use as a stain for pectic substances, though as Luft (1971a) points out botanists have used it "occasionally with a disclaimer as to specificity". Certainly the specificity of such a reagent in its initial interaction with acidic sugar is of major importance. There is biochemical evidence (Utsumi and Oda, 1972) that the staining by ruthenium red of cells depends on the presence of glyco-proteins containing sialic acid at the cell surface. However the complex nature of the interaction of acidic sugars with ruthenium red has been demonstrated by Stoddart et al. (1969), who showed that the hyperchromic effect following the binding of the dye to polygalacturonic acid became a hypochromic effect after titration of all the carboxyl groups, indicating a continued interaction, other than by simple ionic binding, at low pH. Osmium was not present. In 1964, Luft demonstrated that ruthenium red could be adapted with heavy metal staining for use in the electron micro-scope, and has shown its use with a number of animal tissues (Luft, 1971b). He suggests (Luft, 1971b) that this organic dye stains extracellular materials in animal tissues which are probably acidic mucopolysaccharides, and it complements other techniques and has the advantage of being of fine grain, high resolution and good contrast. Further, the method developed by Luft (1971b) does not require conditions of low pH, a distinct advantage compared to other methods described above. This stain has been used by a number of authors including Carr et al. (1970) who examined the surface of the murine peritoneal macrophage with this reagent in the electron microscope, together with a number of the agents described earlier in this section in both the optical and electron microscope. These workers were able to demonstrate with ruthenium red that the murine peritoneal macrophage has a prominent cell coat; this "coat" the authors point out is stained readily by those techniques considered to demonstrate acid mucopolysaccharides (that is ruthenium red, colloidal iron and thorotrast), but less readily by PAS and periodic acid-methenamine-silver methods; techniques regarded as principally demonstrating neutral mucopoly-saccharides. The ruthenium red technique has also been used for investi-gating the changes on the cell surfaces which take place when a cell is transformed with oncogenic virus. Martinez-Palomo and Brailovsky (1968) studying embryonic hamster cells transformed by Adeno-12 and SV40 viruses, found that when compared to normal cells, they showed an increase in thickness of ruthenium red-staining material on the extracellular side of the plasma membrane. This increase in thickness of ruthenium red staining material at the surface of the transformed cells was correlated with their lack of contact inhibition. Martinez-Palomo and Brailovsky (1968) suggested that, as a result of the malignant transformation, the acid mucopolysaccharide at the cell surface increases in amount, and that this result extends the observations of Defendi and Gasic (1963) with Hale stain described above. In a subsequent publication which was extended to two spontaneously transformed cell lines where the staining layer was also found to be increased, Martinez-Palomo *et al.* (1969) quote values (in Å) for the average thickness of the stained layer, and contend that an increase of sialo-mucin is one of the main factors responsible for a cell's loss of contact inhibition. This suggestion will be examined further when the functional significance of membrane heterosaccharides is discussed, though one should here urge a general word of caution on the advisability of extrapolating from such microscopical observations to statements on chemical changes likely to be taking place at the cellular surface. It must be reiterated that it is important that the specificity of any staining reaction used to investigate cell surfaces should be fully understood; in the case of ruthenium red the mechanism of the reaction is seen to be complicated, and although the bulk of available information suggests that ruthenium red has an affinity for acidic poly-saccharides there is evidence that this material will also bind to cephalin. The difficulty of interpreting results from cells which have been grown in a serum-supplemented medium should not be underestimated, for it should be remembered that the denser layer seen with ruthenium red at the cell surface could, at least in part, be serum glycoprotein adsorbed from the culture medium. These general words of caution, it is suggested, apply not only to the specific example of the virally transformed cells and their concomitant surface changes, but to any tissue culture system being so examined. Perhaps the greatest advantage ruthenium red has, as a stain, is the ability to demonstrate material to be present at the cell surface which would not ordinarily be visualized by the conventional heavy metal staining techniques of electron microscopy. To judge from the action of PAS or colloidal iron on the cell surface, this material is likely to contain a significant heterosaccharide component.

In this section the question of surface staining in microscopy has largely centred on results obtained with malignant cells, although it should not be assumed that such data are confined to cancer cells. Here the work of Rambourg and Leblond (1967) and Rambourg (1969) is particularly important. They used periodic acid-silver methenamine, a technique derived from the PAS method (except that the aldehyde groups produced by the periodic acid oxidation are reacted with silver methenamine with the release of free silver), and considered by these authors (Rambourg and Leblond, 1967) to be fairly specific for detection of glycoproteins, together with colloidal thorium as a method for detecting acidic carbohydrates. These investigations were able to confirm earlier light microscopic observations (Rambourg et al., 1966) on the presence of carbohydrate at the surface of some fifty different cell types in the rat. Rambourg et al. (1966) had concluded "not only that the presence of a 'cell coat' is a common feature of vertebrate cells, but also that this coat is rich in carbohydrates". In a further paper Rambourg (1969) examined thin sections of rat tissues, fixed with glutaraldehyde and embedded in glycolmethacrylate and stained with a chromic acid phosphotungstic acid mixture, and observed an intense reaction on the outer surface of the plasma membranes of more than forty-five cell types. In view of the topographical similarity between this staining and that found with PAS, this latter staining technique is considered by Rambourg (1969) to be a means of detecting glycoproteins that contain sialic acid residues. In a recent review Rambourg (1971) points out that the anionic stain, phosphotungstic acid, has been criticized as a stain for polysaccharide, because it has been suggested that there is no direct evidence for the involvement of free polysaccharide. However, Rambourg (1971) makes reference to a number of experiments on the action of phosphotungstic acid at low pH with model compounds and on rat tissues which had been subjected to chemical modification, as well as a series of radio-autographic experiments involving the use of ^{3}H-fucose to label glycoprotein. From all of these experiments it is presumed that glycoproteins are indeed involved in the reaction with phosphotungstic acid.

Taken as a whole, the microscopical evidence claimed to indicate the presence of heterosaccharide material at the cell surface is very persuasive, and it would appear from this data that the existence of such materials is a common, and probably universal, feature of animal cell surfaces. It will be seen that histochemical evidence for the presence of sugar-containing macromolecules at the cell periphery has been derived from the fields of light and electron microscopy principally through the use of chemical techniques involving periodate oxidation, as well as the binding of cationic materials to acidic residues. Very often the methods devised for the electron microscope have been adapted from techniques originally used for demonstrating surface heterosaccharides in the light microscope. Although the sum total of microscopic evidence available, especially when taken in

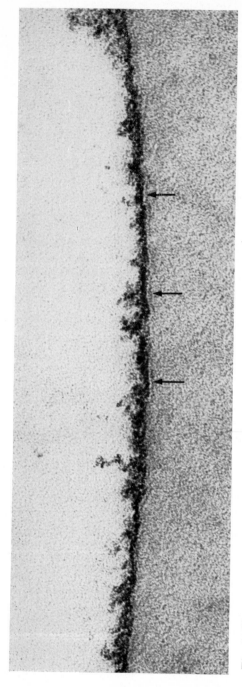

Fig. 5. Electron micrograph of a thin section showing part of the surface of a saline washed human erythrocyte stained with ruthenium red by the technique of Luft (1971b). The inner dense leaflet of the membrane is visible in some places (arrows). The very densely stained surface layer is considered to be the site of the glycosylated portion of membrane glycoproteins. ×240,000 (from an unpublished study by A. M. Glauert and G. M. W. Cook).

conjunction with the results of electrophoresis experiments, strongly supports the contention that sugars are important constituents at the cell surface, nonetheless the question of the specificity of the reagents used is of major consideration and a number of the techniques used suffer from the criticism that their specificity is not sufficiently precise. For example, hydroxyamino acids may give a false positive for carbohydrate in the PAS reaction, though, as Rambourg (1971) points out, this is only likely in those cases where prolonged periodate oxidation has been used. Ruthenium red reacts with a range of molecules including carbohydrates and lipids (Luft, 1971b) and again a knowledge of the specificity of the method is of considerable importance when it is intended to use such a compound to identify a particular class of molecule in cellular structures. Admittedly this cationic dye may encounter principally carbohydrate rich macromolecules at the intact cellular surface, but Luft (1971b) rightly points out that "considerably more work is required to establish the specificity of the ruthenium red reaction for acidic polysaccharides". It has been shown (Glauert and Cook, unpublished observation) that in the case of the saline washed human erythrocyte, a thick layer of ruthenium red-staining material is demonstrated and the prior treatment of the cells with neuraminidase reduced the thickness of the subsequent stained layer, suggesting that the dye is bound to the sialic acid moieties of the glycoproteins at the cell surface. However in the case of murine lymphocytes, although the saline washed cells possess a ruthenium red-staining layer at their surfaces, neuraminidase treatment does not greatly diminish the staining reaction. Therefore, what may appear as a specific reaction for sialic acid moieties in the case of the human erythrocyte, does not necessarily hold for all cell types. A possible explanation of the above observations may be provided by reference to electrokinetic data from which it has been found that neuraminidase-sensitive sialic acid residues account for a much greater proportion of the negative s surface changes in the case of the human erythrocyte than is the case for the murine lymphocyte (Cook and Jacobson, 1968). That neuraminidase treated lymphocytes still possess an appreciable anionogenic charge might suggest that ruthenium red had a specificity for polyanions in general; a further complication is that human erythrocytes fixed for extended periods with formaldehyde, which reacts only with cationogenic groups, remain unstained after treatment with ruthenium red.

B. Methods Involving the Use of Lectins as Histochemical Reagents

1. APPLICATION IN FLUORESCENCE MICROSCOPY

Recently, the problem regarding the devising of a stain with specific binding properties for saccharide moieties for use in microscopy has been greatly aided by the use of lectins. Lectins are plant agglutinins, which are proteins that have specific binding properties for sugar ligands. A number of different lectins have been isolated and their specificity for saccharide moieties determined. Concanavalin A, a lectin isolated from Jack beans (*Canavalia ensiformis*) and having a specificity towards α-D-gluco-pyranosyl and α-D-mannopyranosyl residues, has aroused considerable interest in investigating the chemistry of cell surfaces. This lectin has been tagged with fluorescein isothiocyanate, using mild conditions of about pH 7, by Wayne Smith and Hollers (1970) and used to examine lympho-cytes of mouse, guinea pig and man. These workers found that fluorescent staining of the membrane of the "tail" of motile lymphocytes was apparent after 10 minutes of treatment and of the cytoplasm of the tail region after half an hour. The staining could be inhibited by α-methyl-D-manno-pyranoside and α-methyl-D-glucopyranoside as well as unconjugated lectin. It is likely that the fluorescence seen in the cytoplasm is the result of pinocytosis, stimulated by the binding of the lectin to glycoprotein receptors on the cell surface. Indeed cyanide and antimycin, used at concentrations known to inhibit pinocytosis in macrophages, prevented cytoplasmic fluorescence but not the fluorescence at the membrane. Concanavalin A tagged with fluorescein isothiocyanate was also investigated by Allen *et al.* (1971) using murine peritoneal macrophages. In this system staining shown to be specific by the use of methylglycosides as haptenic inhibitors, was confined to the cell surface when performed in the presence of potassium cyanide or antimycin. These latter authors were able to show by means of Ouchterlony gel diffusion against shellfish glycogen, that the labelling of the lectin with fluorescein isothiocyanate did not modify the specific carbohydrate binding properties of the agglutinin. Both the above papers indicate that carbohydrate residues which are either α-D-mannopyranosyl or α-D-glucopyranosyl are present in the cell types discussed.

In addition to the use of fluorescein conjugates of concanavalin A in fluorescence microscopy, the technique of immunofluorescence for detecting cell bound lectin, and consequently the demonstration of the presence of certain sugar residues, has been described (Mallucci, 1971). In this method anti-concanavalin A serum was raised in golden hamsters and was used, together with commercial fluorescein conjugate of anti-hamster globulin, to detect, by the indirect method, lectin bound at the surface of a number of normal and virally transformed cells of mice. It was noted (Mallucci, 1971) that when this technique was used with macro-phages, it was necessary to pre-adsorb the fluorescein conjugate of anti-hamster reagent before addition to the system in order to prevent non-specific fluorescence occurring during the assay. This procedure was not necessary when the other cell types were examined. This study (Mallucci, 1971), besides being useful for comparing the lectin-binding properties of the various cell types, was particularly interesting because the method was used to study the reappearance of lectin-binding sites after trypsin treatment. No detectable binding of concanavalin A was found after treating the cells with 0·2% trypsin; the reappearance of lectin-binding took some 8–10 hours to return to the control levels.

Fluorescence microscopy of lectins has not been confined to the use of concanavalin A. Cronin *et al.* (1970) have conjugated another lectin, wheat germ agglutinin, with fluorescein isothiocyanate and shown an increase in the number of cells staining when Chinese hamster fibroblasts were passaged after transformation with Schmidt-Ruppin strain of Rous sarcoma virus; while Fox *et al.* (1971) using fluorescein tagged wheat germ agglutinin have found that normal 3T3 cells have lectin sites exposed during mitosis. The

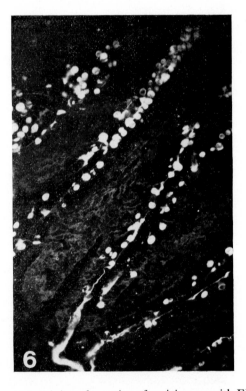

Fig. 6. Fluorescent staining of a section of rat jejunum with FITC labelled concanavalin A. There is intense fluorescence of goblet cells lining the crypts and fluorescence is also visible in collagen and reticulin, some of the latter probably representing the surface carbohydrate of other cells. The intense fluorescence of goblet cells arises from the staining of mucin (see Chapters 3 and 5) within their cytoplasm. (Photograph by courtesy of Dr. J. A. Kiernan) ×170.

possible significance of these results will be discussed later (see Chapter 6) but they are included as an example of the use of a lectin, other than concanavalin A, in the demonstration of saccharide residues of the cell surface by fluorescence microscopy.

Recently the use of fluorescent labelled lectins has been extended to the investigation of paraffin sections of variously fixed animal tissues. Stoddart and Kiernan (1973a) used fluorescein-labelled concanavalin A to locate the α-D-arabinopyranoside configuration in a wide range of paraformaldehyde fixed tissues of the rat, and compared their results with conventional staining procedures. Many cellular surfaces gave some staining, while some components of the intracellular matrix (especially the glucosyl groups of collagen) and some mucins showed very strong staining. The specificity of the method was confirmed by selective abolition of fluorescence with competitive inhibitors and EDTA.

In a parallel study (Kiernan and Stoddart, 1973), aprotinin, a small protein of bovine lung, was used as a similar specific stain for sugars containing carboxyl groups (Stoddart and Kiernan, 1973b) and again showed staining of cellular surfaces, certain matrix materials and mucins. Recent unpublished work by Collins, Jacobson and Stoddart has given strong indications that such fluorescent-labelled reagents can be used to detect differences between the surfaces of normal and malignant cells in certain types of leukaemia in mice and humans, both in smears and paraffin sections, provided that suitable fixations are used. In this work, methanol or glutaraldehyde gave better results than formalin fixation. Both concanavalin A and aprotinin occasionally stain nuclear structures, especially chromosomes during late prophase to anaphase.

2. Application in Electron Microscopy

In the case of the electron microscope, Bernhard and Avrameas (1971) have made use of the lectin concanavalin A to visualize carbohydrate-containing cellular components, especially those present in the cell "coat". Noting that concanavalin A has more than one active site these authors detected the presence of cell-bound lectin, and hence certain cell-bound saccharide moieties, by binding a glycoprotein, horse-radish peroxidase, to the free valency sites on the agglutinin. The catalytic activity of the peroxidase was finally revealed by the diaminobenzidine method. Control experiments, in which α-methyl-mannopyranoside was used as a specific haptenic inhibitor, demonstrated the specificity of the technique. With this technique, Bernhard and Avrameas (1971) were able to employ concanavalin A successfully for demonstrating cellular carbohydrate components. When the concanavalin A pretreatment was carried out on living cells, the final electron-dense reaction product was limited exclusively to the cell surface or within pinocytotic vesicles close to the cell periphery.

Penetration into the cytoplasm could be achieved by prolonged treatment of fixed cells with lectin when Bernhard and Avrameas (1971) were able to detect specific and strongly positive staining in the area of the Golgi apparatus and in the perinuclear space; though in the latter case this may

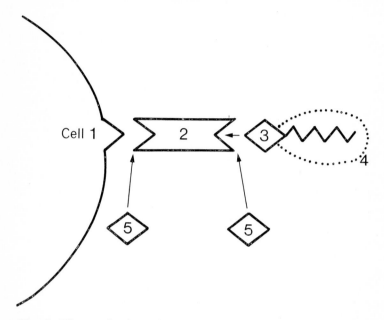

Fig. 7. The mechanism of the concanavalin A-peroxidase reaction for the visualization of cellular carbohydrates by the method of Bernhard and Avrameas (1971). 1, Cell with appropriate carbohydrate residue available at cell periphery. 2, Concanavalin A shown diagrammatic purposes as being in a divalent form. 3, Peroxidase. 4, Electron-dense reaction product after diaminobenzidine. 5, Appropriate competitive methyl glycoside for use in the control experiments. Redrawn from the scheme proposed by Bernhard and Avrameas (1971).

be due to non-specific trapping of the lectin. The reaction with the Golgi area is not surprising in view of the importance of this structure in the biosynthesis of glycoproteins. The results of the technique of Bernhard and Avrameas (1971) greatly strengthens the contention that sugars are important cell-surface constituents, and the authors made the point that "the use in electron microscopy of well defined lectins to visualize carbohydrates of a given specificity is tempting, particularly for further explanation of the cell coat". Their technique has been used with Ehrlich ascites carcinoma cells and ascitic form of murine L1210 leukaemia (Roth *et al.*, 1972).

In addition to the use of peroxidase for detecting concanavalin A bound at the cell surface, Smith and Revel (1972) have made use of haemocyanin (*Busycon canaliculatum*) in a similar manner. The distribution of sugar residues on the surfaces of various cell types, including the rabbit peritoneal polymorphonuclear leucocyte, was studied by binding concanavalin A to the cells present on glass coverslips. The lectin was then visualized by attaching haemocyanin, presumably via its sugar residues, to the bound lectin. The haemocyanin-concanavalin A complex was stabilized on the cell surface by glutaraldehyde treatment, and the individual haemocyanin molecules could then be recognized (haemocyanin is a molecule of distinctive shape and uniform size, being 350Å in diameter, and appearing in the replicas as rectangles and circles, the side and end-on views of the molecule) and counted in shadow cast replicas, prepared with platinum-palladium (80 : 20) and coated with carbon. To ascertain that their method was specific, Smith and Revel (1972) used α-methyl-glucoside as a haptenic inhibitor. The advantage of the method (Smith and Revel, 1972), in common with the use of ferritin labelled lectins (see following paragraphs), is that it enables one to map the distribution of lectin sites over the cell surfaces. For as Smith and Revel (1972) point out, the use of thin sections (as used by Bernhard and Avrameas, 1971) does not allow mapping of the cell surface unless one resorts to serial sectioning and reconstruction procedures. Besides providing a method for mapping the distribution of lectin receptor sites in three-dimensional replicas of intact cells at good resolution with the transmission electron microscope, their work is important in drawing attention to the inhomogeneous distribution of binding sites for concanavalin A on the cell surface. In the case of the rabbit polymorphonuclear leucocyte, the concanavalin A binding sites were found to occupy a characteristic band near the central bulging zone of the cell; in addition, patches of binding sites were often seen in the central zone of the cell, while very few marker molecules were seen on the thin peripheral zone.

In addition to these studies, methods using concanavalin A (to which electron-dense material has been attached by covalent linkage) have also been successfully developed for use as a specific saccharide stain for electron microscopy. Nicolson and Singer (1971) reported the use of the lectin, to which horse-spleen ferritin has been coupled by the use of glutaraldehyde, to examine membrane preparations of human and rabbit erythrocytes in the form of ghosts prepared by lysis in water. In some of the preparations, where the upper membrane of the whole ghost became torn and folded back, exposing the inner surface of the plasma membrane, it was possible to study the staining of ferritin-concanavalin A conjugates with respect to both surfaces of the plasmalemma. Nicolson and Singer (1971) found a very striking difference between the two faces of the membrane after treatment with conjugated lectin. Concanavalin A bound specifically and exclusively to the outer surface of the human and rabbit erythrocyte membranes, more binding being detected in the latter case. The inner membrane could be exclusively labelled by the use of ferritin-labelled anti-spectrin,[3] and in order to show that spectrin did not obstruct the binding of conjugated concanavalin A to the inner surface, membranes from which spectrin had been removed by treatment with EDTA and mercaptoethanol were prepared, and distribution of bound lectin again shown to be limited to the outer surface of the membrane. In the latter case membrane-staining with anti-spectrin was absent. This investigation not only demonstrated very elegantly the presence of certain saccharide moieties in the plasma membrane of the erythrocytes of two different species, but revealed very clearly that there is an asymmetry of staining, the saccharide moieties being confined to the outer surface of the cell membrane. These authors mention unpublished work in which ferritin conjugates of ricin, a lectin with a specificity for galactosyl residues, showed the same asymmetric staining pattern. This asymmetry led Nicolson and Singer (1971) to conclude "that this asymmetry is a general phenomenon for the oligosaccharides of plasma membranes of eucaryotic cells, and to suggest that it is the result of a specific mechanism for membrane biogenesis". Certainly this work adds greatly to the store of evidence favouring the presence of heterosaccharides at cell surfaces, and supports the results made some six years previously by Benedetti and Emmelot (1967) with colloidal iron-stained liver plasma membranes in which a similar asymmetric distribution of sugar moieties (sialic acid) was demonstrated.

Stobo and Rosenthal (1972) have also reported on the use of conjugated lectins as specific saccharide stains for use in microscopy. These workers prepared concanavalin A complexes of ferritin and cytochrome peroxidase, using glutaraldehyde coupling, examined the conjugates against murine thymocytes, and were able to show that these materials were ideal reagents for visualizing surface carbohydrate receptors. In both the electron microscope and the light microscope the cytochrome peroxidase was visualized by the diamino benzedine reaction. In the preparation of their conjugates, care was taken by Stobo and Rosenthal (1972) to protect the carbohydrate-binding sites with a suitable saccharide ligand, α-methyl-D-glucopyranoside, in the cross-linking step so that the biological activity was preserved. Further, these investigators used baker's yeast as a source of cytochrome peroxidase, a preparation which, in contrast to horseradish peroxidase, contains no detectable carbohydrate, and therefore can be used without the possibility of non-covalent binding between the label and the saccharide binding sites on the lectin. Although the work of Stobo and

---

[3] Spectrin is a protein of uncertain homogeneity isolated from erythrocytes in the presence of mercaptoethanol, saline and EDTA, and by some other procedures (Marchesi and Steers, 1968).

Rosenthal (1971) substantiated the presence of carbohydrate moieties at the cell surface as shown by others, using suitable lectin conjugates, their work is particularly important in that these authors were able to show by means of the DNA-synthetic response that their ferritin conjugate possessed biological activity similar to the unmodified agglutinin. The use of cyto-chrome peroxidase is particularly appropriate in studies where a lectin-peroxidase conjugate is used for detecting surface carbohydrate, because in the case of the horseradish peroxidase, a glycoprotein, problems of precipitation are encountered in the coupling procedure (Stobo and Rosenthal, 1972), due to non-covalent binding of the lectin to the complex. In the case of the work of Bernhard and Avrameas (1971), the peroxidase-lectin complex was subsequently formed at the cell surface.

C. Terminology

Taking into consideration the use of lectins in light and electron microscopy, together with the results of earlier chemical techniques, the histochemical evidence for the presence of heterosaccharides at the animal cell surface can be regarded as being firmly established. In the course of this chapter, heterosaccharide materials have been variously described as "mucin" or "sialomucin", and in the literature "sialomucopolysaccharide" and other terms are commonly encountered. Where such terms have been used, the authors have only intended to refer to the stainable entity by describing it in terms used by the original investigators, without necessarily endorsing their usage. Undoubtedly the majority of heterosaccharide[4] materials demonstrable by histochemical techniques are chemically glycoproteins, a number of which contain sialic acid, and it is unfortunate that such loose terminology as "sialomucin" and "sialomucopolysaccharide" has been used in the histochemical literature. In addition to glycoproteins, glyco-lipids are known constituents of the plasma membrane, and in some cell types mucopolysaccharides or glycosaminoglycans are occasionally found to be associated with the cell surface. In the case of the glycolipids, their exact location in the cell periphery has not been entirely elucidated, though they are likely to be near to the external surface since a number of glyco-lipids possess an antigenic expression in the intact cell; hence their contri-bution to histochemical studies on the cell surface must be considered. However, as pointed out by Neutra and Leblond (1966), glycolipids which are mostly soluble in organic solvents, will, in a number of cases, be largely removed from tissues in histological processing. In addition to the above comments on terminology, it is perhaps appropriate to consider here the use of the terms such as "glycocalyx" (Bennett, 1963) and "cell coat" as a

---

[4] Heterosaccharide is used here as a general term for glycoproteins, glycolipids and glycosaminoglycans. The term "complex carbohydrate" has been used by Neutra and Leblond (1966) in a synonymous manner (see Chapter 5).

description for the carbohydrate rich layer present at the cell surface. These terms have no doubt arisen from histochemical evidence of the type discussed above, and Rambourg and Leblond (1967) appear to favour the view that the "cell coat" is an external layer covering the plasma membrane. While "cell coat" may be an ideal usage from a descriptive point of view, this term has been criticized (Cook, 1968), and it has been argued (Cook, 1971) that it should not be thought to imply that molecules assigned by means of electron microscopy to this region of the cell are extra-membranous. Arguments have been advanced (Cook, 1968 and Chapter 1) for glycoproteins and glycolipids as integral parts of the plasma membrane, and indeed there is more recent evidence (Bretscher, 1971) to suggest that glycoprotein molecules may span the width of the plasma membrane. More recently, Rambourg (1971) would prefer to include the plasma membrane and the "cell coat" within the concept of a larger functional complex which Revel and Ito (1967) referred to as the "greater membrane".

In summary, it is stressed that while the term "cell coat" may once have been quite suitable for describing these molecules at the cell periphery by microscopical techniques, it is now misleading since recent advances indicate that such molecules do not constitute a layer apart from the cell membrane but are an integral part of that organelle.

III. Immunochemistry and Lectins

A. Blood-Group Substances

A discussion of the evidence which demonstrates the presence of carbohydrate-containing macromolecules at the cell surface would not be complete without reference to the large body of immunological data available. The carbohydrate nature of blood-group substances has been known for a considerable number of years, though the presence of glycosubstances in membrane structure has been completely ignored until recently. It seems strange that so little correlation was made between the results of immunochemistry and membrane biophysics. This may be partly explained when one considers that the majority of investigations on serologically active molecules had been performed on material derived from sources other than the erythrocyte, the cell whose plasma membrane has been used as a classical model for cell membranes. However, it had been pointed out (Watkins and Morgan, 1952; Morgan and Watkins, 1959) that there is a large quantity of evidence to suggest that only part of the molecules of group specific substances are responsible for their specific serological properties. Further, it would seem reasonable to suppose that the specific fragments of these complex molecules are the same as those specific centres present in the red cell ultrastructure. Indeed, it had been indicated (Watkins, 1955) in the 1950s by the inactivation of water-soluble H blood group mucoid with crude extracts of *Trichomonas foetus*, and also by the fall in the agglutination titre of red cells after incubation with a partially purified enzyme from this extract, that the serological specific structure of the water-soluble H substance is chemically similar to the receptors present in the cell surface of the erythrocyte. It is therefore surprising that the role glycoproteins and glycolipids play in membrane structure has only recently been accomodated in models of plasma membranes. Electrokinetic and microscopical evidence in favour of carbohydrate at the cell surface has been discussed earlier in this chapter, not because the results obtained with these techniques establish any chronological priority over the results of immuno-chemistry (indeed the reverse is a more correct view of the matter), but rather because the results of such physical techniques as cell electrophoresis had always been interpreted in terms of lipoproteins and the classical models of plasma membrane structure. It was only when this interpretation was shown to be invalid that a direct challenge to the classical models could be made. Moreover, cell electrophoresis has the advantage of demonstrating the position of ionogenic materials in the membrane, and it was this tech-nique, used in the absence of chemical fixatives, which showed that carbo-hydrate-containing materials occupied a peripheral position in the mem-brane. This peripheral location of carbohydrate-rich molecules was sub-stantiated by the results of electron microscopy, and together these methods finally established the inadequacies of the bi-molecular leaflet model in providing an explanation not only of cell surface, but also cell membrane structure.

It is not proposed to go into the details of the chemistry of blood group antigens here, the blood group substances have been reviewed in detail elsewhere (Watkins, 1966) and the structure of blood group active materials in plasma membranes is considered in Chapter 3.

B. Lectins

1. GENERAL PROPERTIES

Reference has already been made of the use in microscopy of groups of naturally occurring haemagglutinins, often found in the seeds of leguminous (and other) plants and termed lectins, for the study of the cell surface. In dealing with immunological evidence in favour of the presence of hetero-saccharides at the cell surface, it is appropriate to discuss these carbo-hydrate-binding proteins of plants; indeed the first information on the carbohydrate nature of blood group specificity was obtained with lectins (Watkins and Morgan, 1952). These materials, which have been reviewed recently (Sharon and Lis, 1972), may be divided into two main classes: those which have a specificity directed towards specific blood group substances (for a review, see Bird, 1959), and those which will agglutinate erythrocytes regardless of blood group type, though this need not imply great chemical differences. The latter group of lectins, though being "non-specific" as regards their reaction to human blood group specific structures, nevertheless have a defined specificity towards a particular saccharide residue or short sequence of carbohydrates present in a glycoprotein. These lectins, which occur in a far greater abundance than the type-specific lectins, have been increasingly used in the last few years as probes for investigating the characteristics of cell surfaces, especially the changes which take place at the cellular periphery when the cell undergoes malignant transformation. Although the properties of lectins have been known for a considerable number of years, indeed the study of lectins commenced before the end of the last century (see Sharon and Lis, 1972), it is only over the last five or six years that they have been directed towards an intensive study of cell surfaces; the impetus for these investigations has no doubt been stimulated by the observations that certain lectins are able preferentially to agglutinate tumour cells.

2. INTERACTION OF LECTINS WITH MALIGNANT CELLS

a. *Wheat germ agglutinin*

In 1963, Aub *et al.* described an impurity present in a wheat germ lipase preparation which acted as a preferential agglutinin for tumour cells. The cell-agglutinating and the lipolytic activities appeared to be separable: the substance causing the agglutination was active in the superantants of wheat germ lipase heated up to 65°, even though the lipase activity decreased markedly over this range and, indeed, was abolished at 65°. From dialysis experiments, the clumping factor was shown to be a large molecule, possibly mucopolysaccharide in nature. Aub *et al.* (1963) suggested that they might be dealing with a plant agglutinin, but pointed out that they had been unable to find a precedent in the literature of such a molecule (with a specific reactivity towards tumour cells). It should be pointed out that Aub *et al.* (1963) were not the first to demonstrate that wheat germ lipase preparations had an apparently selective action on tumour cells, for in 1961 Ambrose *et al.* had shown that of two proteolytic enzymes, trypsin and elastase, acid phosphatase and wheat germ lipase tested for their effect on normal and tumour cell growth, only wheat germ lipase showed a selective action on the tumour cells. However, Ambrose *et al.* (1961) interpreted their results in terms of the lipase's reacting enzymatically with the ester linkage between the fatty acid chain and the glycerol residue of a lipid molecule, and suggested that there was a distinct possibility that a change in the lipid component of the cell is involved in carcinogenesis. This interpretation probably reflects on cell surface structure. The "clumping factor" (Aub *et al.*, 1963) or agglutinin was subsequently isolated (Burger and Goldberg, 1967) and characterized as a glycoprotein,

Table I   Some of the More Commonly Used Lectins in Cell Surface Studies

Some properties of purified lectins with panagglutinating properties

Source	Molecular Weight	Subunits	Sugar Specificity	Number of Binding Sites
Canavalia ensiformis (Jack bean)	55,000	2	α-D-Man α-D-Glc	2
Glycine max (Soybean)	110,000		D-Gal NAc	2
Lens esculenta, also known as Lens culinaris (common lentil)	42,000 −69,000	2	α-D-Man	2
Triticum vulgaris (wheat)	26,000		(D-Glc NAc)₂	
Ricinus communis (castor bean) I II	120,000 60,000		β-D-Gal β-D-Gal. D-Gal NAc	

Specificity of some blood group specific lectins

Source	Specificity Human Blood Type	Structure Capable of Reacting	Comments
Dolichos biflorus (horse gram)	A	α-D-Gal NAc	Predominantly A_1 specific. Affinity for A_1 cells over five hundred fold greater than for A_2 cells.
Lotus tetragonolobus (asparagus pea)	H(O)	α-L-Fuc	Three separate agglutinins have been isolated from this source, all with specificity for α-L-Fuc.
Vicia graminea	N	$\text{Gal} \xrightarrow[\beta]{1-3} \text{Gal NAc} \xrightarrow{\alpha}$	Seed anti N lectins cross-react weakly with M-cells.
Iberis amara	M	$\text{Gal} \xrightarrow{\beta} \overset{\displaystyle \text{NANA}}{\underset{}{\text{Gal NAc}}} \xrightarrow[1-3]{\alpha \mid 2-6} \xrightarrow{\alpha}$	

Data compiled from Bird (1959), Sharon and Lis (1972), Uhlenbruck and Dahr (1971).

and that the specific surface site that interacts with the agglutinin contains
N-acetylglucosamine. More recent evidence obtained by Allen and
Neuberger (1972) indicates that wheat germ agglutinin is not a glycoprotein. This lectin has found increasing use, especially when examining
virally transformed cells, as a powerful tool for studying the chemistry of
the cell surface.

b. *Concanavalin A*

Other lectins, in particular the agglutinin from Jack beans (*Canavalia
ensiformis*), concanavalin A, have also been widely used in this field.
Special attention has been focused on lectins since they apparently have
the ability to agglutinate virally transformed cells, as opposed to normal
cells (Burger, 1969; Inbar and Sachs, 1969a). No doubt it is hoped that this
differential agglutinability of normal and transformed cells may be
connected with some surface alteration, which may itself be responsible for
the characteristic properties of the malignant cells. It should, however, be
borne in mind that the use of an agglutination assay for monitoring changes
at the cell surface may not be without problems of interpretation. This
problem has been discussed by Jansons and Burger (1971) who indicate
that agglutination may not be a simple one-step phenomenon with a simple
binding of the agglutinin to the carbohydrate receptor. Apart from this, the
monosaccharide which shows the best haptenic inhibition may not necessarily be the residue present in the oligosaccharide of the surface receptor.
As an example of this latter problem Jansons and Burger (1971) quote the
finding of Kornfeld and Kornfeld (1970) that the carbohydrate moiety
present in the phytohaemagglutinin receptor site of red blood cells does not
contain the haptenic inhibitor N-acetylgalactosamine. This latter point is of
particular interest here since it relates to the question of whether the binding
of lectins to the cell surface can be taken as direct evidence for the presence
of a particular saccharide residue at that surface. Undoubtedly the binding
of lectins to the cell surface suitably controlled with haptenic inhibitors
may be regarded as a further demonstration of the presence of heterosaccharide materials at this site, although the rigorous proof of this contention would entail the isolation and characterization of the respective
receptor sites, an area in which a significant amount of successful effort has
recently been expended (see Chapter 3).

c. *Mechanism of Differential Agglutinin of Transformed Cells*

Apparently virally transformed cells are more agglutinable by several
lectins than are non-malignant cells. Following such a transformation the
cells undergo a loss of contact inhibition of growth, and this change has
been correlated with the appearance of certain agglutinin receptor sites on
the surface of the tumour cell, not available on the normal cellular surface

but which could be exposed by gentle proteolytic enzyme digestion (Burger, 1969). Certainly, brief digestion with proteolytic enzyme of the normal cell surface can bring about a transient release from growth control (Burger, 1970) which makes the above correlation attractive, but enzymatic treatment may bring about more complex changes than are envisaged in a simple uncovering of various sites on the membrane. An alternative mechanism envisages the agglutinin sites as being in a "semi cryptic" form, that is to say, they are available for agglutinin binding but agglutination is prevented by the presence of other surface structures which can be removed by digestion with a proteolytic enzyme.

The more recent findings of Cline and Livingston (1971) and Ozanne and Sambrook (1971), using radioactively labelled concanavalin A, as well as

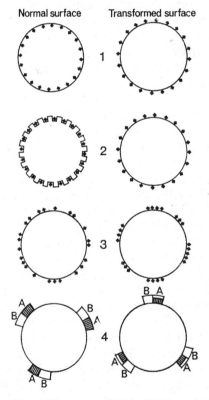

Fig. 8. The possible explanations for the increased agglutinability of transformed cells by lectins. 1, Cryptic site model. 2, Semi-cryptic site model. 3, The redistribution mechanism. 4, The possible involvement of metabolic sites in concanavalin A agglutination after the model of Inbar *et al.* (1971). Site B binds the lectin, while site A determines agglutination. The blank area = active; The hatched area = inactive. Site A may be activated by trypsin. (From Kemp *et al.*, 1973.)

wheat germ agglutinin, and of Sela *et al.* (1971), using labelled soybean
agglutinin, are of special interest and show that there is no difference in the
amount of lectin bound to normal and transformed cells. As such, these
findings indicate that the apparent susceptibility of tumour cells to
agglutination by lectins cannot be explained simply by differences in the
number of lectin sites available at the surface of the transformed cells. In
earlier work by Inbar and Sachs (1969b) with ^{63}Ni-labelled concanavalin A,
a positive correlation between the ability of cells to be agglutinated by
concanavalin A and their ability to bind ^{63}Ni-labelled lectin was found.
However, the more recent papers on binding studies indicate that ^{63}Ni-
labelled preparations are unsatisfactory since varying results were obtained
with different batches of lectin (Ozanne and Sambrook, 1971). Cline and
Livingston (1971) also experienced difficulties with ^{63}Ni-labelled concanavalin A, for example the molecular weight as judged by gel filtration is
lower than that of native lectin. Apparently ^{125}I-labelled or ^{3}H-labelled
(as acetyl group) preparations, where the radioactive label is covalently
linked to the lectin molecule, do not suffer from the problems associated
with the ^{63}Ni product and give consistent results in these other workers'
hands.

More recently, alternative suggestions have been advanced to explain the
differential agglutinability by lectins of transformed and normal cells.
Nicolson (1971) has produced electron microscopic evidence in favour of a
different surface distribution of lectin binding sites in normal and transformed cells. Using ferritin labelled concanavalin A, it was found, on the
surface of mouse 3T3 cells, that the lectin was bound in an essentially
random distribution, while the same line of cells transformed with SV40
virus showed a clustered binding pattern with ferratin labelled concanavalin A[5]. Nicolson (1971) proposes that it is the more clustered distribution of the lectin sites which is responsible for the increased agglutinability
of the transformed cells by concanavalin A, and that therefore there is
no need to invoke the "cryptic" or "semi cryptic" agglutinin sites model
to explain the phenomenon. In the case of the normal cell, Nicolson
(1971 and 1972) has found that treatment with trypsin induces changes
in the distribution of binding sites for concanavalin A, in which certain
regions of the enzyme-treated cell surface possess a clustered arrangement
of lectin binding sites which are very similar to that found in the transformed cell. This proteolysis-induced change in binding-site distribution
is apparently reversible over a period of six hours, with a return of the
properties associated with the normal cell (Nicolson, 1972).

In addition to the above suggestion, a rather more involved mechanism
has been put forward in the case of concanavalin A by Inbar *et al.* (1971),

---

[5] This is a secondary effect. (Nicolson, 1973; Rosenblith *et al.*, 1973.)

who suggest that the binding sites on cell surfaces have two components, namely one which binds the lectin and another which determines agglutination. The former is not temperature sensitive and is active to the same extent in both normal and transformed cells, whereas the component which determines agglutination is temperature sensitive and is in an active form only in the transformed cell. The agglutination of transformed cells takes place at 24° but can be suppressed by lowering the temperature to 4°. With normal cells, trypsin treatment activates the agglutination component and then the cells are agglutinable both at 24° and 4°. Indeed, in the case of the transformed cell after trypsin treatment, the cells retain their ability to agglutinate even at the lower temperature. In both normal and transformed cells the agglutination at 4° requires 500 μg of lectin/ml; using a concentration 5 μg/ml agglutination was only perceived at 24°. These workers (Inbar et al., 1971) consider that the component which determines agglutination may be associated with a metabolic activity, and suggest that an enzyme might alter the surface charge of the cell, although no evidence was given to support this latter contention.

It should be added that the temperature sensitive component which determines agglutination is specific for concanavalin A, as agglutination by wheat germ agglutinin and soybean agglutinin are not temperature sensitive.

The mechanisms suggested above as explanations of the differential agglutinability of normal and transformed cells by lectins have been dealt with in some detail, for as such they demonstrate that heterosaccharides are an important element of the cellular periphery. The use of lectins in the study of cell surfaces, however, would probably not have taken place had it not been for the impetus given by their use in cancer research. Without such an impetus, the use of lectins might well have been confined to blood typing.

d. *Purification*

The use of lectins in cancer studies has made it necessary to obtain these materials in purified form. This need, though desirable, is not essential for blood typing, since there are very few cases where more than one specificity is present in the same extract. Purification may not always be reflected in an increasing activity of the preparation. For example, on purification the lectin from asparagus pea (*Lotus tetragonolobus*) agglutinates human erythrocytes (type *O* cells) at a concentration of 38 μg/ml. However, if crude extract derived from the same seeds is added at a dilution which would not in itself be sufficient to produce agglutination, a much lower concentration of lectin (2 μg/ml) is sufficient to produce agglutination (Yariv et al., 1967). Such lectins may be classified as behaving as though they were "incomplete" (Bird, 1959; Sharon and Lis, 1972). It would be interesting to determine whether the apparent absence of L-fucose-binding sites on the surfaces of a number of normal and transformed cells of hamster, rat and mouse (Inbar *et al.*, 1972) claimed from binding experiments performed with ^{3}H labelled fucose-binding protein, could be confirmed when subagglutinating doses of crude extract are added to such a system. Here it is interesting to note that commercial preparations of concanavalin A have been shown (Uchida and Matsumoto, 1972) to contain varying amounts of demetalized material which would have an effect on the binding capacity of the preparation. However, full activity can be restored by the addition of the appropriate divalent ions. The presence of metal ions is a requirement of a number of lectins for full carbohydrate binding activity.

Usually the preparation of lectins involves the extraction of the seeds in an appropriate buffer. If purified material is required, the extract can then be subjected to a combination of ammonium sulphate fractionation and chromatographic techniques, and in the latter procedure materials such as hydroxyl apatite (Lis *et al.*, 1966) have been used. Recently, techniques of affinity chromatography which utilize the carbohydrate-binding properties of the lectins have been used increasingly in this field. In these techniques, the partially purified plant extract is passed through a column of an appropriate adsorbent, the residual non-specific protein is removed with buffer and the lectin eluted in buffer containing the appropriate competitive inhibitor. Concanavalin A and lectins from lentil (*Lens culinaris* also known as *Lens esculenta*) have been purified (Howard *et al.*, 1971) by use of cross-linked dextran gels (Sephadex: Pharmacia), while agarose gels have been employed in the preparation of Ricinus communis agglutinin (Nicolson and Blaustein, 1972). In the case of fucose-binding proteins of *Lotus tetragonolobus* affinity chromatography on agarose amino caproylfucosamine, which is available commercially (Miles-Yeda), has been described (Inbar *et al.*, 1972). While a conjugate of agarose and *N*-aminocaproyl-D-galactopyranosylamine has been used for the purification of soybean agglutinin (Gordon *et al.*, 1972).

The obtaining of a highly purified lectin with optimal carbohydrate-binding properties is not without problems in its application to cell surface chemistry. Molecular weights reported for a particular lectin may vary greatly, possibly because these proteins are composed of subunits which undergo association-dissociation reactions depending upon the conditions used. Concanavalin A possesses a subunit[6] molecule weight of 27,000

---

[6] Besides the intact subunit, Wang *et al.* (1971) report the presence of three naturally occurring fragments of the subunit. Similar results have been reported by Abe *et al.* (1971) who noted that the ratio of intact subunit to naturally occurring fragments varied amongst a number of preparations of concanavalin A examined. Cunningham *et al.* (1972), noting that it has been suggested that these naturally occurring fragments are formed as a result of proteolytic digestion before or during the isolation of the lectin, showed that trypsin or chymotrypsin treatment of the subunit did not produce these fragments.

(Wang *et al.*, 1971). At pH values lower than 6·0 this lectin exists in solution as a dimer of molecular weight 55,000, while at pH values greater than 7·0 sedimentation equilibrium studies indicate that it exists as a species with a molecular weight of 110,000 (Kalb and Lustig, 1966). The results of low resolution X-ray diffraction studies on crystals grown at pH 6·8 indicate that the lectin is present as a tetramer of crystallographically equivalent subunits of 27,000 molecular weight (Quiocho *et al.*, 1971). Each subunit is capable of binding one saccharide residue, and in the case of the tetrameric concanavalin A four sugar molecules can be found in symmetry equivalent positions (Becker *et al.*, 1971). The distances between these sites on different dimers of the same tetramer are 24·6 Å and 35·1 Å which Becker *et al.* (1971) point out, though short, are not inconsistent with possible distances of closest approach of cells in the agglutination reaction.

IV. Conclusion

In this chapter the main lines of evidence in favour of the presence of macromolecules that contain carbohydrate at the cell surface have been discussed. The discussion has centred on work performed on animal cells, though the occurrence of surface carbohydrate in other eukaryotic cells should not be overlooked and will recieve special attention in Chapter 4.

Certainly the evidence for heterosaccharides as important constituents of cell surfaces can no longer be doubted, and the challenging question must surely be that of their biological function.

References

ABE, Y., IWABUCHI, M. and ISHII, S. (1971). *Biochem. biophys. Res. Commun.* **45**, 1271–1278.

ADA, G. L. and FRENCH, E. L. (1959). *Nature, Lond.* **183**, 1740–1741.

ALLEN, A. K. and NEUBERGER, A. (1972). *Biochem. J.* **130**, 35P.

ALLEN, J. M., COOK, G. M. W. and POOLE, A. R. (1971). *Expl. Cell Res.* **68**, 466–471.

AMBROSE, E. J., DUDGEON, J. A., EASTY, D. M. and EASTY, G. C. (1961). *Expl. Cell Res.* **24**, 220–227.

ANDERSON, S. G. (1948). *Aust. J. exp. Biol. med. Sci.* **26**, 347–354.

AUB, J. C., TIESLAU, C. and LANKESTER, A. (1963). *Proc. natn. Acad. Sci. U.S.A.* **50**, 613–619.

BALLENTINE, R. and PARPART, A. K. (1940). *J. cell. Comp. Physiol.* **16**, 49–54.

BANGHAM, A. D., PETHICA, B. A. and SEAMAN, G. V. F. (1958). *Biochem. J.* **69**, 12–19.

BATEMAN, J. B., ZELLNER, A., DAVIS, M. S. and McCAFFREY, P. A. (1956). *Archs. Biochem. Biophys.* **60**, 384–391.

BECKER, J. W., REEKE, G. N. Jnr. and EDELMAN, G. M. (1971). *J. biol. Chem.* **246**, 6123–6125.

BENEDETTI, E. L. and EMMELOT, P. (1967). *J. Cell Sci.* **2**, 499–512.

BENKE, O. (1968). *J. Ultrastruct. Res.* **24**, 51–69.

BENNETT, H. S. (1963). *J. Histochem. Cytochem.* **11**, 14–23.

BERNHARD, W. and AVRAMEAS, S. (1971). *Expl. Cell Res.* **64**, 232–236.

BIRD, G. W. G. (1959). *Br. med. Bull.* **15**, 165–168.

BRETSCHER, M. S. (1971). *Nature New Biology* **231**, 229–232.

BURGER, M. M. (1969). *Proc. natn. Acad. Sci. U.S.A.* **62**, 994–1001.

BURGER, M. M. (1970). *Nature, Lond.* **227**, 170–171.

BURGER, M. M. and GOLDBERG, A. R. (1967). *Proc. natn. Acad. Sci. U.S.A.* **57**, 359–366.

BURNET, F. M. and STONE, J. D. (1947). *Aust. J. exp. Biol. med. Sci.* **25**, 227–233.

BUZZELL, A. and HANIG, M. (1958). *Adv. Virus Res.* **5**, 289–346.

CARR, I., EVERSON, G., RANKIN, A. and RUTHERFORD, J. (1970). *Z. Zellforsch. mikrosk. Anat.* **105**, 339–349

CLINE, M. J. and LIVINGSTON, D. C. (1971). *Nature New Biology* **232**, 155–156.

COOK, G. M. W. (1968). *Biol. Rev. Cambridge Phil. Soc.* **43**, 363–391.

COOK, G. M. W. (1971). *A. Rev. Pl. Physiol.* **22**, 97–120.

COOK, G. M. W. and JACOBSON, W. (1968). *Biochem. J.* **107**, 549–557.

COOK, G. M. W., HEARD, D. H. and SEAMAN, G. V. F. (1960). *Nature, Lond.* **188**, 1011–1012.

COOK, G. M. W., HEARD, D. H. and SEAMAN, G. V. F. (1961). *Nature, Lond.* **191**, 44–47.

COOK, G. M. W., HEARD, D. H. and SEAMAN, G. V. F. (1962). *Expl. Cell Res.* **28**, 27–38.

COOK, G. M. W., SEAMAN, G. V. F. and WEISS, L. (1963). *Cancer Res.* **23**, 1813–1818.

CRONIN, A. P., BIDDLE, F. and SANDERS, F. K. (1970). *Cytobios* **7**, 225–231.

CUNNINGHAM, B. A., WANG, J. L., PFLUMM, M. N. and EDELMAN, G. M. (1972). *Biochemistry, N.Y.* **11**, 3233–3239.

CURTAIN, C. C. (1953). *Aust. J. exp. Biol. med. Sci.* **31**, 623–630.

DEFENDI, V. and GASIC, G. (1963). *J. cell. comp. Physiol.* **62**, 23–31.

EMMELOT, P. and BOS, C. J. (1966). *Biochim. biophys. Acta* **115**, 244–247.

ENGSTROM, A. and FINEAN, J. B. (1958). *In* "Biological Ultrastructure", p. 236, Academic Press, New York.

EYLAR, E. H., MADOFF, M. A., BRODY, O. V. and ONCLEY, J. L. (1962). *J. biol. Chem.* **237**, 1992–2000.

FLETCHER, J. M., GREENFIELD, B. F., HARDY, C. J., SCARGILL, D. and WOODHEAD, J. L. (1961). *J. chem. Soc.* pp. 2000–2006.

FORRESTER, J. A., AMBROSE, E. J. and MACPHERSON, I. A. (1962). *Nature, Lond.* **196**, 1068-1070.

FOX, T. O., SHEPPARD, J. R. and BURGER, M. M. (1971). *Proc. natn. Acad. Sci. U.S.A.* **68**, 244–247.

FUHRMANN, G. F., GRAZER, E., KUBLER, W., RUEFF, F. and RUHENSTROTH-BAUER, G. (1962). *Z. Naturf.* **17b**, 610–613.

FURCHGOTT, R. F. and PONDER, E. (1941). *J. gen. Physiol.* **24**, 447–457.

GASIC, G. and BERWICK, L. (1963). *J. Cell Biol.* **19**, 223–228.

GASIC, G. and GASIC, T. (1962a). *Proc. natn. Acad. Sci. U.S.A.* **48**, 1172–1177.

2 CARBOHYDRATES OF ANIMAL CELL MEMBRANES

GASIC, G. and GASIC, T. (1962b). Nature, Lond. 196, 170.
GASIC, G. and GASIC, T. (1963). Proc. Soc. Exp. Biol. Med. 114, 660–663.
GORDON, J. A., BLUMBERG, S., LIS, H. and SHARON, N. (1972). FEBS Letters 24, 193–196.
GOTTSCHALK, A. (1957). Biochim. biophys. Acta 23, 645–646.
HANIG, M. (1948). Proc. Soc. Exp. Biol. Med. 68, 385–392.
HAYDON, D. A. (1961). Biochim. biophys. Acta 50, 450–457.
HEARD, D. H. and SEAMAN, G. V. F. (1960). J. gen. Physiol. 43, 635–654.
HEARD, D. H. and SEAMAN, G. V. F. (1961). Biochim. biophys. Acta 53, 366–374.
HIRST, G. K. (1942a). J. exp. Med. 75, 49–64.
HIRST, G. K. (1942b). J. exp. Med. 76, 195–209.
HIRST, G. K. (1948). J. exp. Med. 87, 301–314.
HOWARD, I. K., SAGE, H. J., STEIN, M. D., YOUNG, N. M., LEON, M. A. and DYCKES, D. F. (1971). J. biol. Chem. 246, 1590–1601.
INBAR, M. and SACHS, L. (1969a). Proc. natn. Acad. Sci. U.S.A. 63, 1418–1425.
INBAR, M. and SACHS, L. (1969b). Nature, Lond. 223, 710–712.
INBAR, M., BEN-BASSAT, H. and SACHS, L. (1971). Proc. natn. Acad. Sci. U.S.A. 68, 2748–2751.
INBAR, M., VLODAVSKY, I. and SACHS, L. (1972). Biochim. biophys. Acta 255, 703–708.
JANSONS, V. K. and BURGER, M. M. (1971). In "Glycoproteins of Blood Cells and Plasma" (G. A. Jamieson and T. J. Greenwalt, eds), Lippincott, Philadelphia and Toronto, pp. 267–279.
KALB, A. J. and LUSTIG, A. (1966). Biochim. biophys. Acta 168, 366–367.
KEMP, R. B., LLOYD, C. W. and COOK, G. M. W. (1973). "Progress in Surface and Membrane Science" (J. F. Danielli, M. D. Rosenberg and D. A. Cadenhead, eds), 7, 271–318, Academic Press, London.
KIERNAN, J. A. and STODDART, R. W. (1973). Histochemie 34, 77–84.
KLENK, E. (1958). In "The Chemistry and Biology of Mucopolysaccharides" (G. E. W. Wolstenholme and M. O'Connor, eds), p. 311, J. and A. Churchill, London.
KLENK, E. and LEMPFRID, H. (1957). Hoppe-Seyler's Z. Physiol. Chem. 307, 278–283.
KLENK, E. and UHLENBRUCK, G. (1958). Hoppe-Seyler's Z. Physiol. Chem. 311, 227–233.
KORNFELD, R. and KORNFELD, S. (1970). J. biol. Chem. 245, 2536–2545.
KRAEMER, P. M. (1968). Biochim. biophys. Acta 167, 205–208.
LIS, H., SHARON, N. and KATCHALSKI, E. (1966). J. biol. Chem. 241, 684–689.
LUFT, J. H. (1964). J. Cell Biol. 23, 54A–55A.
LUFT, J. H. (1971a). Anat. Rec. 171, 347–368.
LUFT, J. H. (1971b). Anat. Rec. 171, 369–416.
MALLUCCI, L. (1971). Nature New Biology 233, 241–244.
MANGIN, L. (1893). C. r. hebd. Séanc. Acad. Sci., Paris 116, 653–656.
MARCHESI, V. T. and STEERS, E. (1968). Science, N.Y. 159, 203–204.
MARIKOVSKY, Y. and DANON, D. (1969). J. Cell Biol. 43, 1–7.
MARTINEZ-PALOMO, A. (1970). Int. Rev. Cytol. 29, 29–75.
MARTINEZ-PALOMO, A. and BRAILOVSKY, C. (1968). Virology 34, 379–382.

Martinez-Palomo, A., Brailovsky, C. and Bernhard, W. (1969). *Cancer Res.* **29**, 925–937.

McCrea, J. F. (1947). *Aust. J. exp. Biol. med. Sci.* **25**, 127–136.

Morgan, W. T. J. and Watkins, W. M. (1959). *Br. Med. Bull.* **15**, 109–112.

Morton, J. A. and Pickles, M. M. (1947). *Nature, Lond.* **159**, 779.

Neutra, M. and Leblond, C. P. (1966). *J. Cell Biol.* **30**, 119–136.

Nicolson, G. L. (1971). *Nature New Biology* **233**, 244–246.

Nicolson, G. L. (1972). *Nature New Biology* **239**, 193–197.

Nicolson, G. L. (1973). *Nature New Biology* **243**, 218–220.

Nicolson, G. L. and Blaustein, J. (1972). *Biochim. biophys. Acta* **266**, 543–547.

Nicolson, G. L. and Singer, S. J. (1971). *Proc. natn. Acad. Sci. U.S.A.* **68**, 942–945.

Northrop, J. H. (1926). *J. gen. Physiol.* **9**, 497–502.

Ozanne, B. and Sambrook, J. (1971). *Nature New Biology* **232**, 156–160.

Parsons, D. F. and Subjeck, J. R. (1972). *Biochim. biophys. Acta* **265**, 85–113.

Pickles, M. M. (1946). *Nature, Lond.* **158**, 880.

Ponder, E. (1951). *Blood* **6**, 350–356.

Pondman, K. V. and Mastenbroek, G. G. A. (1954). *Vox Sang.* **4**, 98–107.

Quiocho, F. A., Reeke, G. N. Jnr., Becker, J. W., Lipscomb, W. N. and Edelman, G. M. (1971). *Proc. natn. Acad. Sci. U.S.A.* **68**, 1853–1857.

Rambourg, A. (1969). *J. Microscopie* **8**, 325–342.

Rambourg, A. (1971). *Int. Rev. Cytol.* **31**, 57–114.

Rambourg, A. and Leblond, C. P. (1967). *J. Cell Biol.* **32**, 27–53.

Rambourg, A., Neutra, M. and Leblond, C. P. (1966). *Anat. Rec.* **154**, 41–72.

Revel, J. P. (1964). *J. Microscopie* **3**, 535–544.

Revel, J. P. and Ito, S. (1967). *In* "The Specificity of Cell Surfaces" (B. D. Davies and L. Warren eds), pp. 211–234, Prentice-Hall, New Jersey.

Rosenblith, J. Z., Ukena, T. E., Yin, H. H., Berlin, R. D. and Karnovsky, M. J. (1973). *Proc. natn. Acad. Sci. U.S.A.* **70**, 1625–1629.

Roth, J., Meyer, H. W., Bolck, F. and Stiller, D. (1972). *Expl. Path.* **6**, 189–192.

Seaman, G. V. F. and Cook, G. M. W. (1965). *In* "Cell Electrophoresis" (E. J. Ambrose ed.), pp. 48–65, J. and A. Churchill, London.

Seaman, G. V. F. and Heard, D. H. (1960). *J. gen. Physiol.* **44**, 251–268.

Sela, B. A., Lis, H., Sharon, N. and Sachs, L. (1971). *Biochim. biophys. Acta* **249**, 546–568.

Sharon, N. and Lis, H. (1972). *Science, N.Y.* **177**, 949–959.

Simon-Reuss, I., Cook, G. M. W., Seaman, G. V. F. and Heard, D. H. (1964). *Cancer Res.* **24**, 2038–2043.

Smith, S. B. and Revel, J. P. (1972). *Devl. Biol.* **27**, 434–441.

Stobo, J. D. and Rosenthal, A. S. (1972). *Expl. Cell Res.* **70**, 443–447.

Stoddart, R. W. and Kiernan, J. A. (1973a). *Histochemie* **33**, 87–94.

Stoddart, R. W. and Kiernan, J. A. (1973b). *Histochemie* **34**, 275–280.

Stoddart, R. W., Spires, I. P. C. and Tipton, K. F. (1969). *Biochem. J.* **114**, 863–870.

Stone, J. D. and Ada, G. L. (1952). *Br. J. exp. Path.* **33**, 428–439.

Uchida, T. and Matsumoto, T. (1972). *Biochim. biophys. Acta* **257**, 230–234.

UHLENBRUCK, G. and DAHR, W. (1971). *Vox Sang.* **21**, 338–351.
UTSUMI, K. and ODA, T. (1972). *In* "Histochemistry and Cytochemistry 1972" (T. Takeuchi, K. Ogawa and S. Fujita, eds), Proceedings of the 4th International Congress of Histochemistry and Cytochemistry, Kyoto, Japan. Japan Society of Histochemistry and Cytochemistry, Kyoto.
WALLACH, D. F. H. and ESANDI, M. V. DEP. (1964). *Biochim. biophys. Acta* **83**, 363–366.
WALLACH, D. F. H. and EYLAR, E. H. (1961). *Biochim. biophys. Acta* **52**, 594–596.
WANG, J. L., CUNNINGHAM, B. A. and EDELMAN, G. M. (1971). *Proc. natn. Acad. Sci. U.S.A.* **68**, 1130–1134.
WATKINS, W. M. (1955). *In* "Vth International Congress of Blood Transfusion", Communication 306, Croutzet, Paris.
WATKINS, W. M. (1966). *Science, N.Y.* **152**, 172–181.
WATKINS, W. M. and MORGAN, W. T. J. (1952). *Nature, Lond.* **169**, 825–826.
WAYNE SMITH, C. and HOLLERS, J. C. (1970). *J. Reticuloendothelial Soc.* **8**, 458–464.
WINKLER, K. C. and BUNGENBERG DE JONG, H. G. (1941). *Archs. néerl. Physiol.* **25**, 431–466.
YARIV, J., KALB, A. J. and KATCHALSKI, E. (1967). *Nature, Lond.* **215**, 890–891.

3

Structures of Mucopolysaccharides, Glycolipids and Glycoproteins of Animal Cell Surfaces: Fungal Cell Walls

I. Surfaces of Animal Cells

A. Introduction

The carbohydrates of the surfaces of animal cells can be considered to fall into two groups: those that are components of the "intercellular matrix", and those which are integral parts of the plasmalemma. The former can be said to have an ephemeral association with the plasmalemma during their biosynthesis or secretion, while the latter remain a definite part of the plasmalemma. In reality the distinction is not absolute, but it forms a convenient practical division, since those matrix polysaccharides and glycoproteins that are known in detail are almost all the products of cells specialized for their synthesis, which elaborate extensive matrices. Hence very little of these materials is in contact with, or near to, the plasmalemma. They do, nevertheless, act as part of a permeability barrier to the cells within the matrix and confer certain mechanical properties upon those tissues that contain large amounts of them, such as cartilage. Many of the same general observations that apply to the plant cell wall (see Chapters 4 and 6) can be applied to the matrix, with regard to its influence upon further differentiation of the cells within, and to the difficulties inherent in the ascription of functions to individual components of the whole structure. It should also be borne in mind that those so-called matrix polysaccharides, mucopolysaccharides and glycoproteins may be simply enormously over-produced in certain specialized tissues, while in other cell types they may still be present as quite minor components of the plasmalemma.

Many of the structural, physical-chemical and biosynthetic properties of several matrix substances have been reviewed in considerable detail in recent years (Balazs, 1970; Spiro, 1972; Yosizawa, 1972), as has been the biochemistry of the glycolipids (Kiss, 1970). Accordingly, this chapter will be devoted somewhat to the glycoproteins of the cellular surface, the problems of their isolation and analysis and the partial elucidation of the structure of the best known of them—those of the human erythrocyte.

B. Nomenclature of Sugars

The basic terminology used in the description of sugars and of their linkages is summarized in Fig. 1. A sugar is described as a pentose, hexose, etc. upon the basis of the number of carbon atoms it contains, linked together in its ring and major side-chains. The ring may be five or six-membered; the former is termed a furanose and the latter a pyranose. Both contain an oxygen atom. In solution simple sugars (monosaccharides) exist as equilibrium mixtures of ring and open-chain forms, but in polymers they occur only as their furanose or pyranose forms. A sugar is called an aldose or a ketose, depending upon whether its open-chain form would have an aldehyde or a keto group. In what follows in this chapter aldoses are far more common than ketoses. Sugars form "glycosidic" bonds via C1 of an aldose or C2 etc. of a ketose. These bonds can have α or β configurations and are termed "anomers", and the anomeric configurations of oligosaccharides are fixed (Fig. 1). Both pyranosides and furanosides exist in a variety of interconvertible conformations (Bentley, 1972), though it is seldom known which are assumed by the molecules to be discussed below, especially in membranes.

Sugars are described as D- or L- and this refers to their absolute configuration. The D- and L- isomers are mirror images of each other, about a plane at right-angles to the plane of the page (as drawn) and to the mean plane of the sugar ring (Fig. 1).

The term deoxy-sugar implies a sugar which lacks an hydroxyl group on a carbon atom of its structure which would "normally" bear one. Often it is replaced by an amino or acetyl-amino group, as in N-acetylglucosamine (2-deoxy-2acetamido glucose). Uronic acids are sugars in which the terminal —CH_2OH group is oxidized to a carboxyl group (all here are uronic acids of hexoses, so C6 is oxidized). For recent reviews of aspects of this see Balazs (1970), Hunt (1970), Spiro (1972), Yosizawa (1972) and Gallop *et al.* (1972).

C. Carbohydrate-Containing Molecules of the Matrix etc.

1. HOMOPOLYSACCHARIDES

a. *Chitin*

Chitin is one of the most abundant organic chemicals on earth, and its annual synthesis is estimated to be second only to that of cellulose. It is a component of the exoskeleton of very many invertebrates (but not Echinoderms, and Tunicates), and of the walls of many fungi and some green algae. Like cellulose (q.v., Chapter 4), chitin is normally isolated after exhaustive extraction of other, more soluble, materials by a variety of

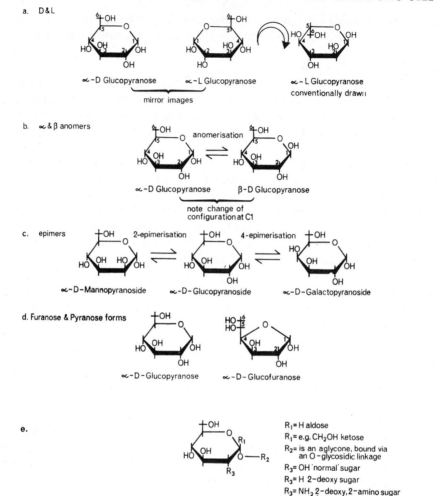

Fig. 1. Nomenclature of sugars etc. Most of the sugars found at the surfaces of eukaryotic cells are aldoses (i.e. they have an aldehyde group at carbon atom 1, that is C1, in their open-chain form). Sialic acid is a ketose (i.e. has a potential keto-group) and forms glycosidic links via C2, not C1. The D- and L-sugars are mirror images of each other, but are usually drawn with the same orientation of the sugar ring. Anomers differ in their configuration at the glycosidic carbon atom and epimers about other carbon atoms. The sugar rings can be bent and folded in several ways: see the review of Bentley, 1972. In an oligo- or polysaccharide the sugar residue with its glycosidic carbon atom unsubstituted, if an aldose, has reducing properties and this end of the polymer chain is called the "reducing end", of which there is only one. Several "non-reducing ends" can occur if the chain branches.

severely degradative procedures and is itself a degradation product. Though homogeneous with respect to its repeating unit, it shows some polydispersity of apparent molecular weight. The procedures generally used in its isolation include extractions with hot alkali and mineral acid, which probably do not degrade the polysaccharide chains too severely, but are certainly likely to cleave more labile bonds between the chitin and other macromolecules (especially proteins), to which it is almost certainly linked *in vivo* (Hunt, 1970). In the majority of exoskeleton (cuticular)

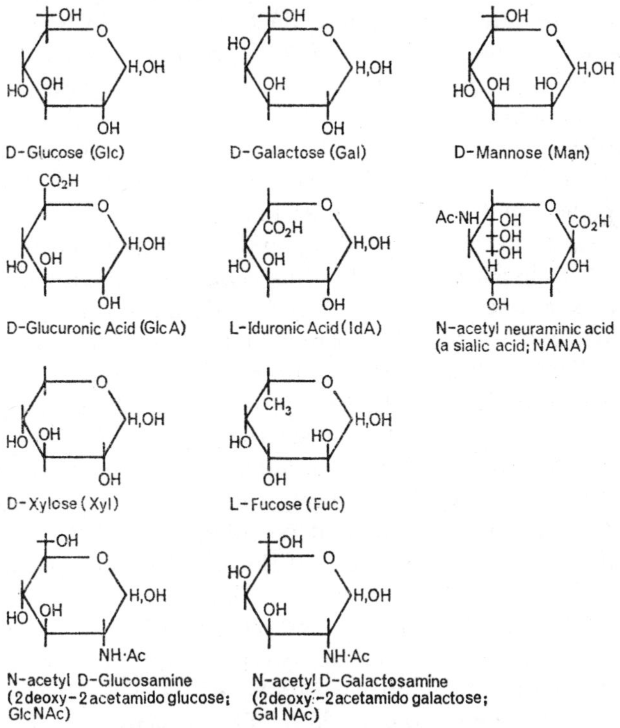

Fig. 2. Sugars common at surfaces of animal cells.

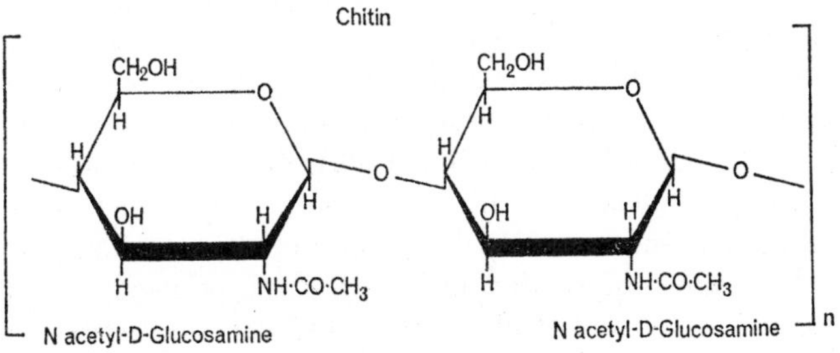

Fig. 3. Chitin: repeating unit.

materials that have been studied, chitin accounts for only 30–40% of the dry weight, and most of the rest is protein. In the *Crustacea* the chitin content is higher (up to about 75%), but these are atypical (though widely used as a source of chitin for chemical analysis). The chitin contents of fungi and algae are lower and variable between species, though about 2·5–5% by weight is a fairly common content for the higher fungi.

Chitin is particularly insoluble in aqueous solutions, though salts that have ions of large radius of hydration do assist its partial dissolution, if present in high concentrations. Thus lithium salts (and especially the thiocyanate) are quite effective, but the "solutions" obtained are colloids, rather than true solutions. Tetrammino-cupric salts do not dissolve chitin, nor, in general, do organic solvents (though chitin is sparingly soluble in anhydrous ammonia).

The evidence for the chemical structure of the chitin chains has been summarized elsewhere (e.g., Brimacombe and Webber, 1964), and is broadly of the same general type as that for the structure of cellulose (q.v., Chapter 4), in that it depends upon the demonstration and character-ization of the disaccharide and oligosaccharide derivatives. The poly-saccharide is a linear chain of 2-acetamido-2deoxy-βD-glucopyranosyl residues (i.e., β-D-*N*-acetylglucosaminyl residues) which are linked via C4 (Fig. 3). Almost all samples of chitin contain some free glucosaminosyl residues, which are likely to be largely, if not entirely, an artefact of the mineral acid-extraction step during isolation. Falk *et al.* (1966) reported a totally *N*-acetylated form of chitin (chitan) from diatoms which shows a different three-dimensional structure from other chitins and possibly indicated that it really is a different polymer.

One of the most important aspects of chitin is that it has an exceptionally well known three-dimensional structure. The generally accepted structure of the unit cell is that derived by Carlström (1962) from X-ray analysis. Using a chitin of *Homarus americanus* (Lobster) he demonstrated an orthorhombic unit cell of dimensions a = 4·76 Å, b = 10·28 Å and c = 18·84 Å. The fibre repeat was 10·28 Å. This structure is folded and the chains asymmetrical. A point of special interest was that the unit cell contained two chains of saccharosyl residues running in opposed directions. A very similar structure, with the same fibre repeat distance, is proposed for cellulose and this gives rise to serious problems in the consideration of the mechanism of biosynthesis of these polymers (for cellulose see Chapters 4 and 5). The studies of Pearson *et al.* (1960) and Marchessault *et al.* (1960) upon the infra-red spectra of chitin are compatible with this model, and both inter- and intra-molecular hydrogen bonds must stabilize the structure.

Not much is known of the detailed biosynthesis of chitin. The precursor in *Neurospora* (Glaser and Brown, 1957) and probably elsewhere (Carey and Wyatt, 1960; Candy and Kilby, 1962) is UDP-*N*-acetylglucosamine, and chitosan primers appear to be required. It would be of great interest to know more of structures of the plasmalemmata of those cells that produce chitin in all types of organism that contain chitinous materials, and to compare this with cellulose-synthesizing systems (the features and problems of which are discussed in Chapters 4 and 5). In particular, the hypodermis of crustaceans might prove suitable for a detailed ultrastructural and biochemical study, since it is known to contain UDP-N-acetylglucosamine and is fairly easily obtainable in reasonable amounts. Possibly, the use of fluorescent-labelled lectins (Stoddart and Kiernan, 1973a; Kiernan and Stoddart, 1973) and of heavy-metal labelled lectins could be applied to the localization of sites of new chitin synthesis, and to the analysis of patterns of chitin deposition and the organization of such patterns.

b. *Animal Celluloses*

Celluloses, essentially similar to those of plants, have been reported from a variety of animal sources, both invertebrate and vertebrate. Among the invertebrates cellulose is known from a number of ascidians (tunicates), but

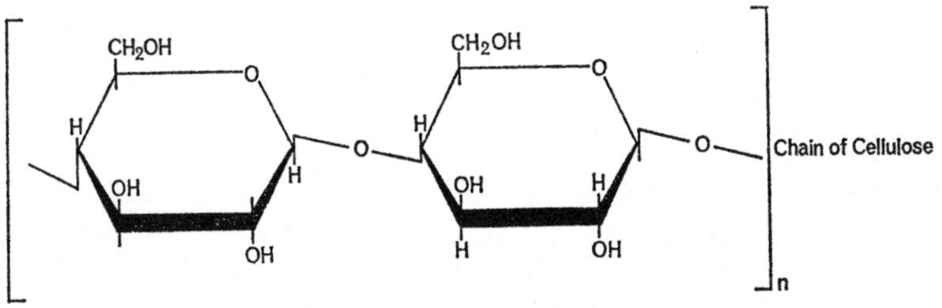

Fig. 4. Cellulose: repeating unit.

not, with certainty, from other invertebrates. These ascidian celluloses are often called tunicins and are associated with the tests of the organisms, in which they are probably always admixed with acidic mucopolysaccharides. Quite extensive chemical studies have been made of them and they appear to be identical in basic structure with the cellulose of plants and bacteria. Hunt (1970) has reviewed these studies. It is very likely that they are synthesized by mechanisms similar to those of plants, and the same general problems of biosynthesis apply, as those discussed in Chapters 4 and 5 and above, for chitin.

In 1958 Hall *et al.* reported the presence of cellulose in aortic tissues and dermis of humans, and the same group subsequently extended their studies to compare tunicate and mammalian celluloses (Hall and Saxl, 1960, 1961; Hall *et al.*, 1960), and found them very similar.

2. GLYCOSAMINOGLYCANS

a. *Chondroitin*

Chondroitin is an acidic polysaccharide isolable from bovine cornea and reported from the skin of a squid (Anno *et al.*, 1964), and very closely related to the chondroitin sulphates and dermatan sulphate. It consists of a repeating unit of (1,4)-*O*-β-D-glucuronosyl (1,3)-2-acetamido-2-deoxy-β-D-galactose, with both sugars in the pyranose form (Fig. 5). It is possible that chondroitin represents the precursor of the chondroitin sulphates, but the relation *in vivo* is still unclear, as is the relation of chondroitin to the surfaces of the cells that produced it (Fig. 5).

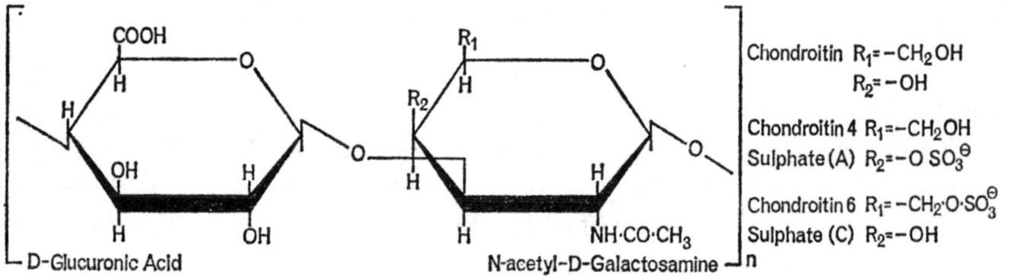

Chondroitin $R_1 = -CH_2OH$
$R_2 = -OH$

Chondroitin 4 $R_1 = -CH_2OH$
Sulphate (A) $R_2 = -O\ SO_3^{\ominus}$

Chondroitin 6 $R_1 = -CH_2 \cdot O \cdot SO_3^{\ominus}$
Sulphate (C) $R_2 = -OH$

D-Glucuronic Acid   N-acetyl-D-Galactosamine

Fig. 5. Chondroitin sulphates A and C: repeating units.

b. *The Chondroitin Sulphates*

The chondroitin sulphates (Fig. 6) are a group of sulphated mucopolysaccharides found in a wide variety of connective tissues from many animal sources. They have been reported in the *Annelida, Arachnida, Mollusca* and vertebrates and will probably be shown in many other phyla in the future. In particular, they are very well known in specialized mammalian connective tissues and chemical studies have largely been directed to materials from this source.

All the chondroitin sulphates are laid down with collagen and other glycoproteins and usually with lipids, in assembly of the complex matrices

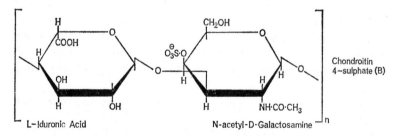

Chondroitin 4−sulphate (B)

L-Iduronic Acid   N-acetyl-D-Galactosamine

Fig. 6. Chondroitin sulphate B: repeating unit.

of connective tissues such as cartilage. It is not known how far they are ordered by other components of the system, and how far they organize the deposition of other substances, in either case the pattern of organization may ultimately derive from the activities and properties of the surfaces of those cells that are producing the matrix. Also, these mucopolysaccharides will constitute an important part of the peripheral microenvironment of the cells that gave rise to them.

(i) *Chondroitin Sulphate A:* Chondroitin sulphate A is of very wide distribution and has the same structure as chondroitin, with a sulphate group substituting at C4 of the *N*-acetylgalactosaminosyl residue (Fig. 5). Brimacombe and Webber (1964) have reviewed the chemical evidence for this structure in detail. *In vivo* the polysaccharide is linked to protein via a glucuronosyl-(galactosyl)$_2$-xylosyl-*O*-serine chain (Lindahl and Rodén, 1966; Rodén and Smith, 1966).

(ii) *Chondroitin Sulphate B or Dermatan Sulphate:* This polysaccharide has been isolated from a variety of sources including skin (Meyer and Chaffee, 1941), lung (Marbet and Winterstein, 1951), heart valves (Deiss and Leon, 1955), umbilical cord (Fransson, 1968) and elsewhere in mammals. It is closely similar to the A sulphate of chondroitin, but differs in having an iduronosyl group in the repeating unit, in place of the glucoronosyl group of chondroitin (Fig. 6). Like chondroitin sulphate A, dermatan sulphate appears to be attached to protein *in vivo* via galactosyl and xylosyl groups, *O*-glycosidically linked to serine (Bella and Danishevsky, 1968; Fransson, 1968). It is notable that heparin is known to have very similar types of linkage.

(iii) *Chondroitin Sulphate C:* This has chondroitin as the basic structure but substituted with sulphate at C6 of the *N*-acetylgalactosaminosyl groups (Fig. 6), in contrast to chondroitin sulphate A. Like chondroitin sulphate A it is very widespread and shows the same linkage to protein (Fransson, 1968).

(iv) *Chondroitin Sulphate D:* A chondroitin sulphate from the cartilage of shark has been named chondroitin sulphate D and is closely similar to chondroitin sulphate C, but more heavily sulphated, with some sulphate groups on the glucuronosyl residue.

(v) *Chondroitin Sulphates of Invertebrates:* Most of the partial structures reported for these resemble those of vertebrates very closely and Hunt (1970) has summarized the field in his book. It is very likely that future studies will reveal several new forms of chondroitin sulphate both in invertebrates and vertebrates, but a proper understanding of their functional importance seems a more distant prospect.

c. *Keratan Sulphate or Keratosulphate*

This polysaccharide occurs in mammalian skeletal tissues, such as inter-vertebral discs, in cornea and in aorta. It is probably quite widespread. The repeating unit is $(1,3)O$-β-D galactopyranosyl $(1,4)$-2-acetamido-2-deoxy-6-O-sulpho-β-D-glucopyranosyl (Fig. 7). Removal of the sulphate yields a product with some cross-reactivity to sera directed against A and B blood-group substances, and it is claimed that fucose is present in the

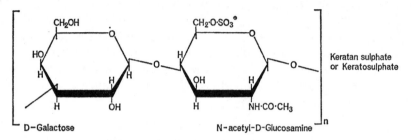

Fig. 7. Keratan sulphate: repeating unit.

polysaccharide to a small extent. There is some evidence for a partial sulphation of the galactosyl residues at C6. Clearly the presence of adequate sulphation is likely to be important in such a molecule, lest it react with circulating antibody. Similar polysaccharides are known in invertebrates (Hunt, 1970).

d. *Hyaluronic Acid*

Hyaluronic acid is a very widespread component of many tissues and seems to be present in the interstitial "spaces" between the plasmalemmata of a great variety of cells. It occurs in the walls of blood vessels, in vitreous humor, synovial fluid and is even produced by some micro-organisms. Brimacombe and Webber (1964) discuss the problems of the isolation of the material in some detail, and also the chemical evidence for its structure.

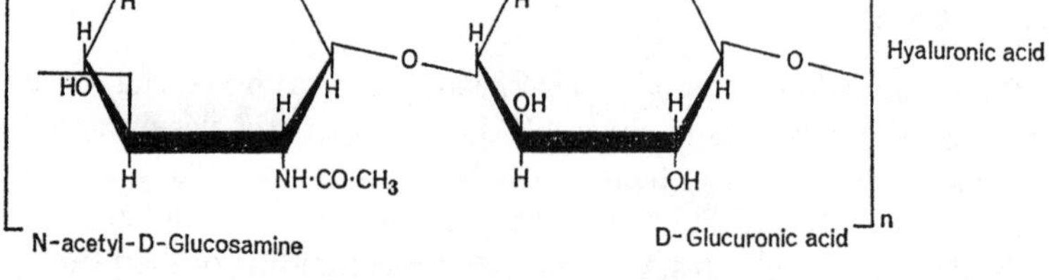

Fig. 8. Hyaluronic Acid: repeating unit.

In vivo it is almost certainly covalently linked to protein. The structure is a chain of alternate *N*-acetylglucosaminosyl and glucuronosyl residues, linked via C1 and C4 of the uronosyl group and C1 and C3 of the amino sugar, both links being of β configuration (Fig. 8). The reducing terminal is an *N*-acetylglucosaminosyl group and the non-reducing terminal is glucuronic acid.

e. *Glucan Sulphates*

Many invertebrates contain sulphated polysaccharides of types other than those described above, including various sulphated forms of cellulose. In some cases these molecules seem to replace the chondroitin sulphates and in others they occur with them, though they are more usually components of secretions, and so fall outside the scope of this chapter. However, the possibility of their presence in the periphery of many invertebrate cells is quite real and will have to be taken into account in any studies of plasmalemmata derived from cells of tissues that produce or contain them. An excellent account of their chemistry has been given by Hunt (1970).

3. OTHER HETEROSACCHARIDES

a. *Polysaccharide Phosphates*

A small number of polysaccharide phosphates have been reported among yeasts and invertebrates. Among the yeasts the wall mannans are sometimes phosphorylated, and can justifiably be considered to fall within the terms of reference of this book. None of the polysaccharide phosphates of invertebrates has yet been shown to be intimately associated with the cellular periphery, but the possibility that some are must be borne in mind.

b. *Jelly-Coat Materials*

The majority of animals produce an envelope, rich in polysaccharide, with which to surround their eggs. The structures of these capsules can be complex and involve the association of components from the ovum itself and from surrounding, secretory tissue. In some organisms the capsule is elaborated into a thick, gelatinous coat and sufficient quantities can be obtained to allow of chemical analysis. A variety of different types of polysaccharide and glycoprotein have been described, but their origins are seldom clear, so that it is difficult to describe them with certainty as cell-surface carbohydrates, rather than secretory products. Many of the glycoproteins isolated from them are similar to cell-surface materials in the species of sugars that they contain, but also resemble soluble glycoproteins in the ratios of these. The eggs of Echinoderms have received particular attention and in several cases have been shown to possess mucopolysaccharides rich in sugars, especially fucose and galactose, which can be implicated in the fertilization process. A comprehensive review is given by Hunt (1970) in his book.

4. Glycoproteins

a. Collagens

Collagens are a class of fibrous, structural glycoproteins, of closely similar composition which are ubiquitous in the animal kingdom. They form an essential part of the extracellular "skeletons" of all tissues, and in some tissues, such as cartilage or tendon, they can form a substantial part of the dry weight of the whole. Chemically they are distinguished by a very high content of glycine and proline residues, and by the presence of hydroxy-proline.

The structure of collagen is complex and unusual for a protein (see Gallop *et al.*, 1972). The molecules as a whole form long, rigid rods, which are considered to be assembled outside the cell, after the synthesis of individual component chains within the cell and their release through the plasmalemma. Cross-links of several types form between the chains once they are outside the cell, making the collagen strong and insoluble. The fully-formed molecule has a molecular weight of 3×10^5 daltons and an axial ratio of about 270 : 1, with a length of about 2900 Å. It consists of three helically wound chains of about the same molecular weight as each other, so that their degree of polymerization is about 10^3. Each of the chains comprising the triple helix is itself a helix, and the chains are closely similar to each other, but not identical. Denaturation of collagens can liberate free, single chains (called α_1, α_2 etc. depending on their type), aggregates of two chains (called β chains) and aggregates of three chains (called γ chains). This basic structure seems to be universal among the collagens, though there is still dispute about the details of how the three chains are intertwined (Rich and Crick, 1961; Ramchandran 1962 and 1967). It is thought by some that the α-chains are not continuous, poly-peptide chains, but are joined together at intervals by some other bond.

Individual collagen (tropocollagen) molecules are usually linked together with a degree of overlap (about 700 Å) to produce the native collagen fibre, with its typical striated periodicity of near 700 Å, visible by electron microscopy. In some cases the periodicity is lacking and some other quaternary structures can obtain—even one branched collagen is known (spongin B).

All collagens appear to be glycosylated, but to a variable degree. The invertebrates generally show a far greater degree of glycosylation of their collagens than do vertebrates, and more than 15% of the weight of the molecule may be sugar. The vertebrates seldom have more than about one tenth of this. A very large range of sugars have been claimed to be attached to invertebrate collagens, including glucose, galactose, mannose, xylose, arabinose, fucose, glucosamine, galactosamine and uronic acid. Some of these might arise in some instances from other associated polysaccharides, but most of them are likely to be genuinely attached. Hunt (1970) again has reviewed the invertebrate collagens admirably in his book.

In the vertebrates, some considerable study has been made both of the nature of the sugar residues attached, and of the type and site of the linkages by which they are held. In these studies much use has been made of the renal glomerular basement membrane (which can be considered to contain a specialized and exceptionally heavily glycosylated form of collagen), and of the sugar-rich collagens of the eye.

Many workers have now confirmed that much of the carbohydrate of several collagens, both vertebrate and invertebrate in origin, is linked via an O-glycosidic linkage to the hydroxyl group of hydroxylsyl residues of the polypeptide chains. This is an extremely unusual linkage for a glyco-protein, though the collagen of earthworm cuticle, for example, displays the more common O-glycosidic linkage to serine and threonine (to which much of the very large amount of galactose in this collagen is attached).

Collagens of many types yield 2-O-α-D-glucopyranosyl-O-D-galacto-pyranosyl hydroxylysine after alkaline hydrolysis in 2N sodium hydroxide at 105^5C for 24 hours. No larger oligosaccharide residues have yet been found (Katzman and Jeanloz, 1970), though the galactosyl-hydroxylysine does occur. Spiro (1970a) has reported most interesting data to show that the ratio of these two glycopeptides varies with the source of the collagen. Collagens from skin (of rat and calf), bovine achilles tendon, and the cornea and selera of rabbits contained about 55–65% of the disaccharide and 35–45% of the monosaccharide, whereas collagens from glomerular basement membrane and anterior lens capsule (both bovine) contained well over 90% of the disaccharide. Carp ichthyocol, a fish collagen, was of intermediate ratio. The ubiquity of these two structures is remarkable in that the amino acid compositions of collagens vary quite considerably, and Eylar (1965) has suggested that they are involved in the export of the molecules through the plasmalemma. As Spiro (1970a) has pointed out, the substitution of hydroxylysine by sugars may well prevent the formation of cross-links between the polypeptide chains, some of which seem to involve this amino acid. If this is so, the sugars will play an important part in determining tertiary and quaternary structure in the molecule.

Spiro (1970b; 1972) has made extensive studies of a number of basement membranes, which represent in part a specialized form of collagen. In particular, the bovine renal glomerular basement membrane has been used. Again disaccharides of glucose and galactose occurred, attached to hydroxy-lysine residues, but heteropolysaccharide fragments were also found after collagenase-pronase digestion. These contained galactose, mannose, hexosamine, fucose and sialic acids and corresponded to an oligosaccharide of molecular weight greater than 3×10^3 daltons, which was linked via asparagine (i.e., with an N-glycosidic bond) or showed microheterogeneity. It is not clear how this larger saccharide is related to collagen in the membrane, but Spiro (1970b) suggests that it is attached to polypeptides which are linked to the collagen proper via disulphide bridges.

Collagens from the "traditional" sources, such as rat tail, also contain small amounts of mannose, sialic acid, fucose and hexosamine when isolated. Since their carbohydrate content is low it is uncertain whether they are truly attached to the collagen, or are minor contaminants, but it is highly probable that some form of covalent linkage exists here also.

b. *Structural Glycoproteins*

In addition to the clearly definable collagens, described above, matrices of connective tissues contain a range of other glycoproteins. Some of these are undoubtedly plasma glycoproteins, but others do not seem to be and are often tightly linked with collagen, perhaps via disulphide bridges. Important though these substances undoubtedly are, there is little detailed information about either their structures or origins in relation to the plasmalemma. Certainly they can carry antigens which may be cross-reactive with antibodies directed towards cellular surfaces, and to that extent may be considered as likely to be representative of surface materials.

D. Relations Between Matrix and Plasmalemma

The matrix polysaccharides play a most important rôle in maintaining the structure of the microenvironment of the plasmalemmata of most animal cells. They constitute not only a large part of the architecture of the intercellular material, but give it a great deal of its ion-exchange and hydration properties. The mechanical properties of the matrix materials have been very extensively studied, but, as with the plant cell wall, the ascription of particular physical properties to individual components of the matrix is a highly dangerous undertaking.

The biosynthesis of the matrix is a complex process, very closely linked to the properties of the plasmalemma. Collagen is synthesized by a sequence of reactions in which the polypeptide chains are assembled upon polyribosomes, they are released and many of their lysyl residues are hydroxylated, glycosylation and, possibly, partial helix assembly begins and finally assembly is completed after the release of the glycosylated chains through the plasmalemma. Once outside the cell, the collagen is further assembled into the giant molecular lattice of the matrix and slowly becomes further cross-linked and less soluble. Basement membranes appear to be synthesized in an analogous way (Spiro, 1970b). It seems extremely likely, since the deposition of collagen is an orderly process, that the plasmalemma has some rôle in organizing its pattern. Animal celluloses and chitin are probably synthesized in ways very similar to the plant celluloses (q.v.), and in these the plasmalemma is the site of synthesis and has an organizational rôle. A possible mechanism for the regulation of the size of the polymers produced is discussed in Chapter 5.

Revel (1970) and others have produced very good evidence for a function of the Golgi apparatus in the production of the glycosaminoglycans (such as the chondroitin derivatives) for export into the developing matrix. The proteinaceous part of the structure appears to be synthesized in the rough endoplasmic reticulum, before being transferred to the Golgi apparatus for glycosylation prior to its leaving the cell. The site of sulphation is not yet fully certain, but present evidence strongly suggests that it tends to take place in the outermost cisternae of the dictyosomes, that is immediately before export to the exterior. It is not known whether the polysaccharide chains are fully complete before the insertion of sulphate groups begins. How the completed glycosaminoglycans are arranged and inserted into the developing matrix is still unknown, though it is extremely likely that the plasmalemma again plays some part.

E. Carbohydrate-Containing Molecules of the Plasmalemma

Two classes of carbohydrate-containing molecules are recognized as being present in the plasmalemmata of eukaryotic cells, the glycolipids and the glycoproteins. The bulk of our knowledge of them is drawn from studies upon animal, and especially mammalian cells, of which the human erythrocyte has been by far the most popular subject of investigation. Consequently, the remainder of this chapter will be rather biased towards this particular type of cell, and some of what we know of it may eventually be shown to be atypical. That caveat apart, the importance and usefulness of the red cell to this field cannot be stated too strongly.

1. STRUCTURAL STUDIES

a. *Conventional Chemical Methods*

Many techniques have been applied to the structural analysis of glycolipids and glycoproteins, particularly the classical methods of carbohydrate chemistry, including periodate oxidation, methylation, partial hydrolysis and enzymatic degradation. Only an outline of each well established procedure will be given here, and for more detailed information technical compendia, such as those of Gottschalk (1972) and Whistler and Wolfrom (1963–5) should be consulted.

(i) *Periodate oxidations, etc.:* Acidic solutions of periodate, such as sodium metaperiodate below pH 5, will oxidise the 1,2 diol configuration (i.e. will split the bond linking two adjacent carbon atoms, when each is substituted by an hydroxyl group, see Fig. 9). At low pH the reaction liberates two aldehyde groups, but they will not undergo further reaction with the periodate. At higher pH this tends to happen and is called "over-oxidation" or "non-Malapradian oxidation", after the originator of the

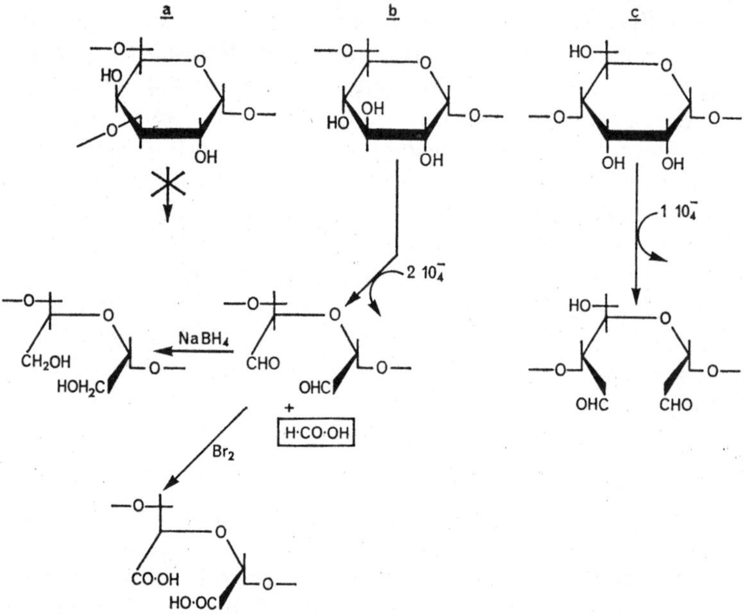

Fig. 9. Periodate Oxidation. The ideal oxidation of three types of linked sugars is shown. Oxidation requires the presence of a 1,2-diol group, at least. Thus a 1,3-linked polymer is unreactive to periodate. A 1,2-diol consumes periodate and a 1,2,3, triol both consumes periodate and liberates formic acid, which can be estimated. The reducing terminal of an oligo- or polysaccharide liberates formaldehyde (if an aldose), which can also be estimated. Thus analysis of the uptake of periodate and the nature and quantities of the products gives information about linkages and sometimes molecular weight. The product of high molecular weight can be oxidized or reduced, hydrolysed and analysed to give information about the original configuration of the saccharides.

periodate method. If the pH falls below about 3, acid hydrolysis may also occur. The oxidation is accompanied by the liberation of formic acid, where the hydroxyl groups form 1,2,3-triols, and by the production of formaldehyde at the free reducing ends (for example in oligosaccharides released by partial hydrolysis of glycolipids). The amount of periodate consumed and products liberated is, therefore, predictable for various structures (see Fig. 9), so making it possible to distinguish different arrange-ments of given sugar residues and to estimate molecular weight by release of formaldehyde. The consumption of periodate and release of products can be followed titrimetrically. Great care must be exercised to ensure that the oxidation of the diols and triols is complete, while no over-oxidation can be allowed. This can be technically very difficult to achieve.

Further information can be extracted from the oligo- or poly-aldehyde also produced, either by reducing it or oxidizing it, and then hydrolysing. Reduction (for example with sodium borohydride) forms a poly-alcohol, the hydrolysis-products of which can be separated by gas–liquid chromatography (GLC), using stero-specific columns to resolve, for example, the erythritol and threitol produced. This can give useful structural information. Oxidation of the poly-aldehyde with bromine-water, followed by hydrolysis, liberates organic acids, such as the two stereoisomers of tartaric acid (sometimes called erythraric and threaric acids, cf. the alcohols). These can be separated again by GLC, by paper chromatography or upon thin layers.

A number of variants of periodate reactions are used for special purposes. A combination of oxidation, reduction and hydrolysis is used in the important Smith degradation of polysaccharides (e.g. Rothfus and Smith, 1963), while more specific oxidizing agents, such as lead tetra-acetate, are occasionally employed.

(ii) *Methylation:* A very informative way of studying the linkages of oligosaccharides is by the methylation of all the hydroxyl groups not occupied by glycosidic, or other links, to form their methyl ethers. After hydrolysis, the various partially methylated products can be separated (upon thin layer, paper or gas–liquid chromatography) and identified, and linkages can then be deduced. The techniques of methylation of sugars are outlined in Table I, and can be applied to oligosaccharides, either free or combined with other materials, and to polysaccharides. Where a dimethyl sulphate (or other) reaction is used at alkaline pH, it is essential to protect any free reducing ends of poly or oligosaccharides by converting them first to their methyl glycosides. This can readily be done by heating them with methanolic hydrogen chloride (i.e. a Fischer–Speier reaction). This glycosidic methyl group is lost again on subsequent hydrolysis. Care must be taken to avoid the loss of the more volatile methylated derivatives during hydrolysis, isolation and characterization.

One of the most difficult aspects of methylation analysis is the synthesis of standard samples of all the various possible partially methylated derivatives of the sugars. To some extent the advent of combined GLC and mass spectroscopy is circumventing this problem.

(iii) *Partial Hydrolysis:* Glycosidic bonds vary greatly in their stability to chemical hydrolysis, and it is common to find much more variation in a hetero-oligosaccharide than in a hetero-oligopeptide of equivalent size. On

Table I. Procedures for Insertion of Methyl Groups into Sugars

Synthesis	Procedure	Application
Methyl glycosides	Heat with Methanol/HCl	Protection of reducing terminal residues
Methyl esters	Expose to diazomethane in ether at ambient temperature	Blockade of carboxyl groups of sialosyl and uronosyl residues
Methyl ethers	i. Dimethyl sulphate at alkaline pH usually several cycles, after protection of the reducing terminal	Exhaustive methylation for structural analysis
	ii. Methyl iodide/silver oxide	Similar to the above: less forcing and hazardous

Methyl glycosides are unstable to aqueous acid and can be visualized, after exposure of chromatograms to strong acid, by treatment with aniline phthalate or phosphate (for thin layer work), or by alkaline silver reagents. Methyl esters and ethers can be detected directly by way of their reducing properties. Methyl ethers are widely separated by gas chromatography and their volatility is a problem in their handling.

the other hand, the specificity of glycosidases is often for one side only of a glycosidic bond, and that bond may occur several times among the small number of sugars present in most oligosaccharides of glycolipids and glycoproteins, so that the use of enzymes in sequencing oligosaccharides differs in nature from the approach in proteins and polypeptides.

The hydrolysis of oligosaccharides under alkaline conditions tends to be complex, and requires the exclusion of oxygen if loss of sugars and formation of tarry products is to be avoided. Hence acid hydrolysis is greatly preferred, and is usually applied first where both forms of hydrolysis are used.

The rate of acid hydrolysis of a glycoside depends partly upon the number of electron-withdrawing or donating groups substituent upon the carbon atoms involved in the bond and, to a much less extent, upon substituents of other atoms nearby. It also depends upon the presence of other groups near the bond which can compete for incoming protons. Hence the nature of the group, at C2 of an aldose, greatly influences the rate of hydrolysis of those glycosides it forms through C1. Glycosides of 2-deoxy-sugars hydrolyse far more rapidly than those of the 2-hydroxylated analogues and these, in turn, are much more labile than the 2-deoxy-2-amino analogues (in each case a factor of about ten times is involved). Thus glycosides of glucosamine are far less easily hydrolysed than those of glucose. If the amino group is blocked by an acetyl group, for example, it does not interact with protons so strongly, so that glycosides of glucosamine are of comparable lability to those of glucose.

Considerations of this type are very important in the design of procedures for partial hydrolyses. If the rate of de-acylation exceeds that of glycosidic hydrolysis in a heterosaccharide containing glycosides of neutral and C2-acetyl-amino sugars, many very resistant bonds will appear. Thus, the condition for optimal partial hydrolysis may be weak acidity, such as 0·1 N-hydrochloric acid, and, for total hydrolysis to determine hexosamine, strong acid hydrolysis with 6 N-hydrochloric acid. This latter would destroy neutral sugars, by converting them to furfural and its derivatives, without destroying hexosamines.

In general, glycosides of sialic acids are very labile to acid hydrolysis and the free acids are, in themselves, fairly unstable. Neutral sugars in the furanoside configuration come next, followed by neutral pyranoses and then the hexosamines. Glycosides of uronic acids are very stable to acid hydrolysis below pH3, but less so above, where the ionized carboxyl group promotes hydrolysis by neighbouring-group participation.

(iv) *Enzymatic hydrolysis:* The majority of glycosidases that are used in the structural analysis of oligosaccharides are *exo*-enzymes, that is to say, they degrade from the exterior of the molecule, normally starting at the non-reducing end. Most are specific for the saccharide on the non-reducing side, for the anomeric configuration and, sometimes, for the position of the glycosidic bond. If glycosidic degradation is performed either by the consecutive application of the enzymes, or by the simultaneous application of a mixture of them, sequences can be derived. In the former case sugars are removed, one at a time, to their limit; in the latter, the time-courses of release of sugars are followed—but this is seldom applicable for more than two or three residues. Branching of oligosaccharides complicates analysis very acutely.

A serious difficulty with the use of glycosidases is their tendency to show different specificities towards substrates of high and low molecular weight. It is essential always to check the nature of the sugars released, with all the attendant problems of micro-analysis that this may imply. *Endo*-glycosidases have found little use in studies upon oligosaccharides, but have been much used by polysaccharide chemists, as a means of obtaining small oligosaccharides.

The chief advantages of enzymatic techniques are their specificity (when established), the mild conditions of their application, their requirement for a minimum of handling of small samples and the information that they give upon anomeric configuration.

b. *Use of Lectins in Analysis of Glycolipids and Glycoproteins*

A new chapter in the development of structural studies upon glycolipids and glycoproteins has been opened by the application of lectins[1] to the problem. These plant proteins which have various binding specificities and affinities for sugars or oligosaccharides, both in the free and bound state, have been discussed in the previous chapter. They can be specific for residues that are peripheral or internal in oligosaccharides and their binding is reversible by high concentrations of the appropriate free sugars. Several are metalloproteins, and they may be di- or poly- valent. Thus, they act as agglutinins towards cells that bear an appropriate surface grouping and, to this extent, are analogous to antibodies. Various animal and fungal sources of carbohydrate-dependent agglutinins are known, but many of these show a rather lesser specificity than do the lectins. Some of the commoner lectins, and their specificities, are shown in Table I of Chapter 2.

Lectins have been applied to the analysis of carbohydrates of cell surfaces in three ways. Fluorescent and heavy-metal labelled lectins have been used to locate sugars at the cell periphery, both in the optical and electron microscopes (see Chapter 2). Unlabelled, soluble lectins have been used to study the chemistry of various surface saccharides *in situ*, especially those that are antigenic (Uhlenbruck, 1973), while the lectins have been used like immunoadsorbents, following attachment to insoluble supports (Lloyd, 1970). Lectins are often divided into those which are "blood-group specific" and those which are not, which is a convenient, but slightly misleading, classification. The two types are not basically different. So-called "blood-group specific" lectins react with oligosaccharides that are found only on cells that bear the appropriate blood-group antigens. The lectin binding-site might be the whole antigen, part of it, or some other saccharide that only occurs with it. No greater degree of specificity is postulated for these lectins than for others, it is simply a reflection of the pattern of occurrence of their ligands.

Where soluble lectins are used, it is possible to see whether treatment with specific glycosidases will abolish the agglutination of cells both by the lectin and by some antibody of interest. If so, then it may be possible to isolate immunologically active glycolipids or glycopeptides by adsorption onto the insoluble form of the lectin. Though studies of this kind are very new, they have already yielded a great deal of valuable information and offer the prospect of immense utility in the future. Despite all this, some care should be exercised in their application, for, though a great number of lectins are now known, very few have been well characterized. It is

[1] The term lectin was originally applied to carbohydrate-binding proteins of plants. This definition is still, strictly, correct, though many authors now use the term for animal and fungal agglutinins. The original definition is used here.

important that the specificities of soluble lectins should be well defined and checked when insoluble or labelled derivatives are prepared.

(i) *Labelled Derivatives of Lectins and Other Agglutinins:* The procedures used in the synthesis of fluorescent or heavy-metal labelled derivatives of lectins are essentially identical with those used for the preparation of labelled antibodies. A general caveat, however, is that lectins are often unstable to the alkaline pH used for labelling antibodies with isothio-cyanates (e.g. of fluorescein or rhodamine) and some (e.g. concanavalin A) can aggregate readily, unless they are below neutrality. A suitable com-promise for some lectins is to label (with a ratio of 2 labels to 1 protein) at about pH 7–7·5 for 18 hours at 15–20°. An excess of a binding sugar can be useful as a protection for the active sites during reaction. Surplus reagent can be removed by dialysis or gel-filtration (though care should be taken to ensure that the lectin does not bind to the gel). Many labelled lectins are stable to freezing and thawing, but it is preferable to store any of uncertain stability at 4° under sterile conditions. In the authors' experience unlabelled lectins are frequently stable as freeze-dried solids if stored at −20°, though the fucose-binding proteins of *Lotus tetragonolobus* are rather unstable upon storage. Not only can the lectins be used as histo-logical reagents and topographical markers of saccharides or cell surfaces (Uhlenbruck *et al.*, 1972; Marchesi *et al.*, 1972), but they represent the only reagents for labelling the non-reducing terminals of oligo- and poly-saccharides. To this extent they offer valuable tools for chemical studies upon glycoproteins and glycopeptides, isolated by microchemical methods such as electrophoresis or electrofocussing in gels (see below). In the authors' laboratory some labelled lectins have been shown to be capable of marking zones of isolated glycoproteins. This is obviously capable of extension to polysaccharides and, possibly, glycolipids, and can certainly be used in some immuno-electrophoretic systems.

(ii) *Adsorbents Derived from Lectins:* Again, as with immunoglobulins, adsorbents are readily prepared by the attachment of several lectins to insoluble supports. A variety of procedures have been described with antibodies, but Agarose beads (such as Sepharose 4B or 6B) and cyanogen bromide are used extensively with the lectins. Activated gel supports are prepared by the treatment of the gel with cyanogen bromide (at about 10 mg/ml of gel) in the presence of alkali. The activated support is then very thoroughly washed and the lectin is allowed to react at pH 7–7·5. When the reaction is complete the adsorbent is washed extensively and the excess unreacted sites must be blocked. If the lectin has an isoelectric point fairly near pH 7, ethanolamine is suitable. In those cases where the lectin possesses an isoelectric point well away from this, it is advisable to block the unreacted sites with a reagent which will tend to oppose the otherwise predominant charge of the lectin. This minimizes non-specific ion-exchange effects. In any case, adsorbents of this type should always be used at a fairly high ionic strength—again to minimize charge effects. Some lectins, at least, are stable in the presence of deoxycholate, so making possible the stabilization of solutions of membrane glycoproteins by this detergent, during adsorptions. An example of this has been the work of Allan *et al.* (1971) upon the adsorption of glycoproteins of lymphocyte plasma membranes to phytohaemagglutinin in the presence of deoxycholate (2%).

c. *Physical Methods*

The recent literature upon the application of nuclear magnetic resonance spectroscopy, infra-red spectroscopy, mass spectroscopy, X-ray crystallography and polarimetry to the chemistry of carbohydrates has been reviewed by Kennedy (1971). Thus far they have had little application to specifically membraneous materials (though some blood-group substances have been studied—but mostly from cyst-fluids), partly because of the quantities involved and partly because of the interpretational problems inherent in analysis of complex heterosaccharides. Likewise the approach to polysaccharide structure by way of computer-based mathematical analysis of models (Rees, 1969) has yet to be applied to heterosaccharides of cellular surfaces. Probably all of these techniques will eventually prove useful in the solution of some structures of membrane-saccharides, and will ultimately lead to the assembly of three-dimensional models of these molecules.

2. Glycolipids of the Plasmalemma

Though a great many glycolipids have been reported and analyzed from all manner of animal, plant and bacterial sources (Kiss, 1970), only a limited number have been demonstrated from circulating blood plasma, and it is quite likely that some are to be associated with the intercellular matrix rather than the plasmalemma. Equally, some circulating glycolipids can exchange with their membrane-bound counterparts.

a. *Isolation and Classification*

Glycolipids from vertebrate sources fall into three main categories and the simplest of these are the cerebrosides, in which the primary hydroxyl group of the sphingosine residue of a ceramide (an *N*-acyl sphingosine) is *O*-glycosidically linked to a mono- or oligosaccharide (Fig. 10). Thus the cerebrosides are the glycosides of ceramides. Where sialosyl groups substitute the oligosaccharide residues of cerebrosides the lipid is termed a ganglioside[2] (Fig. 10), i.e. gangliosides are the sialosyl glycosides of cerebrosides. If a sulphate ester group substitutes a galactosyl residue attached

---

[2] This term used to be restricted to lipids of the central nervous system, but now has more general usage. For terminology see review of Kiss (1970).

to the sphingosine residue of a cerebroside, the lipid is called a sulphatide. In addition to these major groups, a rare type of lipid called a sphingoplasmalogen has been described of the structure shown in Fig. 10.

Glycolipids are generally molecules of molecular weights up to about 2×10^3 and are likely to contain at least 15% by weight of carbohydrate, and often very much more (exceeding 50%). Thus, provided that good sources are available—such as human red cell membranes—it is often possible to isolate sufficient quantities of the lipids for application of the normal techniques of carbohydrate chemistry.

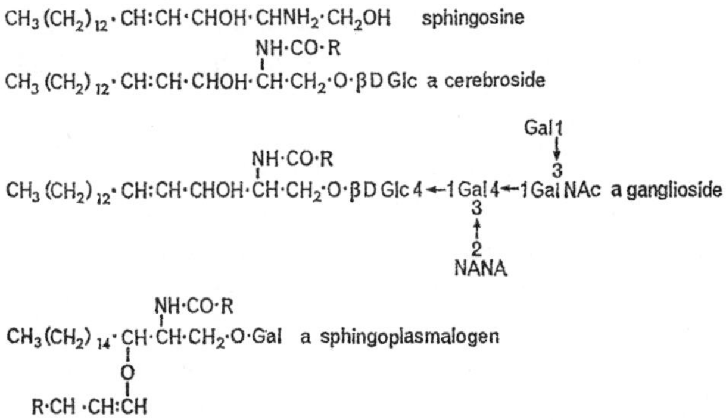

$$CH_3 (CH_2)_{12} \cdot CH{:}CH \cdot CHOH \cdot CHNH_2 \cdot CH_2OH \quad \text{sphingosine}$$

$$\overset{\displaystyle NH \cdot CO \cdot R}{\underset{\displaystyle |}{}}$$
$$CH_3 (CH_2)_{12} \cdot CH{:}CH \cdot CHOH \cdot CH \cdot CH_2 \cdot O \cdot \beta D \, Glc \quad \text{a cerebroside}$$

$$\text{Gal} \, 1$$
$$\downarrow$$
$$\overset{\displaystyle NH \cdot CO \cdot R}{\underset{\displaystyle |}{}} \qquad \qquad 3$$
$$CH_3 (CH_2)_{12} \cdot CH{:}CH \cdot CHOH \cdot CH \cdot CH_2 \cdot O \cdot \beta D \, Glc \, 4 \leftarrow 1 \, Gal \, 4 \leftarrow 1 \, Gal \, NAc \quad \text{a ganglioside}$$
$$3$$
$$\uparrow$$
$$2$$
$$NANA$$

$$\overset{\displaystyle NH \cdot CO \cdot R}{\underset{\displaystyle |}{}}$$
$$CH_3 (CH_2)_{14} \cdot CH \cdot CH \cdot CH_2 \cdot O \cdot Gal \quad \text{a sphingoplasmalogen}$$
$$\underset{\displaystyle |}{O}$$
$$R \cdot CH \cdot CH{:}CH$$

Fig. 10. Sphingosine and its relation to the glycolipids.

Because the glycolipids often contain oligosaccharide residues of five or six sugar residues, their solubility in the solvents normally used for lipid extractions can be very limited. Likewise, many of them are soluble in polar solvents and some of the most heavily glycosylated species are very soluble in water. Hence, there has been a tendency for them to be overlooked during lipid analysis and to appear as contaminants in preparations of membrane glycoproteins, where their water-solubility can be deceptive. It should also be remembered that hydrophobic proteins (especially lipoproteins) can easily be carried over with lipid during extractions with organic solvents.

Chloroform/methanol mixture (2 : 1, by volume) has been the most widely used extractant of lipid from membrane preparations and from crude tissue homogenates, and removes most non-glycosylated lipid fairly well. Most gangliosides distribute largely into the aqueous phase, though the cerebrosides and lower gangliosides occur in both aqueous and organic phases in comparable concentrations. The more heavily the lipid is glycosylated, the less easily it is solubilized in chloroform/methanol. Because some glycolipids tend to partition between organic (e.g. chloroform) and aqueous phases, it is essential to analyze both phases for glycolipids when examining the total glycolipid content of any membrane. Other good solvents for them include pyridine and some of its simpler alkyl derivatives, tetrahydrofuran, N-methyl pyrrolidone, 2-ethoxyethanol, several halogenated alcohols and hexafluoroacetone sesquihydrate. Some of these are also good solvents for most membrane proteins (q.v.).

b. *Fractionation*

The large-scale (i.e. $10-10^4$ mg) separation of glycolipids from each other and from other lipids is usually based upon differential partition in a variety of two-phase systems. Upon the largest scale counter-current distribution has been used (Thierrault, 1963), while column chromatography has been employed, with a variety of stationary (hydrated) phases, including cellulose, rolled paper and silica-silicic acid (see review by Carter *et al.*, 1965). Where the support is a hydrated polysaccharide, specific adsorption effects are likely with the glycolipids, which will tend to retard them. Ion-exchange chromatography is useful for gangliosides and neutral glycolipids can be handled in the same way in the presence of borate ions, when they will behave as their cyclic borate esters.

Though large-scale fractionation has been applied to the separation of glycolipids of the membranes of the red blood cells of several animal species, and to the lipids of brain, which are doubtless mostly of membraneous origin (though seldom proved to be so), small-scale procedures (below 10 mg) are far more widely applicable to problems of membrane chemistry. Here, the technical advances since the review of Carter *et al.* (1965) have been considerable, largely through the widespread application of techniques of thin-layer chromatography. Many systems have been developed, depending upon the particular glycolipids that it is desired to separate (Neskovic *et al.*, 1970), and the majority of these use a layer of silica gel as the support. The organic phase is very variable, and for a novel membrane system it is often necessary to modify systems used for other membranes. The mixture of organic solvents used for the initial extraction of the glycolipids can provide a good basis for the design of a new system and, for example, Brady (1973a and b) uses chloroform-methanol-water (60 : 35 : 8, by volume) to fractionate gangliosides of human and murine origin.

c. *Detection and Analysis*

Though thin-layer chromatography offers the possibility of separating microgram quantities of glycolipids, it poses problems of their subsequent detection and analysis. If silica is used as the layer, it is possible to detect glycolipids by degradation with strong sulphuric acid in the presence of a phenol, so that the derivatives of furfural liberated yields a coloured product with the phenol. Under appropriate conditions such reactions can be made quantitative, either directly with the aid of a chromatogram scanner, or indirectly by elution after light detection, followed by accurate reaction in a test-tube and spectrophotometry. Among reagents that have been used in this way are orcinol (Austin, 1963), resorcinol (Brady, 1973) and chloroxbenzidine (Austin, 1963). Direct charring with sulphuric acid is also possible. In general with membranes, it is preferable to separate radioactive glycolipids, where possible, so that further analysis can be performed without having to degrade some, or all, of the starting material in detection.

Gross analysis of glycolipids is performed upon acid hydrolysates, though in cases where several different types of saccharide (and glycosidic bond) are present, several hydrolyses may be needed, each optimal for the recovery of a particular sugar. If glycolipids are hydrolysed with hydrochloric acid and the hydrolysates are evaporated, some monosaccharides tend to be lost. Spik *et al.* (1969) have shown that this arises from transglycosylations. The content of sphingosine is usually estimated by nitrogen after methanolysis with $1 \cdot 0$ N-sulphuric acid in anhydrous methanol.

The sugars liberated by hydrolysis of glycolipids (or other glycosylated molecules) can be separated and analyzed by a variety of techniques, which can be divided into the two major groups below.

d. *Paper and Thin Layer Chromatography and Electrophoresis*

The separation of sugars by paper chromatography is old and well established. Some useful systems are cited in Table II. Most are used in the descending mode, preferably with the paper pre-washed to remove endogenous sugars. The separated zones can be detected by a wide variety of means, some of which are cited in Table II. In general, loadings of less than 5 μg are insufficient for detection.

Separations of sugars by thin-layer procedures have been quite extensively used, and give similar separations to the paper systems (often with some change of R_F, but not ranking order of zones). The methods of detection used are fundamentally similar, and an improvement in sensitivity down to loadings of about 1 μg is normally feasible.

Electrophoresis offers an excellent technique for the separation of sugars, and is usually carried out in buffers such as $0 \cdot 05$ M-sodium tetraborate (pH $9 \cdot 2$), which form charged derivatives of neutral sugars (Weigel, 1963).

The technique can be applied either upon paper or thin layers, and an initial separation of acidic or basic sugars from their neutral counterparts can be achieved at pH $6 \cdot 5$, in a pyridine acetate buffer. Amino acids can be separated from neutral and acidic sugars at pH 2. The technique is very rapid and because the formation of derivatives such as borate esters depends upon the stereochemistry of the sugars, it gives separations very different in ranking order from those obtained in chromatographic procedures, thereby complementing them. Detection and quantitation are

Table II. Some Chromatographic and Other Procedures for the Separation of Sugars

1. Ethyl acetate/pyridine/water, 8 : 2 : 1 (v/v)	Neutral sugars, including fucose
2. Ethyl acetate/acetic acid/pyridine/water, 5 : 1 : 5 : 3 (v/v) in tank equilibrated with ethyl acetate/pyridine/water 40 : 11 : 6 (v/v)	Uronic acids and amino sugars
3. Ethyl acetate/acetic acid/formic acid/water, 18 : 3 : 1 : 4 (v/v)	Uronic acids and oligosaccharides containing them
4. Butan-1-ol/ethanol/water, 5 : 1 : 4 (v/v)	Fucose, rhamnose, ribose and derivatives: any species that are very mobile in 1.
5. Butanol-1-ol/pyridine/water, 6 : 4 : 3 (v/v)	N-acetyl-amino sugars, sialic acids and derivatives
6. Electrophoresis in 0·05 M sodium tetraborate pH 9.2	Most neutral sugars, amino sugars and uronic acids. Separations are complementary to those (1–5) above

These systems are applicable to paper chromatography and, in most cases, to thin layer separations upon cellulose. Aldoses can be detected with alkaline silver reagents, or aniline phthalate, or phosphate with which neutral hexoses give yellow-brown spots and peutoses pink-purple spots. Uronic acids give orange derivatives, and fucose and rhamnose give brown derivatives. Amino sugars give violet spots with ninhydrin. Oligosaccharides likewise give coloured derivatives and can also be visualized with the periodic acid-Schiff procedure, using para-rosaniline. This gives pink spots with 1,4 and 1,6-linked saccharides, no reaction with those that are 1,3 linked and yellow spots with 1,2 linked saccharides.

carried out as in chromatography. It is surprising and unfortunate that more use has not been made of these techniques in biochemical studies, because of their rapidity and ease of use.

The analysis of isolated glycolipids has been revolutionized by the application of gas–liquid chromatography (GLC). This technique offers great advantages over the older established techniques of paper chromatography in terms of its sensitivity and accuracy. Its chief disadvantages lie in the basic cost of equipment and a variety of technical problems, associated specifically with sugars.

Sugars cannot be introduced into a gas chromatograph directly, because they are not appreciably volatile and char at the operating temperature[3]. Several volatile derivatives have been used. One of the earliest classes of sugar derivative to gain wide acceptance were the trimethylsilyl ethers (Sweeley *et al.*, 1963). Sugars liberated from a glycolipid, or other molecule, by hydrolysis (see below), are usually treated with a mixture trimethyl-chlorosilane and hexamethyldisilazane in anhydrous pyridine, though dimethylformamide and dimethylsulphoxide have also been used. Reaction is rapid, and the analysis is performed upon samples of the reaction mixture injected directly into the chromatograph. Sweeley and Walker (1964) early applied the technique to glycolipids of brain. While having the advantages of rapidity and simplicity in preparation of the derivative, it does lead to the appearance of multiple peaks for each sugar, with consequent difficulties in interpretation of the final trace. Even so, the method is very widely used and is ideal where the multiple peaks of the particular sugars under study can be resolved sufficiently.

Alditol acetates have become extremely popular derivatives of sugars for GLC, since they are fairly easily prepared and yield single peaks. Shaw and Moss (1969) give a useful summary of conditions for their separation. They are prepared by the successive reduction of sugars (e.g. with borohydrides) and acetylation with acetic anhydride and acetyl chloride. More recently alditol trifluoroacetates have also been used (Shapira, 1969) and have the advantages of easy synthesis.

Methylated and partially methylated sugars can be separated by GLC, and this is especially useful when applied to linkage studies (see above). Methylation has most often been carried out by the use of dimethyl sulphate at alkaline pH, preferably under nitrogen. Alternative techniques are to use methyliodide in the presence of silver oxide, or a solution of sodium hydride in dimethylsulphoxide. The last is particularly applicable to the glycolipids, since the whole reaction can be carried out in solution. Methyl-ether derivatives of sugars give only single peaks in the gas chromatograph.

[3] Though this, in itself, has been used by Turner (1969).

The choice of detector in GLC is dictated somewhat by the sugar under study. Electron-capture detection is probably the most popular procedure, and has a linear response facilitating quantitation. The use of mass spectrometers coupled in series with gas chromatographs, is now widespread in many fields and is being applied very usefully to carbohydrate chemistry, in studies such as those of Heyns et al. (1969) and Zinbo and Sherman (1970).

e. *Glycolipids of Blood-Group Specificity and Related Glycolipids of Human Erythrocytes*

Human red cells contain a number of antigenic determinants at their surface. Some of these interact with circulating antibodies of other individuals, others are antigenic only in other species or in a very few humans. The ABO blood-group system is the best known of these. Individuals can possess antigens upon their erythrocytes which are either A or B or both. A person of blood-group A has the A antigen upon his erythrocytes, and the anti-B antibody circulating in his serum. A person of blood-broup B is the converse. Where an individual is classed as AB, both antigens are found upon the red cells, but neither anti-A or anti-B antibodies are present in the serum. The person of blood-group O has neither antigen, but both antibodies. However, the red cells of individuals of group O carry another antigen, called H, which is antigenic in several species other than man. Two subgroups of the A-type can be distinguished serologically, and are called A_1 and A_2, though the molecular basis of their difference is not clear (similarly $A_1 B$ and $A_2 B$ exist). Many humans secrete A, B or H-active substances in fluids such as saliva and semen, where the antigens appear upon soluble glycoproteins, as well as upon the glycolipids of the red-cell membranes of these persons. Such people are called "secretors".

The Lewis blood-group system is closely allied to this. It contains two antigenic determinants, Le^a and Le^b, which occur on the red cells. For detailed descriptions of blood grouping see Race and Sanger (1962) and Bishop and Surgenor (1964).

Recent evidence (Gardas and Kościelak, 1971) indicates that Le^a and Le^b specificities always reside in glycolipids of the red cell, as do A, B and H antigens in non-secretors. Secretors contain another type of A, B and H antigenic material, which is probably glycopeptide in nature (see below).

The earliest experiments upon the chemical basis of the ABO system, which were the first markers to be demonstrated at the surface of a human cell, were carried out on ovarian cyst fluids, which contain these activities. Though the antigens were not, in this case, part of a cellular surface, their structural elucidation was a triumph of analytical biochemistry, and a necessary forerunner of the analysis of the red cell (Kabat, 1956; Watkins, 1966 and 1972; Morgan, 1967). The review of Watkins (1972) gives a detailed account of the history of the analysis of the blood-group active substances of cyst fluids, milk and other sources, and of the enzymes that degrade and modify the antigens and alter their specificity. An account of the genetic basis of these blood groups is given both by Morgan and Watkins (1969) and by Watkins (1972). What follows here is a summary of the basic structural features of the A, B, H, Le^a and Le^b glycolipids and of the mechanisms by which these are generated. For more detailed discussion the reviews above, and that of Pardoe (1971), should be consulted.

The neutral glycolipid fraction of the human red-cell membrane contains a series of glycolipids called globosides, all of which contain a glucosyl residue $\beta 1$, 1 linked to ceramide. Their structures and names are summarized in Fig. 11. Globoside 1 (cytolipin K) is present in substantially

Glc.β.1.1. Ceramide	Glucoceramide
Gal.β.1.4.Glc.β.1.1.Ceramide	Cytolipin H
Gal.∝.1.4.Gal.β.1.4.Glc.β.1.1. Ceramide	Ceramide trihexoside
Gal.NAc.β.1.3.Gal.∝.1.4.Gal.β.1.4.Glc.β.1.1.Ceramide	Globoside 1(Cytolipin K)
NANA∝.2.3.Gal.β.1.4.Glc.β.1.1. Ceramide	Haematoside
Gal.β.1.3.Gal.NAc β.1.4.Gal.β.1.4.Glc.β.1.1.Ceramide	Ganglioside GM1

$$\begin{array}{c} 3 \\ | \\ 2\alpha \\ \text{NANA} \end{array}$$

Fig. 11. Some glycolipids of the erythrocyte (also found in other cell types).

greater amounts than the others (Pardoe, 1971). The simpler of the two gangliodides shown in Fig. 11 (haematoside) is also a normal component of red cell membranes, the other ganglioside (GM1) has been shown in spleen and is suspected in red cells. The neuraminyl group of haematoside is labile to neuraminidase, since it is terminal, in ganglioside (GM1) the sialic acid is resistant to the enzyme, possibly because of steric hinderance. Effects of this sort should be borne in mind when attempting to remove sialosyl groups from cellular surfaces.

The possible structures of those glycolipids of the erythrocyte that possess blood-group activity is summarized in Figs. 12 and 13. The Lewis-active glycolipids have been extracted from tumours, rather than from red cells, in which they are present only in small amounts. A satisfactory explanation of A_1 and A_2 specificities is still awaited. Kościelak (1973) has recently reported a new glycolipid, the possible structure of which is shown in Fig. 14. The sialylation that leads to this glycolipid may compete with the glycosylations leading to the blood-group substances (Fig. 14) and lead to the loss of blood-group antigens upon the white cells in some leukaemias (Kościelak, 1973).

The chemical studies of Kabat, Morgan, Watkins and others have made it possible to advance a mechanism for the genetic control of the synthesis of blood-group substances in secretions and cyst fluids (Morgan and Watkins, 1969), and it seems very probable that the same system applies to the glycolipids of the red cell. In the case of soluble blood-group substances, a common precursor is proposed for all the blood-group determinants and the synthesis of this is under the control of the *H*-gene. The *O*-gene is inactive, while the *A*- and *B*-genes lead to a modification of the H-structure.

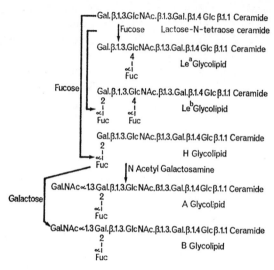

Fig. 12. Proposed Glycolipids of the A, B, H, Le[a] and Le[b] systems, based on chains of type 1 (after Watkins, 1972).

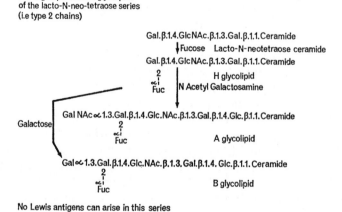

Fig. 13. Proposed Glycolipids of the A, B and H system, based on chains of type 2 (after Watkins, 1972).

The Lewis structures also arise from one of the two possible precursors of the H-structure, under the influence of the *Le* gene (Figs. 12, 13, 15 and 16). The combined activities of *H* and *Le* give rise to Le^b determinants, while Le^a determinants arise from the action of *Le* on one of the classes of H-substance precursor (see Figs. 12, 13, 15 and 16). Secretion is under the

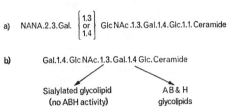

Fig. 14. Sialiglycolipid related to the A, B and H substances (Kościelak, 1973), showing how it probably competes with them for a common precursor.

control of a separate gene (*Se*) which is closely connected with *H*. It is postulated (Watkins, 1958, 1972) that each gene (except *Se*) controls the synthesis of a glycosyl transferase, and the sequential action or non-action of these determines both the phenotype of the red cell (i.e. the antigenic species it contains) and the specificity of the blood-group substances secreted. The secretor gene (*Se*) determines the appearance or non-appearance of all antigens, except Le^a, in the secretions. Because Le^a is determined before the action of the *H*-gene, which acts at the same level as *Le*, it is independent of *Le* and so appears in secretions of all persons in whom the *Le* gene operates. In the red cells, the further elaboration of most Le^a-substances into A, B, H (and Le^b) substances tends to make Le^a too weak to detect. This seems to happen less extensively in secretions.

The α-4 and α-2 fucosyl transferases controlled by the *Le* and *H*-genes respectively, both use guanosine diphosphate fucose (GDPFuc) as the glycosyl donor, as does an α-3 fucosyl transferase, which fucosylates those chains which end in the sequence β-D- Gal 1,4β-GlcNAc. This is under the control of a separate gene called 3-*F* (Watkins, 1972) which is not part of the A, B, O system as such (Fig. 16). The α-*N*-acetylgalactosaminosyl transferase that is controlled by the *A*-gene, and the α-D-galactosyl transferase controlled by the *B*-gene, both use uridine diphosphate derivatives of the sugars.

Though all of these antigens are of immense practical importance in medicine, it is not known what their value is to the individual cells that possess them. The value of their study for the central theme of this book is that they exemplify the way in which oligosaccharides of cell surfaces are assembled, how the pattern of that assembly is genetically determined, and the great complexity and variability of structure that is produced. In no other system do we have such detailed information. One feature which does emerge from the study of these antigens is the great similarity that exists between both the structures of oligosaccharides borne by glycolipids and glycoproteins, and the mechanisms of their assembly. A common control of both is indicated, probably via common or very closely related glycosyl transferases. Quantitatively the blood-group lipids are much less abundant in red cells than many of those glycolipids that are not known to be associated with blood groups. Some other blood groups and lectin-binding receptors are known to be associated with glycolipid, especially the P antigen and the ricin (*Ricinus communis* lectin)—receptor (Pardoe, 1971).

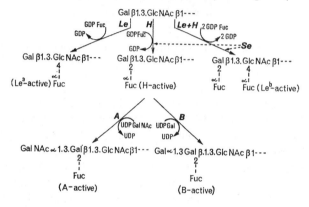

Genetic control of type 1 synthesis

Fig. 15. Genetic control of the synthesis of the A, B, H, Le^a and Le^b specific saccharides of type 1 (after Watkins, 1972).

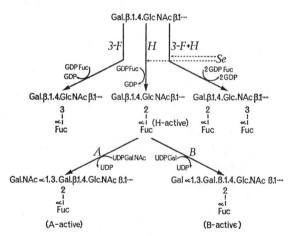

Fig. 16. Genetic control of the synthesis of the A, B and H specific saccharides of type 2 (after Watkins, 1972).

f. *Glycolipids in Chemical Pathology*

Though the study of the role of glycolipids in the determination of blood groups has been the field of the most spectacular advances in their carbohydrate chemistry, it is from the investigation of their pathology that ideas about their biological function are likely to evolve. Two major classes of study have been undertaken. In the first, the central nervous system has been the subject of investigation in relation to a variety of mental disorders, though it is largely presumed that the glycolipids involved are components of cellular surfaces. In the second class of study, the possible rôles of glycolipids in malignant, including virally induced, transformations has been examined.

(i) *Central Nervous System:* A good example of the type of disorder that has attracted biochemical study is Tay-Sachs disease. In this condition the patient shows a progressive, and invariably fatal, lesion of the central nervous system. Onset is often in infancy, but can be as late as early adolescence, and the progress of the disease is over two to four years. Symptoms include psychomoter retardation, intellectual deterioration, seizures, postural abnormalities and, normally, retinal changes. All neurones are eventually affected, but the most obvious lesions are in the cerebrum and diencephalon, causing many of the symptoms above. Lipids accumulate in granules within all the neurones and can cause huge dilatation of the cells. An inborn metabolic error controlled by a recessive gene underlies the disease. For clinical and pathological details of this and similar diseases of glycolipid metabolism see Adams and Sidman (1968).

The lipids of normal human brain are shown in Table III, with those of Tay-Sachs brain beside them. It is clear that the abnormal brain accumulates glycolipids of a shorter oligosaccharide chain length, than those of normal brain, and Okada and O'Brian (1969) have shown a widespread loss of β-D-N-acetyl hexosaminidase in such patients and suggest that this explains the accumulation of ganglioside GM2 which can not be further degraded. More recently the situation has become more complicated since this enzyme and sialidase, which is also supposed to be low in Tay-Sachs disease, have been shown to be inactive towards gangliosides after freezing, but still work on artificial substrates (Brady, 1973b).

A wide range of glycolipid storage diseases are now known, though relatively few of the enzymatic lesions have been identified. Most involve serious injury to the central nervous system, but it is not clear that this always arises from inclusions of lipid within the cells, rather than from changes in neuronal surfaces.

(ii) *Cellular Transformation:* The studies of Brady (1973b) upon the transformation of a mouse cell line with SV40 and polyoma viruses provide a good example of the sorts of modifications that glycolipids can undergo.

Table III. Major Glycolipids of Normal and Tay-Sachs Brains

Structure	Lipid Code (Svennerholm)	Brain Normal	Brain Tay-Sachs
Ceramide. Glc. Gal. NANA	GM3	±	+

Ceramide. Glc. Gal. GalNAc
   |
  NANA
	GM2	+	+++

Ceramide. Glc. Gal. GalNAc. Gal
   |
  NANA
	GMI	++	+

Ceramide. Glc. Gal. GalNAc. Gal. NANA
   |
  NANA
  NANA
	GDla	+++	+

Ceramide. Glc. GalNAc. Gal
   |
  NANA
  NANA
	GD2a	+	±

Ceramide. Glc. Gal. GalNAc. Gal. NANA
   |
  NANA
	GTET	++	±

Normal mouse cells show limitations in their patterns of movement and division above a certain cell-density (contact inhibition of movement and of growth, see Chapter 6), so that their division stops and they do not form multiple layers. After transformation with virus these constraints are lost. The normal cell contains glycolipids analogous to GM3, GM2, GM1, GD1a and GD2a above, with two analogues to each lipid. The transformed cells show little save the GM3 analogues, indicating a blockade in the synthesis of the higher gangliosides. Brady (1972b) has shown a reduction in UDP-*N*-acetylgalactosamine : haematoside *N*-acetylgalactosaminyl transferase activity, which could account for this. Chemical transformation leads to a different picture, in which the lipids corresponding to GM2 predominate, though small amounts of the others are also present (Brady, 1972a). As Brady (1972a) points out, this is somewhat similar to the situation in some hepatomata, where lower gangliosides tend to accumulate relative to their higher homologues. Several other workers have shown similar results. Both Hakomori and Murakami (1968) and Mora *et al.* (1969) found that viral transformation of cells was accompanied by a shortening of glycolipid chains, and Vicker and Edwards (1972) have related this to changes in adhesion following transformation. Hakomori (1970) found that normal cells showed increasing levels of ceramide trihexoside as the density of the cultures increased and that this effect disappeared after transformation with polyoma virus. Robbins and Macpherson (1971) have found similar results with Nil 2 hamster fibroblasts.

In all such studies the association of changes in cellular behaviour with changes in their glycolipids is implied, but not strictly proved—and is very difficult to prove. Since it is known that the control of genetic expression in the ABO blood-group system can influence both glycolipids and secreted glycoproteins (see above), it is possible that oligosaccharides of glycoproteins are also being affected here. If so, it could be the glycoproteins, or both glycolipids and glycoproteins, that regulate (or reflect) cellular behaviour.

Pitesky and Terry (1972) showed that the synthesis of protein and lipid in membranes of *Mycoplasma laidlawii* were largely independent of each other, by applying nutritional constraints upon the availability of lipid. This gave no indication of whether glycosylations of the two were independent. It could be that carbohydrates alone determine some properties of the cellular periphery, irrespective of the aglycone, and are regulated together in glycolipid and glycoprotein.

3. GLYCOPROTEINS OF THE PLASMALEMMA

Following the successful development of the chemistry of the membrane-bound glycolipids in recent years, much attention has been given to the analogous study of the membrane glycoproteins. These studies have received an impetus from the growing awareness that much of the surface carbohydrate of animal cells is protein-bound, (see Chapter 2) and that some properties of the cellular periphery are associated with such macromolecules. The developments of the two fields have much in parallel. As with the glycolipids, the rôle of glycoproteins in blood-group antigenicity has drawn attention to their structure, as has their probable relation to malignant transformation. Consequently, what follows is arranged in a manner analogous to the foregoing section. It should be remembered that the detailed study of the structure of glycoproteins is at a rather earlier stage of development than that of glycolipids, but it already offers even greater rewards in terms of an understanding of the cellular periphery.

a. *Isolation and Characterization of Glycoproteins*

Two types of approach have been made to this problem. One has sought to study the glycopeptides released by controlled proteolysis of the surface of intact cell-membrane preparations. The other approach has attempted to release "intact" glycoproteins from membranes, by a variety of means other than by proteolytic degradation. The former approach has been widely used in studies of the M and N antigens of the human erythrocyte, the latter in comparative studies of virus receptor sites of the erythrocyte and of transformed cells.

From a technical point of view the use of glycopeptides is easier in structural studies, since the soluble glycopeptides released from cell surfaces can be fractionated and analyzed by conventional techniques (such as those collected in Gottschalk's books, 1972), provided that sufficient material is available. The red cell and the blood platelet are readily obtained in bulk and have thus been the first subjects of such studies (see below), but for other cell types micro-analytical and especially radio-chemical methods will have to be used, and still require some further development.

All detailed studies in this field will ultimately require the isolation of individual glycoproteins, before their cleavage into glycopeptides, in order to interpret the relationships of glycopeptides previously analyzed. This is a very much more difficult problem than is that of isolating glycolipids. Membrane proteins are distinctively insoluble, under physiological conditions in the presence of membrane lipids, and are rendered soluble by relatively few procedures. Because of this, not only are they difficult to handle and to isolate, but also it is difficult to set up adequate criteria of their purity. These problems will be considered in some detail, since they are fundamental to an appreciation of the present state of the study of those membrane-proteins that are glycosylated.

b. *Solubilization of Membranes and their Proteins*

The first step necessary for the fractionation of the proteins and glycoproteins of membranes is to convert them into a form in which they can be handled, and this necessitates that they be dissolved in solvents which are compatible with the procedures to be used for their resolution. Essentially two approaches have been used: the differential solubilization of some components of membranes, and the total solubilization of the whole plasma membrane.

(i) *Differential Solubilization:* Successive washing of membrane preparations with aqueous solutions of salts (of increasing ionic strength), chelating agents and, finally, detergents can be used to solubilize various membrane-bound proteins. Such procedures have been used by Rosenberg and Guidotti (1968, 1969) upon the human erythrocyte, and Randall *et al.* (1972) found that the sialic acid of the red cell remained entirely in the least soluble component. In general, the selective extraction of glycoproteins by such procedures has not proved possible, and partition-techniques have to be applied to preparations that have first been totally solubilized (see below).

Another type of approach to differential solubilization is that of Maddy (1964) who has separated the protein and lipid of human red cell ghosts by shaking a suspension of stroma with cold *n*-butanol. About 96% of the proteins pass into the aqueous phase and 4% into the organic phase. The partition of lipids is similar, in the reverse sense. Unfortunately, the total removal of *n*-butanol from the aqueous phase and subsequent handling of the protein can cause it to denature rather easily. This serves to emphasise that not only is it necessary to solubilize membrane glycoproteins, but also there is a need to stabilize the solutions for subsequent manipulations.

(ii) *Total Solubilization:* Several procedures exist for the total solubilization of preparations of cell membranes. Some, such as the use of sodium dodecyl sulphate (SDS) are generally applicable, while in other cases only certain membranes are rendered soluble. Whenever it is desired to make a soluble preparation of a type of membrane not previously studied, a variety of techniques must be evaluated with subsequent handling in mind. Thus the use of SDS would be indicated if a molecular weight-based separation is desired, but subsequent electrofocusing would be precluded (see sections below).

Detergents are the most extensively used solubilizing agents of membranes, and of these SDS is by far the most common. It is particularly used in the fractionation of proteins by molecular weight, in polyacrylamide gels (see below), but has the disadvantage of being extremely difficult to remove from proteins.

It is of interest that detergents of the opposite polarity, such as cetyltrimethyl ammonium bromide, will solubilize cell membranes, without extensive dissociation of the proteins (Triplett *et al.*, 1972). Non-ionic detergents, such as Triton-X 100 or digitonin, are good solubilizing agents and probably do not dissociate membrane proteins as extensively as SDS, even in the presence of mercaptoethanol. They allow a wider range of fractionation procedures to be applied (see below), and will dissolve most membranes.

Chaotropic agents have found some use in the dissolution of membranes. These are compounds which perturb the structure of water and so interfere with hydrophobic bonding. Typically, they are ions with large hydration shells, such as thiocyanate, iodide or lithium ions. In substances such as lithium di-iodosalicylate (Marchesi *et al.*, 1972), detergent and a chaotropic effects are combined. Urea (8 M) and guanidinium salts also interfere with hydrophobic bonding and have been used in physical chemical studies of membrane proteins, but have not been much applied to the isolation of glycoproteins.

Several organic solvents are known to be capable of disrupting membranes and dissolving their proteins. One of the most useful of these is pyridine, which Blumenfeld (1968) has shown to be both a good solvent for human erythrocyte ghosts and leaves the sialoglycoproteins in a water-soluble form. Several pyridine analogues are equally good, or slightly better, for this purpose. The solubilization requires about 33% (by volume) of pyridine at a pH above 8·7, in the cold, and is independent of the pK of the particular derivative of pyridine used (Stoddart, 1970). The pyridine should be dialysed away as quickly as possible so that the pH of the solution is lowered and the danger of β-elimination of the oligosaccharides avoided. Most of the protein precipitates in the sac, but the sialoglycoproteins remain in solution. Zahler (1968) surveyed a range of organic solvents for their ability to dissolve membrane proteins, and reported that methyl-cellosolve, dimethylformamide, concentrated acetic acid and 2-chloroethanol (between pH 2 and 3) were good solvents. Unfortunately, strongly acid solvents are likely to degrade sialoglycoproteins. Stoddart (1970) found that 2,2-dichloroethanol is an exceptionally good solvent for erythrocyte membrane proteins at neutral pH, at concentrations above 40% (by volume). N-methyl pyrrolidone also offered considerable possibilities as a neutral solvent.

c. *Principals of Fractionation of Membrane Proteins*

A plethora of techniques have been applied to the isolation and purification of soluble proteins, but almost all of them exploit various combinations of a very few properties of the macromolecules. A small number of techniques make use of single properties of proteins, at least in principle and approximately in practise.

(i) *Molecular Weight:* The fractionation of proteins by molecular weight is most widely achieved by gel-filtration, and this principle can be applied to membrane proteins in the presence of suitable solubilizing agents, such as detergents (Ne'eman *et al.*, 1972). Anomalous behaviour of glyco-proteins upon polysaccharide gels is not unusual. For the sort of micro-analysis that is becoming increasingly needed where cells are not abundant or easily cultured, techniques of polyacrylamide-gel electrophoresis are useful. Several procedures have been described. Some, like that of Takayama *et al.* (1966) use agents such as urea to keep the proteins in solution. In this method there is an equilibrium of aggregated and dis-aggregated protein, so that very low loadings of protein are required if all the protein is to enter the gel. The band-pattern within the gel is, however, independent of loading. In such procedures the gels are subjected to electrophoresis at a pH at which all the proteins bear the same net charge, and separation primarily results from variations in mobility with molecular size and shape. Even so, it is difficult to be sure that properties such as molecular shape are really equivalent among all membrane proteins under these conditions, especially among those that are glycosylated. For this reason these techniques have tended to lapse from favour. To a large extent such procedures have been replaced by polyacrylamide-gel electrophoresis in the presence of sodium dodecyl sulphate (SDS). SDS is an excellent solubiliz-ing agent for almost all membranes and is known to interact with soluble (non-membraneous) proteins in a stoicheiometric manner, forming rather stable complexes with a very strong predominating negative charge (Reynolds and Tanford, 1970; Nelson, 1971; Ray *et al.*, 1966; Shapiro *et al.*, 1967; Inouye, 1971). Application of a suitable electrical potential gradient to a small cylinder of acrylamide-gel of uniform concentration, at the appropriate pH and in the presence of excess SDS, is considered to cause the proteins to move down the gel in a manner which is determined solely by their size, and hence by their molecular weights. This certainly seems to be true of soluble proteins, but Katzman (1972) has produced good evidence that it is not always true of the proteins and glycoproteins of membranes derived from brain. Though SDS solubilized these molecules, it did not abolish all their interactions, even in the presence of mercapto-ethanol, and their behaviour was highly anomalous. It is now widely recognized that the more heavily glycosylated glycoproteins of cell mem-branes show anomalous relations between molecular weight and mobility, relative to proteins, and Katzman's (1972) study shows that they could, further, be interacting still with other membrane components. Accordingly, every care should be taken in attempting to estimate molecular weights and homogeneity of membrane glycoproteins, and proteins, from measurements in SDS gels. Gels of several different concentrations of polyacrylamide should be seen before any statements about molecular size and homogeneity are made. Where the purity of a single protein or glycoprotein is in question, it is desirable to obtain it by two different procedures and show the identity of the two preparations (Maddy *et al.*, 1972).

Electrophoresis of membrane proteins and glycoproteins in the presence of non-polar detergents, such as Triton X-100, has been little used. It has been found (Stoddart, unpublished) that this detergent will produce a relatively small number of electrophoretically separate proteins when applied to several different membranes which are known to give very many bands with SDS. Probably the Triton X-100 is not fragmenting the membrane as extensively as is SDS, and it gives rise, in all cases so far studied, to one very predominant band, always of about the same mobility. Observations of this type have, in the past, given rise to ideas of "structural protein" in membranes, which are now generally discarded (see Chapter 2), but are important in that they indicate how a sequential dismantling of a membrane might be achieved.

Gel-gradient electrophoresis offers a new and very valuable alternative to these methods. In this technique, a column or slab of polyacrylamide gel is prepared in which there is a gradient of gel-concentration from the top to the bottom, such that a protein applied to the top can migrate into the gel until it reaches that gel-concentration at which it is excluded from further penetration. The electric field serves only to move the proteins. Bands are produced with extremely sharp leading edges and a very narrow bandwidth, so that the peak of any band is higher, relative to the area beneath it, than is the case with conventional electrophoresis (Fig. 17). This makes for greater sensitivity in the detection of minor bands. Gels can be prepared which contain detergents and some of the criticisms of conventional techniques can be overcome. In particular, if any association of the type described by Katzman (1972) is going on and it is reversible (as was his), then in a gradient-gel run to its limit the aggregates will be pulled apart into their subunits, which will be smaller and will penetrate further into the gel until they reach their characteristic exclusion limit. Hence, though molecular weights derived by this technique must be regarded with some care, separations are generally real. The method can also be applied to larger glycopeptides with molecular weights above about 5×10^3 daltons.

(ii) *Isoelectric point:* Where two proteins differ in isoelectric point, their electrophoresis at an intermediate pH will effect a separation. Similarly, a separation will result at a pH near the isoelectric point of either, even if their net charges are of the same polarity. When the pH is remote from their isoelectric points, especially in a restraining lattice such as a polyacrylamide gel, the fractionation becomes increasingly dependent upon properties related to molecular weight (see above). The separation of glycoproteins of cell membranes by electrophoresis at a pH value near their isoelectric point has been little attempted, partly because a knowledge of the isoelectric points is desirable in designing the system. Electrofocusing offers an alternative procedure, where no such information is pre-requisite.

In this technique a pH gradient is set up and stabilized by the presence of suitable ampholytes.[4] The mixture of proteins for fractionation is

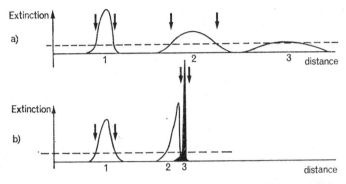

Fig. 17. The principle of gel-gradient electrophoresis (b) compared with SDS-gel electrophoresis (a). In SDS-gel electrophoresis the protein zones move at a rate which depends, in principle, upon their molecular size. An increasing distance of migration leads to a broadening and lowering of the symmetrical peak, which eventually is too diluted to stain up to the visual threshold (dotted line). Hence resolution is not capable of improvement beyond a certain running distance, and the optimal distance of running may vary widely for different proteins under study. In gel-gradient electrophoresis migration is eventually limited by the ability of the protein to enter the gel lattice (again this is related to molecular size). The zones that form are thus very sharp, but asymmetrical, and the sensitivity of the method is very high. The arrows mark bandwidths.

introduced into the gradient, either locally or generally, and an electrical gradient is applied. This is of such a polarity that molecules which are away from their isoelectric pH values tend to move towards those regions that are at these values; there they cease to migrate because they then have no net charge. Thus the separation effected is purely a function of molecular charge, to a good approximation. The method is applicable either in free solution or in polyacrylamide gels. Its major disadvantage is that proteins are often rather insoluble at their isoelectric points and so tend to precipitate, though this is less of a difficulty in gels than in free solution. Gels are a useful microanalytical tool, but bulk separations are usually made in solution in sucrose, or similar, gradients. With membrane proteins the solubility problem is rather acute, and it is necessary to work in non-polar detergents or in 8 M urea. Ionic detergents, such as SDS, are not suitable because they swamp the existing charge on the protein.

[4] An ampholyte is an ion which can bear positive and negative charges simultaneously (a zwitterion) and so will have an isoelectric point of its own, about which it will act as a buffer.

In the authors' laboratory electrofocusing in gels containing Triton X-100 has proved a valuable means of separating membrane proteins and comparing membranes, and this is discussed further below.

A separation similar to that obtained with electrofocusing, but not solely dependent upon charge, can be achieved by isotachophoresis, without the need to operate near the isoelectric points of the species under study. The method depends upon the application of Kohlrausch's Law,[5] and requires that the proteins being separated carry a large part of the total current in the system, unlike electrophoresis. If this condition is met, the proteins form a ranked series of zones in order of their transport numbers, so that the current in each zone is the same. The potential gradient across the system is discontinuous, and each zone has characteristically sharp edges and a "square wave" profile. Zones are contiguous, unless "spacer ions", such as ampholytes, are added. A common counter-ion is required, as is good buffering. When the system, which can be set up in poly-acrylamide gels or in free solution, is at equilibrium, the zones are self-stabilizing and move at a constant speed, giving the procedure its name.

Perhaps because the principles of isotachophoresis have not been readily assimilated and can be difficult to apply, the method has been scarcely used at all with the membrane glycoproteins, though their rather low isoelectric points make them good subjects for the technique. It is not feasible to work at very low or high values of pH, because of the effects of high concentrations of hydrogen ions or hydroxyl ions, so to work near pH 7 with glycoproteins of isoelectric points below or near pH 5 is ideal. Applied as a microanalytical procedure upon polyacrylamide gels, the method gave excellent resolution of the major glycoproteins of the human erythrocyte (Stoddart, unpublished results).

(iii) *Specific Adsorption:* The advent of the insolubilized lectins (see above) has marked the entry of a new range of tools into the armoury of the biochemist, and especially those who study the glycoproteins of membranes.

Where glycoproteins carry specific immunological determinants, selective adsorption, immuno-electrophoresis and allied techniques are feasible. Fluorescent-labelled lectins can be used, to some extent, as stains for specific saccharide residues of glycoproteins in polyacrylamide gels, though the cost of their synthesis limits the use of this technique to rather special cases. Details of the synthesis and use of substances based upon lectins were discussed above.

(iv) *Partition:* Though techniques based upon partition have been immensely valuable in the separation of glycolipids, their use in purifying

---

[5] Kohlrausch's Law can be stated in the form that the total current passing in a solution of an electrolyte is the sum of the independent contributions of all ions present, the measures of which are the product of concentrations and transport numbers.

glycoproteins has been limited. Proteins in general do not withstand exposure to organic solvents very well and tend to denature, especially at interfaces. For many organic solvents the same applies to membrane proteins, but in a few cases proteins do dissolve readily in the solvents, or partition largely into the aqueous phase, without denaturation. Only in such crude, initial steps of fractionations has this class of technique found any practical use.

(v) *Purity and Homogeneity in Membrane Glycoproteins:* A protein is said to be pure with respect to some property, such as an enzyme activity, when no procedure of fractionation will resolve it into a component which has the property and one which does not. For an enzyme the problem is easy, for there is a ready measure of its biological function, but for the membrane glycoproteins this is not so. Their primary functions are not always known, and so fractionation has to depend upon chemical criteria of purity, and in a few cases upon properties which are of a secondary nature, such as blood-group antigenicity or viral and lectin-binding. It is, of course, quite possible that similar antigens will occur upon very different glycoproteins. In addition, different biological specificities may reside on the same glycoproteins, and the possible variation in structure in a glycoprotein in which several different oligosaccharides are attached at many different sites is enormous. It is not yet clear how far such microheterogeneity exists in glycoproteins of membranes, but it is to be expected and probably varies between classes of glycoprotein.

(vi) *The Comparison of Membranes:* A great deal of interest has developed in the possible changes that may occur in surface glycoproteins when cells differentiate, become malignant, undergo viral transformation or divide. These are considered in detail below and in subsequent chapters, but there is a common approach embodied in all such studies, which is the making of comparisons between cells that are in some way related. In a few cases comparisons have been made between glycopeptides, rather than whole glycoproteins (for example, Warren *et al.*, 1972), but in most cases a whole membrane preparation is solubilized and compared with its normal counterpart by one of the microanalytical techniques described above. In this section consideration will be given to the sort of variations between glycoproteins that can be expected, and brief assessment will be made of the ability of these analytical techniques to detect them.

The differences between two cells might lie in the total absence of a glycoprotein or glycoproteins from one, or in an alteration of the relative abundances of the glycoproteins. If many of the glycoproteins are closely similar in molecular weight, such differences will not be easily resolved by SDS-gel electrophoresis and may lie within the normal range of variation of the gels. Even the total absence of a protein can be overlooked in a multiplicity of bands, some of which contain several proteins. For such a study gel electrofocusing is preferable because of its high resolution, and because few membrane proteins have identical isoelectric points.

If the difference between the glycoproteins of two cells lies in a change of the size of the oligosaccharides, it is possible that a sufficient change of molecular weight will occur to be detectable upon SDS-gels. If not, it may still be detectable in gel-gradient systems. A change in isoelectric point will not necessarily occur, so electrofocusing will not be the preferred method here.

When the differences between two glycoproteins is in their charge, because of changes in amino-acid sequences or charged-sugar content, electrofocusing will be a particularly sensitive detector, whereas molecular weight-based methods will be inapplicable.

Changes in the sequences of oligosaccharides, without change of molecular weight or charge, are likely to be reflected in alterations of antigenicity and reactivity towards lectins. Only when the changes are in the non-polar parts of polypeptide chains is it likely that no procedure will detect the difference, and for these careful amino-acid analysis is required.

None of the analytical techniques above will reveal modifications in the topographical distribution of oligosaccharides upon cell surfaces, though they may hint at them. For this the electron microscope and labelled lectins or antibodies must be used (see Chapter 2).

d. Glycoproteins of the Erythrocyte Membrane

As with the glycolipids, the erythrocyte glycoproteins have been a common subject of study, because of the ready availability of the cell in bulk, the simplicity of its structure and its possession of clinically interesting antigens. Some caution must be exercised in generalizing too far from our knowledge of the erythrocyte. It is a highly differentiated, specialized cell, which is peculiar in its lack of common membrane-linked properties, such as adhesiveness, motility or the ability to ingest. Its plasmalemma may, therefore, in many respects be quite atypical, and is likely to be simpler than those of the cells of many other tissues. Even so, it is a valuable test piece on which to develop analytical methods, and yields information which is important in itself.

Studies of glycoproteins of erythrocytes have had several explicit and implied aims. They have sought to determine how many classes of such glycoproteins there are, whether each class is homogeneous or contains several species, how the classes differ from each other and whether one type predominates. The immunochemical approach has tried to elucidate the structures of antigens (especially M and N, see below), and of viral receptor sites, and to relate the two. The study of lectin-binding sites has now been added to these. Attempts have been made to relate the glycoproteins to gross membrane structure (Bretscher, 1971a and b: Chapter 1), and a synthesis of the information yielded from all these attacks should allow something near to a chemical description of a whole erythrocyte glycoprotein to be made fairly soon.

(i) *The M and N Antigens:* The M and N antigens are associated specifically with the protein portion of the erythrocyte membrane, and were first detected by immunization of rabbits with human red cells, followed by titration of the cells with antisera (Landsteiner and Levine, 1927a and b). Two distinct classes of cells, and of antisera, were discovered, which were not immunologically cross-reactive. A third class of cell was found which was reactive towards all the antisera, and a third class of antiserum was reactive towards all cells. Analysis by sequential adsorption of antibodies enable the cells to be typed as M, MN or N. The sera were anti-M, anti-N or mixed (i.e. anti-M and anti-N). The separate P system of blood groups was found at the same time (see Watkins, 1964). This is lipid-associated (Pardoe, 1971) and contains carbohydrate (Watkins, 1964). Genetic control of the MN system is considered to be via two allelic genes, *M* and *N*, giving three genotypes *MM*, *MN* and *NN*. The inheritance is independent of other blood group systems. Some subgrouping has been added (Race and Sanger, 1962), but is not yet explicable at a molecular level. A few humans produce antibodies directed against M or N cells, so the system is of some clincial interest. M and N antigens are known upon several other cell types.

The M and N antigens are removed from the surfaces of erythrocytes by treatment with proteolytic enzymes, such as trypsin or pronase. The antigenicity of the fragments released appears to be greatest in those of largest size, perhaps indicating a need for a considerable degree of tertiary structure in the antigen. If erythrocytes are treated with neuraminidase, other than very briefly, they lose both M and N specificities (Springer and Ansell, 1958; Mäkelä and Cantell, 1958; Klenk and Uhlenbruck, 1960), which suggests either that sialic acid is part of the antigenic site or that it is needed to maintain some specific configuration of the antigen. More recently it has been found that briefer or milder degradation with neuraminidase, weak acid or trypsin will abolish M antigenicity, but makes the N antigen appear instead (Yokoyama and Trams, 1962; Baranowski and Lisowska, 1963; Uhlenbruck, 1969; Huprikar and Springer, 1970). Neuraminidase and acid could be removing some sialosyl residues, needed for M specificity, revealing thereby the N antigen. This could suggest that N is the precursor of M. The effect of trypsin is not so easy to understand on this basis, unless it exposes otherwise hidden N antigens buried within the membrane. If the antigenicity of these substances is very dependent on configuration of both saccharide and protein, then the two types of effect are quite understandable. It may be that the *M*-gene controls the addition of a sialosyl group to an oligosaccharide, which lacks it in a configuration where M antigenicity is expressed and N antigenicity is concealed, even though that sialosyl group itself is not part of the antigen. Some sialic acid is required for the expression of N antigenicity. Huprikar and Springer (1970) found a requirement for a branched structure, terminating in a β-D-galactosyl residue and containing an N-acetylneuraminyl residue, for the activity of the N antigen.

In 1967 Lisowska and Morawiecki showed that blockade of amino groups in MN glycoprotein led to a loss of both M and N antigenicity. Presumably the amino groups are in the polypeptide portion of the molecule, since no free amino groups of amino sugars are known to be present. This implies that the polypeptide plays some part in the expression of other antigens, either by being a part of the antigens themselves, or by maintaining their configuration.

(ii) *Other Blood-Group Antigens:* Recent studies of A, B and H antigens have shown that they also can be associated with a glycopeptide or glycoprotein in the erythrocyte (Whittemore et al., 1969; Gardas and Kościelak, 1971; Kościelak, 1973) and that in secretors (see above) there is too little glycolipid to account for all the antigenic sites. A class of remarkably heavily glycosylated materials, containing only a minority of lipid and protein and possessing immense A, B or H antigenic activity, has been isolated by Kościelak and his group (Kościelak, 1973) and is discussed below. The genetics of the ABO system was summarized in the section upon glycolipids (above).

Dzierkowa-Borodej et al. (1970) have reported that anti-I antibodies will react with glycoproteins of human red-cell membranes and the F antigen is known to be labile to neuraminidase (Hatheway et al., 1969). Among other antigens which are supposed to be associated with membrane proteins are those of the S; s; U; u; Tn; T and Duffy systems (Pardoe, 1971). Relations among these, and their relation to the M, N system and various sites have been discussed in detail by Pardoe et al. (1971). Some structural implications of this will be considered below. A recent series of studies by Roelcke and his colleagues (Ebert et al., 1971; Ebert et al., 1972; Merz and Roelcke, 1971; Roelcke et al., 1971), have shown a rôle for sialic acid in the Pr_1; Pr_2 system and its relation to M and N antigens. In particular Ebert et al. (1972) showed that treatment with periodate caused a loss of M, N and Pr_1 antigens and the appearance of Pr_2, indicating a possible biosynthetic relation between Pr_1 and Pr_2. The chemical significances of these studies will be discussed below.

Sialic acid has been implicated in erythrocyte antigens in species other than man, for example Spooner and Maddy (1971) describe a sialoglyprotein antigen in the ox.

(iii) *Viral Binding Sites:* When influenza virus binds to the erythrocyte it first causes aggregation, then a disaggregation which is not reversible upon addition of more virus. As was explained in Chapter 2, this is now known to be a process in which the virus first binds to site that contains sialic acid, then cleaves the sialic acid from the rest of the oligosaccharide by means of a neuraminidase, causing viral release and loss of agglutination, with an irreversible change in agglutinability by virus. The exploration of the influenza virus agglutinating site has been conducted both upon the receptor-site of the red cell, and upon cross reactive sites of soluble glyco-proteins. Gottschalk *et al.* (1972) have reviewed the field in detail. The viral binding site in red cells is closely associated with the same macromolecules that contains the M and N antigens, and its structure and relation to these is discussed below.

Though this is the best known glycoprotein-based viral receptor site many others are likely to be similar (Winzler, 1972), and so this may be considered as a useful model.

(iv) *Glycoproteins of the Whole Erythrocyte Membrane:* If erythrocyte ghosts are dissolved in SDS/mercaptoethanol and fractionated by acryl-amide-gel electrophoresis, a large number of zones are resolved which stain for protein. Quite a lot of these also stain weakly in the periodate Schiff procedure for carbohydrate, but weak staining of this sort can be an artefact, so that it is not clear how many minor components are present.

A recent study of the tritiation of periodate-oxidized erythrocyte membranes and glycoproteins (Zvilichovsky *et al.*, 1972) with sodium borohydride (^{3}H), showed that most of the label entered oligosaccharides and appeared especially in derivatives of sialic acid. Thus most periodate-generated aldehyde groups must arise in carbohydrate, suggesting that most periodate-Schiff positive material in gels is saccharide in nature. Strong staining is found in the very mobile zone that represents glycolipid, and in another zone that appears in 5% gels at a position corresponding to an apparent molecular weight of 10–11×10^4 daltons (Lenard, 1970a and b; Bretscher, 1971a and b). Care should be taken, as Bretscher has empha-sized, to guard against proteolytic degradation from endogenous proteinases in the erythrocyte. As far as possible, similar precautions should be taken for glycosidases. With aged ghosts or red cells it is quite common to find blurring of the glycoprotein zones and the appearance of new and diffuse Schiff-positive zones of lower molecular weight. Antigens also change on ageing (Greenwalt *et al.*, 1971). The best preparations sometimes show a strongly staining narrow doublet of zones, rather than a single zone, at the position of the major Schiff-positive staining above, and these also stain with ruthenium red (Stoddart, unpublished). Isotachophoresis of a pyridine

Table IV. Composition of the MN-Active Glycoprotein of Human Erythrocyte Membranes (Glycophorin)

Sugar composition (g%)

Sugar	Analysis After Winzler (1972)	Analysis After Marchesi et al. (1972)
Glucose	0·3	—
Galactose	13·2	10–12
Mannose	2·6	—
N-Acetylglucosamine	6·5	6
N-Acetylgalactosamine	13·2	12
Fucose	1·2	—
Sialic Acid	27·8	25
Total Sugars	64·8	53–55
Amino Acids	37·5	40

Amino Acid composition (residues per molecule) after Marchesi et al. (1972)

Amino Acid	Residues from Peptides	from Protein	Amino Acid	Residues from Peptides	from Protein
Ser	28	28	Leu	15	13
Thr	22	22	Ile	14	14
Asp	15	12	Val	16	17
Glu	21	20	Ala	12	11
			Gly	12	13
Lys	9	8	Phe	3	3
Arg	8	9	Tyr	6	6
			Pro	16	15
Met	4	4	His	5	8
CySH	—	—	Total Residues	206	203

extract also reveals this doublet of zones. Both this region of the gel, and the fast-moving lipid zone, contain sialic acid. It should be noted that any entirely 1,3 linked oligosaccharide would be unreactive with periodate, and so would escape detection with the periodate-Schiff method. Such links are quite numerous in the A,B,H and Lewis substances, for example, so completely 1,3 linked oligosaccharides are quite possible.

The major periodate-Schiff positive species, of apparent molecular weight of $10–11 \times 10^4$ daltons (in 5% acrylamide gels) has the MN antigens and the influenza-virus binding site (Winzler, 1972). Its true molecular weight is uncertain. Marchesi (cited by Winzler, 1972) has found apparent molecular weight values of $5 \cdot 5–11 \cdot 0 \times 10^4$, depending upon the gel concentration, and Bretscher (in Jamieson and Greenwalt, 1971 p. 107) has shown similar results. However, Kathan et al. (1961) and Kathan and Winzler (1963) found a molecular weight of $3 \cdot 1 \times 10^4$ daltons by sedimentation-diffusion of a phenol-extracted preparation. Blumenfeld also found these in pyridine extracts (see Jamieson and Greenwalt, 1971 p. 111; Zvilichovsky et al., 1971). Morawiecki (1964) showed that the material is composed of subunits which tend to aggregate, but separate in detergent. Gross analysis of several preparations has indicated a content of about 60% carbohydrate, by weight (Winzler, 1972), and Marchesi (see Jamieson and Greenwalt, 1971, p. 111) considers that the protein contains 208 amino acid residues (Marchesi et al., 1971) and 60% carbohydrate, by weight, giving a molecular weight of $5 \cdot 5 \times 10^4$ daltons. The question still remains unsettled and emphasizes the difficulty of dealing with highly glycosylated molecules, and the limitations of many common techniques for measuring molecular size. There seems general agreement, however, on a carbohydrate content of 60–65% for this glycoprotein and a very low content of lipid. The protein contains large amounts of serine, threonine and glutamic acid and is low in aromatic amino acids, though it has a considerable number of hydrophobic amino acid residues. The approximate sugar composition is given in Table IV.

Weicker (1968) has reported a distinct xylose-containing glycoprotein, separable from the above by phenol partition.

Another type of glycoprotein or glycopeptide of human erythrocyte membranes is that found by Gardas and Kościelak (1971) by the use of the lectins of *Ulex europeus* and *Lotus tetragonolobus*. They obtained evidence for glycoproteins of A,B and H specificity in the erythrocytes of secretors of these antigens. Further studies (Kościelak, 1973) have made possible the isolation of molecules of this class. They were isolated by flotation of the erythrocytes of secretors at a toluene-water interface, extraction of the residues with ethanol and a butanol partition. The butanol-saturated aqueous phases were concentrated and fractionated on CM- and DEAE-celluloses, before a final de-lipidation with chloroform/methanol. Yields of

Table V. Carbohydrate Compositions of Some Glycopeptides and Glycoproteins of M and N Blood Group Activities (figures as % by weight)

Enzyme or Extractant	Pronase						Trypsin		
Author(s)	Cook and Eylar (1965)		Ohkuma and Shinohara (1967)			Winzler et al. (1967)	Jackson and Seaman (1972)		
Peptide No.							API	LAPI	LAPII
Blood Group Specificity	MM	NN	MM	NN	MN	MM and NN	MM and NN		
Hexose (total)	21·6	13·1	18·9	19·1	21·5	24·4	15·36	11·42	7·71
Galactose	21·6	13·1					13·6	9·02	6·58
Glucose							1·76	0·49	1·13
Mannose							—	1·91	—
Hexosamine (total)	28·8	22·4	14·6	15·5	15·2	20·6	16·47	14·92	3·84
N-Acetylgalactosamine						13·5	11·08	10·15	1·80
N-Acetylglucosamine						7·1	5·39	4·77	2·04
Sialic Acid	18·3	38·4	34·3	35·0	32·5	33·8	29·89	25·76	26·71
Fucose						1·8			
Ratios relative to Hexosamine									
Total Hexose	0·75	0·58	1·30	1·23	1·43	1·18	0·93	0·77	2·05
Total Hexosamine	1·00	1·00	1·00	1·00	1·00	1·00	1·00	1·00	1·00
Sialic Acid	0·66	1·71	2·35	2·26	2·14	1·64	1·81	2·26	6·95
Fucose						0·09			

	Ficin			Phenol					diiododosalicylate/phenol
	Ebert et al. (1971)			Kathan and Adamany (1967)			Springer et al. (1966)	Winzler (1972)	Marchesi et al. (1972)
Author(s)	1	2	3	MM	NN	MN	NN		
Peptide No.	1	2	3						
Blood Group Specificity	MM and NN			MM	NN	MN	NN	MM and NN not specified	not specified
Hexose (total)	28·7	28·4	17·8	14·5	14·2	14·2	15·0	16·1	10–12
Galactose	12·2	18·0	13·1				11·1	13·2	10–12
Glucose	14·5	6·2	2·3				0·3	0·3	
Mannose	2·0	4·2	2·4				5·4	2·6	
Hexosamine (total)	14·5	28·4	21·5	14·7	14·1	14·6	9·8	19·7	18
N-Acetylgalactosamine	6·2	17·5	14·3				4·07–5·7	13·2	12
N-Acetylglucosamine	18·3	10·9	7·2				3·07–6·3	6·5	6
Sialic Acid	21·19	22·35	21·60	24·0	24·1	24·4	12·4–16·2 (as Me-pentose)	27·8	25
Fucose	1·3	0·9	1·5	1·4	1·6	1·4	0·7–0·8	1·2	
Ratios relative to Hexosamine									
Total Hexose	1·98	1·00	0·83	0·99	1·07	0·97	1·53	0·82	0·56
Total Hexosamine	1·00	1·00	1·00	1·00	1·00	1·00	1·00	1·00	1·00
Sialic Acid	1·46	0·79	1·05	1·63	1·71	1·67	1·27–1·65	1·41	1·39
Fucose	0·09	0·03	0·07	0·10	0·14	0·10	0·07–0·08	0·06	

The table is calculated from data contained in the following papers:

Cook, G. M. W. and Eylar, E. H. (1965). *Biochim. biophys. Acta* **101**, 57–66.

Ebert, W., Metz, J., Wiecker, H. and Roelcke, D. (1971). *Hoppe-Seyler's Z. Physiol. Chem.* **352**, 1309–1318.

Jackson, L. J. and Seaman, G. V. F. (1972). *Biochemistry* **11**, 44–49.

Kathan, R. H. and Adamany, A. (1967). *J. biol. Chem.* **242**, 1716–1722.

Marchesi, V. T., Tillack, T. W., Jackson, R. L., Segrest, J. P. and Scott, R. E. (1972). *Proc. natn. Acad. Sci. U.S.A.* **69**, 1445–1449.

Ohkuma, S. and Shinohara, T. (1967). *Biochim. biophys. Acta* **147**, 169–171.

Springer, G. F., Nagai, Y. and Tegtmeyer, H. (1966). *Biochemistry* **5**, 3254–3272.

Winzler, R. J. (1972). *In* "Glycoproteins: their composition, structure and function" (Gottschalk, A. ed.), part B, pp. 1268–1293, Elsevier, Amsterdam, London and New York.

Winzler, R. J., Harris, E. D., Pekas, D. J., Johnson, C. A. and Weber, P. (1967). *Biochemistry* **6**, 2195–2201.

the pure products were very small, but the materials were extremely antigenic. They were excluded from Sephadex G200, even in the presence of 5% Triton X-100 plus 2% SDS, and so were considered to be large. The lectins of *Lotus tetragonolobus* and *Dolichos biflorus* reacted with them. Lewis antigens were absent.

The amino acid composition of all these substances were similar and resembled whole stroma. They were quite low in serine and threonine, and the total polypeptide content was only about 7%, by weight. About 1% each of sphingosine and fatty acid remained, so this could still be a new class of glycolipid. The galactosyl contents were high (36–42%) as were glucosaminyl groups (29–33%), which would be acetylated *in vivo*. Galactosamine was associated only with the A-specific substance, while small amounts (1–3%) of sialic acid and glucose were found in each, and the latter was resistant to separation. Fucose accounted for 10–13% by weight. The lack of Lewis antigens was felt to be indicative of the presence of type II chains (see section on glycolipids). Liotta *et al.* (1972) have isolated rather similar materials, but with a higher lipid content than those of Kościelak (1973).

Recently Weiss *et al.* (1971) reported the isolation of a most curious glycopeptide from human red cells, which, though too small, perhaps, to be included here *sensui stricto*, is important because it contains an S-glycosidic link. Its structure is $(Glc)_3$.Cys.Glu.Gly.His.Asp.His.Gly.Ala. and the only similar peptide is that isolated by the same group from urine. No other S-glycoside is yet known in membranes.

(v) *Glycopeptides from the Red Cell Glycoproteins:* Digestion of the erythrocyte, or glycoproteins isolated from it with proteolytic enzymes, has been used as a first step in several structural studies. Cook and Eylar (1965) degraded stroma from MN specific cells with pronase and fractionated the digest. They were able to isolate crude fractions with M and N specificities, and showed differences in their carbohydrate composition (Table V) and elution characteristics on DEAE-Sephadex. The glycopeptides had molecular weights of about 1×10^4 daltons and were very soluble in water. A glycopeptide of similar size (Winzler, 1972) is released from the whole red cell with trypsin. Its composition is given in Table V. Tryptic degradation of stroma leads to the release of a soluble glycopeptide containing about 80%, by weight, of carbohydrate and 20% of amino acids, while the insoluble residue contains only about 3% of its weight as carbohydrate. The soluble glycopeptide is greatly enriched in serine and threonine, relative to the residue and starting material, and is somewhat richer in aspartic acid. It is depleted in hydrophobic amino acids (Winzler, 1972). It would, thus, appear that the glycoprotein is probably asymmetric, with hydrophilic and hydrophobic ends.

Treatment of the soluble glycopeptide or the whole glycoprotein with alkaline borohydride releases 30–40%, by weight, of the sugar (Kathan and Adamany, 1967; Winzler *et al.*, 1967). The products of the reaction and their labelling pattern showed that the reaction was a β-elimination of *O*-glycosidically-linked oligosaccharides. (Reactions of this type are well known in soluble glycoproteins; Marshall, 1972). Weber and Winzler (1969) found that the alkaline cleavage and reduction is accompanied by a loss of galactosamine and an equivalent loss of serine and threonine, with the appearance of alanine and aminobutyric acid (the reduced derivatives

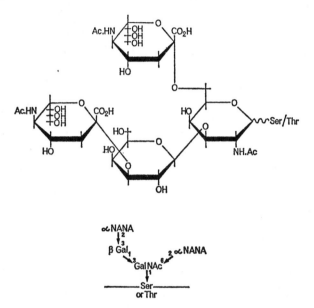

Fig. 18. Structure of the alkali-labile saccharide of the human erythrocyte membrane described by Winzler (1972). The structure is shown in ring form (Haworth formula) and in the simpler convention used in Figs. 12 and 20.

of these). The oligosaccharides released were small, and Thomas and Winzler (1969a and b) and Adamany and Kathan (1969) analysed the structure of the most abundant one by enzymatic degradation and by periodate oxidation. The molecule of tetrasaccharide contained two residues of *N*-acetylneuraminic acid, one of galactose and one of reduced *N*-acetyl-galactosamine for (reduced) reducing terminal. Neuraminidase released, both of the sialosyl residues and the disaccharide remaining, released galactose with β-galactosidase, leaving *N*-acetylgalactosaminitol (^{3}H). Periodate oxidation of the tetrasaccharide did not degrade galactose, showing that it must be substituted at C3, and it could only be linked to sialic acid (N-acetylneuraminic acid). Borohydride reduction of the periodate-oxidased product yielded threosaminitol, which is only possible if the other sialic acid residue is linked to C6 of the N-acetylgalactosamine group. The whole structure is shown in Fig. 18. It should be emphasized that though this structure is common to M, N and MN erythrocytes, there is no direct evidence that it forms a part of either of these antigenic determinants. The data of Cook and Eylar (1965) would indicate that perhaps it does not. It is involved in viral binding.

A study by Kornfeld and Kornfeld (1969, 1970) of the phytohaemagglutinin receptor site of the human erythrocyte has demonstrated another class of glycopeptide. It was found that a soluble glycopeptide, released from the surface of erythrocytes by trypsin, would inhibit haemagglutination by phytohaemagglutinin and also abolish its mitogenic effect on lymphocytes. Alkaline borohydride was used to β-eliminate other oligosaccharides, since the phytohaemagglutinin receptor was found to be alkali stable. Proteolytic degradation with Pronase, gel-filtration and ion-exchange chromatography on DEAE cellulose were used to purify the active oligosaccharide. The final product had a molecular weight of about 2×10^3 daltons. Its binding specificity was associated with a galactosyl residue of the oligosaccharide, with the galactosyl group substituted with an N-acetylneuraminyl group in some cases (Kornfeld and Kornfeld, 1969). Glycosidases were used to determine the sequence of the oligosaccharide (Kornfeld and Kornfeld, 1970) and the possible and proposed structures derived are shown in Fig. 19. It is a branched structure and contains the alkali-stable N-glycosidic bond between N-acetylglucosamine and asparagine.

Winzler (1972) has reported rather similar studies upon the structure of an alkaline borohydride-resistant glycopeptide, derived from a pronase digest. It has a molecular weight of 3×10^3 daltons by gel filtration, and was degraded with specific glycosidases to give the possible partial structure shown in Fig. 20. Its structure is very similar to that proposed by Kornfeld and Kornfeld (1969, 1970). In a study of the Pr_1 and Pr_2 antigens, Ebert et al. (1971) used a digest with the proteolytic enzyme ficin to release glycopeptides from the surface of erythrocytes, and they were able to associate the Pr_2 and I antigens with glycoproteins (sugars were estimated as alditol acetates). In the same year Merz and Roelcke (1971) showed that Pr_1 and Pr_2 antigens were not affected by acylation of amino groups. They also showed that the trifluoroacetyl group could be used to block amino groups, and could be used to abolish M and N antigenicity reversibly. Roelcke et al. (1971) isolated glycoproteins from phenol/saline extracts of erythrocytes, and digested them with ficin to yield four glycopeptide fractions. Each of fractions 1–3 contained about 20%, by weight, sialic acid; fraction 4 had none. Fractions 1–3 showed M and N antigenicity, especially fraction 2. I-antigenicity was peculiar to fraction 1, Pr_1 activity to fraction 2 and Pr_2 activity to fraction 3. No antigens were found in fraction 4.

Presant and Kornfeld (1972) have reported a study of the binding site for the mushroom (*Agaricus bisporus*) agglutinin on erythrocytes and lymphocytes. Normal lymphocytes had about twice as many receptor sites for the agglutinin as did erythrocytes, while lymphocytes from patients with chronic lymphocytic leukaemia had only about one fifth of the normal

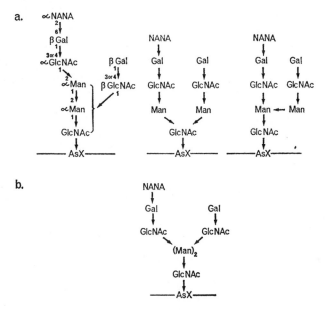

Fig.19. Structure of an alkali-stable saccharide of the human erythrocyte membrane described by Kornfeld and Kornfeld (1970). Possible structures are shown in (a) (only in (a) have the linkages and anomeric configurations been shown, for simplicity). The postulated structure is shown in (b).

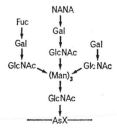

Fig. 20. Structure of an alkali-labile saccharide of the human erythrocyte membrane described by Winzler (1972). Note the similarity to the structure found by the Kornfelds (Fig. 19).

number. The ratio between the numbers of phytohaemagglutinin (*Phaseolus vulgaris* lectin) and mushroom haemagglutinin sites differed in the two cell types and this was presumed to indicate structural differences in their oligosaccharides. A tryptic digest released a glycopeptide from erythrocytes, which was found to increase in binding potency towards the mushroom agglutinin after removal of sialic acid. All the binding of glycopeptide was attributable to species having the structure galactose $\beta1,3$-N-acetylgalactosaminyl serine (or threonine), thus all were O-glycosides.

These studies serve to emphasise an important point. If a lectin is said to be blood group specific it does not mean that it has the same binding specificity as the antibody used for typing. It means that it has only been found to be reactive with cells of that particular type, and this means no more than to say that its particular ligand is rather rare and segregates with

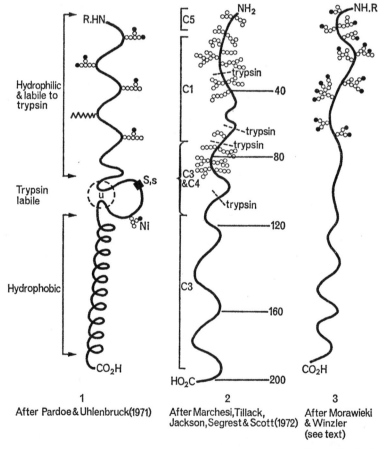

1
After Pardoe & Uhlenbruck (1971)

2
After Marchesi, Tillack,
Jackson, Segrest & Scott (1972)

3
After Morawieki
& Winzler
(see text)

[*see opposite page*

a blood group. The chemical specificity of the lectin need not be as great as that of a lectin which agglutinates very many cell types, but happens to have a very abundant ligand. Hence the "anti-N" lectin of *Vicia graminea* does not react precisely with the antigen of N-substance, since it will also agglutinate horse erythrocytes which lack the antigen (Lisowska, 1963).

(vi) *Models of the Major Glycoproteins:* Morawieki (1964) and Winzler (1969a and b; 1970; 1972) have advanced models for the structure of the major erythrocyte glycoprotein. The two models are very similar, and that of Winzler is shown in Fig. 21. It envisages a hydrophobic "tail" by which the molecule is held into the plasmalemma and which bears the C-terminal of the polypeptide (Marchesi *et al.*, 1971). The carbohydrate is supposed to lie near the (substituted) N-terminal and to form a hydrophilic head. A recent investigation by Marchesi *et al.* (1972) supports this model. After isolating the glycoprotein with the lithium di iodosalicylate procedure, they cleaved the molecule with cyanogen bromide and separated five fragments, most of which came from the N-terminal half of the polypeptide. An asymmetric distribution of hydrophobic residues was found, and a model (Fig. 21) very similar to that above was put forward. Marchesi and his colleagues have correlated this glycoprotein with "bumps" seen in the surface of the erythrocyte membrane upon freeze-etching (Marchesi *et al.*,

---

Fig. 21. Proposals for the structure of the major glycoproteins (or "glycophorin") of the human erythrocyte membrane. The model of Pardoe and Uhlenbruck (1971) in 1, is based upon immunological studies of the location of saccharide antigens and related structures. An alpha-helical structure was put forward for the non-antigenic, hydrophobic "tail" of the molecule, but this is not essential. The M and N antigens are thought to lie near the very hydrophilic N-terminal. The model of Marchesi *et al.* (1972), in 2, is based upon chemical analysis of peptides released by treatment of the protein with cyanogen bromide (the C-labelled peptides in the figure). Several sites of tryptic cleavage were identified in the hydrophilic portion of the molecule. The site nearest the C-terminal was accessible only after disruption of the membrane. An unsubstituted N-terminal was proposed and the C-terminal was supposed to be buried deep within the membrane and not to be accessible at its inner face (as part of the polypeptide chain could be). The original model of Morawieki (1964) and of Winzler (1969a and b, 1970, 1972) in 3, proposed a protein with a hydrophilic "head" and a hydrophobic "tail", with a substituted N-terminal and most of the sugar quite near this. The three models correspond very well and are, therefore, probably correct. There is some dispute over the substitution of the N-terminal and Tanner and Boxer (1972) do not find it, unlike Marchesi *et al.* (1972), but like Morawieki (1964) and Winzler (1969a and b, 1970, 1972). Segrest *et al.* (1972) have evidence for long sequences of hydrophobic residues near the C-terminal.

1971, 1972). This was discussed in Chapter 2. Pardoe *et al.* (1971) have used such a model to explain the complexities of the rare subgroups on the M and N system, and associated systems of blood grouping (Fig. 21).

Essentially this model proposes that M and N antigens lie near to the N-terminal of the polypeptide, and that rare variants (such as the Mk—non-antigenic in man, weakly antigenic in rabbit) occur in the same region. The myxoviral receptor lies further towards the C-terminal, but still on the N-terminal side of the protease-susceptible area. The U and S antigenic systems occur in the protease-susceptible area and an α-helical hydrophobic "tail" of the protein is suggested, though there is little evidence for helicity in this molecule (Jirgensons and Springer, 1968). Thus, knowledge of the detailed molecular architecture of this glycoprotein is still scant and almost nothing is known of the others present in the erythrocyte membrane. Even so, the erythrocyte remains the best known cell in respect of its surface carbohydrate content and structure.

e. *Surface Carbohydrates of Other Cell Types*

(i) *The Blood Platelet :* Like the erythrocyte, the blood platelet has a layer of ruthenium-red positive material at its surface (Hovig, 1965), has the myxo virus receptor site (Pepper and Jamieson, 1968; Barber *et al.*, 1971, cited in Jamieson *et al.*, 1971), and releases glycopeptides that contain sialic acid when treated with a variety of proteolytic enzymes (Pepper and Jamieson, 1970; Jamieson *et al.*, 1971). Studies of the tryptic digest of the surface revealed several glycopeptides of which two could be isolated in reasonable purity. The larger of these glycopeptides had a molecular weight of 12×10^4 daltons, which is much greater than for any erythrocyte sialoglycopeptide, and Jamieson *et al.* (1971) have called it a macroglycopeptide. It behaved as a single species upon electrophoresis in the presence or absence of SDS, and was particularly rich in sialic acid, galactose, N-acetylglucosamine and N-acetylgalactosamine. Its contents of serine, threonine and proline were high. Of the other glycopeptides isolated, one of moderate size ($2 \cdot 25 \times 10^4$ daltons) and a small, heterogeneous one of molecular weight $0 \cdot 5 \times 10^4$ daltons, were markedly different in composition from the macroglycopeptide. The small glycopeptide resembled glycopeptides derived from the plasma fractions in its composition. The platelet macroglycopeptide appears to be of an organ-specific character (Jamieson *et al.*, 1971).

Jamieson *et al.* (1971) describe procedures for the preparation of platelet membrane fractions and report that the effect of digesting these with trypsin is very similar to that upon the whole platelet. Again the macroglycopeptide was found. Intact ghosts release what appears to be chondroitin 4 or 6 sulphate (Jamieson *et al.*, 1971; Olsson and Gardell, 1967), whereas membranes do not, suggesting that this may arise from within the platelet during enzymatic treatment. In a study of the platelets of porcine blood, Mullinger and Manley (1968) found that a galactose-rich sialogly-copeptide was released after treatment with pronase. Two glycosamino-glycans were described, as Anderson and Odell (1958) found for rat blood platelets. In all these platelets the carbohydrate was largely bound by alkali stable links, presumably N-glycosides.

(ii) *The Liver Cell:* The association of sialic acid with the surface of the hepatocyte is now well established (Evans, 1970; Simon *et al.*, 1970; Shimizu and Funakoshi, 1970; Smith and Walborg, 1972). The studies of Smith and Walborg (1972) are a good example of the application of analysis of glycoproteins to the problem of malignancy. They isolated surface sialoglycopeptides from the surfaces of AS-30D rat ascites hepatoma cells, before and after progression of the tumour. A change in the morphology of the tumour towards smaller cell masses accompanied an increase in its malignancy during passage. This was matched by a fall in the content of sialic acid, labile to neuraminidase, between the 107th and 109th trans-plantation generations. Several glycopeptides were released from the cells by papain and these, unusually for cell-surface materials, contained glucose. A fall in glucose content with increasing passage was the most marked effect. In a related investigation, Wray and Walborg (1971) studied the binding sites for wheat germ agglutinin and concanavalin A in the Novikoff tumour cell. A crude mixture of sialoglycopeptides released by Pronase was fractionated by gel-filtration and ion-exchange chromatography, and the wheat germ agglutinin binding site was localized in one of these sialo-glycopeptides.

(iii) *Brain and Retina:* There is now considerable evidence for the presence of a large amount of sialic acid in the region of synapses (for example Dekirmenjian *et al.*, 1969), while the glycoprotein of retinal rods has been studied in some detail (Heller, 1969; Heller and Lawrence, 1970) and is associated with the visual pigment. Studies by Stoddart and Kiernan (1973a and b) and Kiernan and Stoddart (1973), using fluorescent-labelled lectins, provide evidence for the presence of carbohydrate at the surfaces of several kinds of cell in the brain, and show some interesting variations between cell types.

(iv) *Other Cell Types:* Many other cell types (Winzler, 1972) have been shown to possess carbohydrate-enriched surfaces, lectin binding sites or antigens which may well be saccharide in nature, but few of these studies have yet enabled partial structures to be advanced.

In an investigation of the surface of the Ehrlich ascites tumour cell, Langley and Ambrose (1967) isolated a "sialomucopeptide" which contained nearly equimolar amounts of galactosamine and sialic acid. Almost all the sialic acid of the cells was accessible to neuraminidase, and at pH 12·9 a disaccharide was lost from the cell surface. An important class of study is that of which the work of Glossman and Neville (1971) is an example. These authors reported a comparison of the membranes of liver cells, kidney brush border cells and erythrocytes of the rat. Analysis by electrophoresis in SDS-acrylamide gels showed a considerable similarity between the band patterns, but each membrane had its own characteristic pattern of glycoproteins, as revealed by periodate-Schiff staining. All the most strongly Schiff-positive bands contained sialic acid. Comparative studies of this sort may well lead to an understanding of tissue and tumour antigenicity, and of the specificity of interactions between cell types, in molecular terms. It seems an increasingly safe prediction that carbohydrates will be an essential part of any such description (see also Jansons and Burger, 1971).

II. Surfaces of Fungal Cells

Though the fungi are often classified with the plants, they represent a very different type of eukaryotic cell. In their contents of chitin and polypeptide-linked mannan they are unlike most plants, but have some resemblances to animal cells. They are, therefore, included here rather than in Chapter 4.

Like plant cells (see Chapter 4), fungal cells show clearly definable walls, which are intimately associated with the plasmalemma beneath. Little is known about the detailed molecular architecture of the walls of most species, though the wall of the yeast *Saccharomyces cerevisiae* has been examined in some detail. Since the fungal wall is so closely associated with the plasmalemma, it is not possible to assert that the cell membrane contains any distinctive saccharides which are not part of the wall and the two histologically separate structures are best considered to be a biochemical unity.

Despite a degree of formal similarity to the plant cell wall, the fungal wall is chemically very different. Many species contain chitin and the presence of mannan-protein complexes is a widespread feature. The wall of *S. cerevisiae* forms a good example.

The attachment of protein to polysaccharide in the cell wall of this yeast was shown by Northcote and Horne (1952) who found that the composition of the wall was glucan 29%, mannan 31%, protein 13%, lipid 8·5% and ash 3% all by weight. A number of procedures have been reported for the solubilization of these saccharides, including proteolysis (Eddy, 1958), degradation with strong alkali (Kessler and Nickerson, 1959) and extraction with anhydrous ethylenediamine (Korn and Northcote, 1960). Several investigators found traces (1–3%) of glucosamine in the wall, and this was generally considered to be derived from chitin (Falcone and Nickerson, 1956; Eddy, 1958).

Following the work of Falcone and Nickerson (1956), and Eddy (1958), which indicated the existence of protein-carbohydrate linkages, Korn and

Northcote (1960) attempted to isolate the wall polysaccharide under mild conditions. Cell walls were prepared by breaking the cells with glass beads in a vibratory mill and isolating them by differential centrifugation. The walls, which probably contained some plasmalemma-derived material, were extracted with anhydrous ethylenediamine at 37° for three days. The solid remaining was called fraction C. The supernatant was concentrated and precipitated with methanol, resuspended in water and dialysed against water. The supernatant after centrifugation was fraction A and the pellet fraction B. Fraction A contained a mannan-protein complex, fraction B a mannan-glucan-protein complex, while fraction C resembled very swollen native cell walls (though no trace of the two distinct membranes of the cell surfaces could be detected in it by electron microscopy). A molecular weight was obtained for the mannan (5.9×10^4 daltons) prepared according to Haworth, Heath and Peat (1941), and a little protein and glucosamine was found in it (as well as in the glucan, prepared according to Bell and Northcote, 1950). Not more than 9% of the total glucosamine of the wall could be explained as chitin.

The problem was re-examined by Sentendreu and Northcote in 1968. Procedures similar to those of Korn and Northcote (1960) were followed, but the fraction A (above) was gel-filtered to give glycopeptide A_2. When glycopeptide A_2 was treated with 2 N-sodium hydroxide at 98° for 1 hour, and then precipitated with ethanol, it yielded a mannan of high molecular weight which contained traces of amino acids, glucosamine and phosphate. The same glycopeptide on β-elimination, with 0.1 N-sodium hydroxide for 7 days, gave a glycopeptide of high molecular weight consisting of a mannan with amino acids covalently linked to it. Mannose and oligosaccharides were released. Likewise, a mannan containing a little amino acid, glucosamine and phosphate could be released from the whole yeast, or isolated walls, with 0.5 N-sodium hydroxide and purified by a series of precipitations.

Glycopeptide A_2 had a molecular weight of 7.6×10^4 daltons and contained 4% by weight of amino acids. It contained over 40% of serine and threonine, lacked cystine and methionine, and contained glucosamine. The β-elimination led to the appearance of degradation derivatives of serine and threonine, and the existence of O-glycosidic links of mannose to these amino acids was established. The high molecular weight mannan component remained attached to peptide via an alkali-stable linkage, and this peptide contained equimolar amounts of aspartic acid and glucosamine, after treatment with 2 N-sodium hydroxide. The presence of mannose-6-phosphate in the structure was proved, and a periodate oxidation of the polymer, followed by reduction and hydrolysis, yielded glycerol, but no erythrytol. Thus no mannose monomers were present linked only by 1,4-bonds. The glucosamine present was in two forms, one associated with the mannan and resistant to *Helix pomatia* enzyme, the other, possibly, associated with chitin (Bacon *et al.*, 1966). It seemed probable that the polysaccharide was highly branched near the site of its attachment to the polypeptide.

Brock (1965) has described the isolation and partial characterization of a mannan-protein complex of two   strains of the yeast *Hansenula wingei*, which is related to the sex-factor of this yeast. The sex-factor is connected with agglutinability of mixtures of different strains of the yeast (see Chapter 6). It was felt that the sex-factor could well be cytoplasmically located, and it was shown to be susceptible to irreversible inhibition of its agglutinating properties by 2-mercaptoethanol and periodate at ambient temperature. Periodate was ineffective in the cold. Brock (1965) suggested that disulphide bonds were required for agglutination, but the periodate effect would not be surprising in a highly branched and sterically hindered polymer.

When the plasmalemma of *S. cerevisiae* was studied by freeze-etching, areas of particles in hexagonal patterns were seen, joined to the wall by fibres (Northcote, 1969), demonstrating the intimate connection between the two structures.

If the yeast cell wall is used as a model for the fungi, it is clear that their walls are profoundly different from those of plants (Chapter 4) and from the surfaces of animal cells, yet have some resemblances to both, especially the latter, and to the surfaces of prokaryotic cells.

References

ADAMANY, A. M. and KATHAN, R. H. (1969). *Biochem. biophys. Res. Commun* **37**, 171–178.

ADAMS, R. D. and SIDMAN, R. L. (1968). "Introduction to Neuropathology", McGraw-Hill, New York, Toronto, Sydney, London.

ALLAN, D., AUGER, J. and CRUMPTON, M. J. (1971). *Expl. Cell Res.* **66**, 362–368.

ANDERSON, B. and ODELL, T. T. (1958). *Proc. Soc. exp. Biol. Med.* **99**, 765–768.

ANNO, K., SENO, N. and KAWAI, Y. (1964). Sixth Int. Congr. Biochem. New York. *Abs.* **4-6**, 502.

AUSTIN, J. H. (1963). *J. Neurochem.* **10**, 921–930.

BACON, J. S. D., DAVIDSON, E. D., JONES, D. and TAYLOR, I. F. (1966). *Biochem. J.* **101**, 36c.

BALAZS, E. A. (Ed.) (1970). "Chemistry and Molecular Biology of the Inter-cellular Matrix", 3 Vols, Academic Press, London and New York.

BARANOWSKI, T. and LISOWSKA, E. (1963). *Arch. Immunol. Terapii Doswiado-zalnej* **11**, 609–617.

BELL, D. J. and NORTHCOTE, D. H. (1950). *J. chem. Soc.* pp. 1944–1947.

BELLA, A. and DANISHEVSKY, I. (1968). *J. biol. Chem.* **243**, 2660–2664.

BENTLEY, R. (1972). *A. Rev. Biochem.* **41**, 953–996.

BISHOP, C. and SURGENOR, D. M. (1964). "The Red Blood Cell: a compre-hensive treatise", Academic Press, New York and London.

BLUMENFELD, O. O. (1968). *Biochem. biophys. Res. Commun.* **30**, 200–205.

BRADY, R. O. (1973a). *In* "Membrane Mediated Information: Function and Biosynthesis of Membrane Lipids and Glycoproteins", Oxford International Symposium 1972, (P. W. Kent, ed.), Deutscher Akademischer Austauschdienst, in press.

BRADY, R. O. (1973b). *In* "Membrane Mediated Information: Function and Biosynthesis of Membrane Lipids and Glycoproteins", Oxford International Symposium 1972, (P. W. Kent, ed.), Deutscher Akademischer Austauschdienst, in press.

BRETSCHER, M. S. (1971a). *Nature New Biology* **231**, 229–232.

BRETSCHER, M. S. (1971b). *J. molec. Biol.* **59**, 351–357.

BRIMACOMBE, J. S. and WEBBER, J. M. (1964). "Mucopolysaccharides" B.B.A. Library, Vol. 6, Elsevier, Amsterdam, London, New York.

BROCK, T. D. (1965). *Proc. natn. Acad. Sci. U.S.A.* **54**, 1104–1112.

CANDY, D. J. and KILBY, B. A. (1962). *J. exp. Biol.* **39**, 129–140.

CAREY, F. G. and WYATT, G. R. (1960). *Biochim. biophys. Acta* **41**, 178–179.

CARLSTRÖM, D. (1962). *Biochim. biophys. Acta* **59**, 361–364.

CARTER, H. E., JOHNSON, P. and WEBER, E. J. (1965). *A. Rev. Biochem.* **34**, 109–142.

COOK, G. M. W. and EYLAR, E. H. (1965). *Biochim. biophys. Acta* **101**, 57–66.

DEISS, W. P. and LEON, A. S. (1955). *J. biol. Chem.* **215**, 685–689.

DEKIRMENJIAN, H., BRUNNGRABER, E. G., JOHNSON, M. L. and LARRAMENDI, L. M. H. (1969). *Expl. Brain Res.* **8**, 97.

DZIERZKOWA-BORODEJ, W., LISOWSKA, E. and SEYFUCDOWA, H. (1970). *Life Sci.* **9**, 111–120.

EBERT, W., METZ, J., WEICKER, H. and ROELCKE, D. (1971). *Hoppe-Seyl. Z. physiol. Chem.* **352**, 1309–1318.

EBERT, W., METZ, J. and ROELCKE, D. (1972). *Eur. J. Biochem.* **27**, 470–472.

EDDY, A. A. (1958). *Proc. R. Soc. B.* **149**, 425–440.

EVANS, W. H. (1970). *Biochim. biophys. Acta* **211**, 578–581.

EYLAR, E. H. (1965). *J. theor. Biol.* **10**, 89–113.

FALCONE, G. and NICKERSON, W. J. (1956). *Science, N.Y.* **124**, 272–273.

FALK, M., SMITH, D. G., McLACHLAN, J. and McINNES, A. G. (1966). *Can. J. Chem.* **44**, 2269–2281.

FRÄNSSON, L. Å. (1968). *Biochim. biophys. Acta.* **156**, 311–316.

GALLOP, P. M., BLUMENFELD, O. O. and SIEFTER, S. (1972). *A. Rev. Biochem.* **41**, 617–672.

GARDAS, A. and KOŚCIELAK, J. (1971). *Vox Sang.* **20**, 137–149.

GLASER, L. and BROWN, D. H. (1957). *J. biol. Chem.* **228**, 729–742.

GLOSSMANN, H. and NEVILLE, D. M. JNR. (1971). *J. biol. Chem.* **246**, 6339–6346.

GOTTSCHALK, A. (1972). *In* "Glycoproteins, their composition, structure, and functions" (A. Gottschalk, ed.), two vols. (A and B), Elsevier, Amsterdam, London and New York.

GOTTSCHALK, A., BELYAVIN, G. and BIDDLE, F. (1972). *In* "Glycoproteins, their composition, structure, and functions" (A. Gottschalk, ed.), part B, pp. 1082–1096, Elsevier, Amsterdam, London and New York.

GREENWALT, T. J., STEANE, E. A. and PINE, N. E. (1971). *In* "Glycoproteins of Blood Cells and Plasma" (G. A. Jamieson and T. J. Greenwalt, eds), pp. 235–244, Lippincott, Philadelphia and Toronto.

HAKOMORI, S. (1970). *Proc. natn. Acad. Sci. U.S.A.* **67**, 1741–1747.

HAKOMORI, S. and MURAKAMI, W. T. (1968). *Proc. natn. Acad. Sci. U.S.A.* **54**, 254–261.

HALL, D. A. and SAXL, H. (1960). *Nature, Lond.* **187**, 547–550.

HALL, D. A. and SAXL, H. (1961). *Proc. R. Soc.* B. **155**, 202–217.

HALL, D. A., LLOYD, P. F. and SAXL, H. (1958). *Nature, Lond.* **181**, 470–472.

HALL, D. A., HAPPEY, F., LLOYD, P. F. and SAXL, H. (1960). *Proc. R. Soc.* B. **151**, 497–516.

HATHEWAY, C. L., WESELI, D., LUDWICK, T. and HINES, H. (1969). *Vox Sang.* **17**, 204–216.

HAWORTH, W. N., HEATH, R. L. and PEAT, S. (1941). *J. chem. Soc.* pp. 833–842.

HELLER, J. (1969). *Biochemistry, N.Y.* **8**, 675–679.

HELLER, J. and LAWRENCE, M. (1970). *Biochemistry, N.Y.* **8**, 864–869.

HEYNS, K., SPERLING, K. R. and GRÜTZMACHER, H. F. (1969). *Carbohydrate Res.* **9**, 79–97.

HOVIG, T. (1965). *Thromb. Diath. Haemorrh.* **13**, 84–113.

HUNT, S. (1970). "Polysaccharide-Protein Complexes in Invertebrates", Academic Press, London.

HUPRIKAR, S. and SPRINGER, G. F. (1970). *In* "Blood and Tissue Antigens" (E. Aminoff, ed.), Academic Press, New York.

INOUYE, M. (1971). *J. biol. Chem.* **246**, 4834–4838.

JAMIESON, G. A. and GREENWALT, T. J. (1971). "Glycoproteins of Blood Cells and Plasma" (G. A. Jamieson and T. J. Greenwalt, eds), Lippincott Philadelphia and Toronto.

JAMIESON, G. A., BARBER, A. J. and URBAN, C. L. (1971). *In* "Glycoproteins of Blood Cells and Plasma" (G. A. Jamieson and T. J. Greenwalt, eds), pp. 219–234, Lippincott, Philadelphia and Toronto.

JANSONS, V. K. and BURGER, M. M. (1971). *In* "Glycoproteins of Blood Cells and Plasma" (G. A. Jamieson and T. J. Greenwalt, eds), pp. 267–279, Lippincott, Philadelphia and Toronto.

JIRGENSONS, G. and SPRINGER, G. F. (1968). *Science, N.Y.* **162**, 365–367.

KABAT, E. A. (1956). "Blood Group Substances: their chemistry and immunology", Academic Press, New York and London.

KATHAN, R. H. and ADAMANY, A. M. (1967). *J. biol. Chem.* **242**, 1716–1722.

KATHAN, R. H. and WINZLER, R. J. (1963). *J. biol. Chem.* **238**, 21–29.

KATHAN, R. H., WINZLER, R. J. and JOHNSON, C. A. (1961). *J. exp. Med.* **113**, 37–45.

KATZMAN, R. L. (1972). *Biochim. biophys. Acta* **266**, 269–272.

KATZMAN, R. L. and JEANLOZ, R. W. (1970). *In* "Chemistry and Molecular Biology of the Intercellular Matrix", (E. A. Balazs, ed.), Vol. 1, pp. 217–228, Academic Press, London and New York.

KENNEDY, J. F. (1971). *In* "Carbohydrate Chemistry", **4**, 149–166, Chemical Society Specialist Periodical Reports.

KESSLER, G. and NICKERSON, W. J. (1959). *J. biol. Chem.* **234**, 2281–2285.

KIERNAN, J. A. and STODDART, R. W. (1973). *Histochemie* **34**, 77–84.

KISS, J. (1970). *Adv. Carb. chem. and Biochem.* **24**, 382–433.

KLENK, E. and UHLENBRUCK, G. (1960). *Z. physiol. Chem.* **319**, 151–160.

KORN, E. D. and NORTHCOTE, D. H. (1960). *Biochem. J.* **75**, 12–17.

KORNFELD, S. and KORNFELD, R. (1969). *Proc. natn. Acad. Sci. U.S.A.* **63**, 1439–1446.

KORNFELD, R. and KORNFELD, S. (1970). *J. biol. Chem.* **245**, 2536–2545.

KOŚCIELAK, J. (1973). *In* "Membrane Mediated Information: Function and Biosynthesis of Membrane Lipids and Glycoproteins", Oxford International Symposium 1972, (P. H. Kent, ed.), Deutscher Akademischer Austausch-dienst, in press.

LANDSTEINER, K. and LEVINE, P. (1972a). *Proc. Soc. Exp. Biol. Med.* **24**, 600–602.

LANDSTEINER, K. and LEVINE, P. (1927b). *Proc. Soc. Exp. Biol. Med.* **24**, 941–942.

LANGLEY, O. K. and AMBROSE, E. J. (1967). *Biochem. J.* **102**, 367–372.

LENARD, J. (1970a). *Biochemistry, N.Y.* **9**, 1129–1132.

LENARD, J. (1970b). *Biochemistry, N.Y.* **9**, 5037–5040.

LINDAHL, U. and RODÉN, L. (1966). *J. biol. Chem.* **241**, 2113–2119.

LIOTTA, I., QUINTILIANI, M., QUINTILIANI, L., BUZZONETTI, A. and GUILIANI, E. (1972). *Vox Sang.* **22**, 171–182.

LISOWSKA, E. (1963). *Nature, Lond.* **198**, 865–866.

LISOWSKA, E. and MORAWIECKI, A. (1967). *Eur. J. Biochem.* **3**, 237–241.

LLOYD, K. O. (1970). *Archs. Biochem. Biophys.* **137**, 460–468.

MADDY, A. H. (1964). *Biochim. biophys. Acta* **88**, 448–449.

MADDY, A. H., DUNN, M. J. and KELLY, P. G. (1972). *Biochim. biophys. Acta* **288**, 263–276.

MÄKELÄ, O. and CANTELL, K. (1958). *Ann. med. exp. Biol. Fenniae* (Helsinki) **36**, 366–374.

MARBET, R. and WINTERSTEIN, A. (1951). *Helv. chim. Acta* **34**, 2311–2320.

MARCHESI, V. T., TILLACK, T. W., JACKSON, R. L., SEGREST, J. P. and SCOTT, R. E. (1972). *Proc. natn. Acad. Sci. U.S.A.* **69**, 1445–1449.

MARCHESI, V. T., TILLACK, T. W. and SCOTT, R. E. (1971). *In* "Glycoproteins of Blood Cells and Plasma", (G. A. Jamieson and T. J. Greenwalt, eds), pp. 106–113, Lippincott, Philadelphia and Toronto.

MARCHESSAULT, R. H., PEARSON, F. G. and LIANG, C. Y. (1960). *Biochim. biophys. Acta* **45**, 499–507.

MARSHALL, R. D. (1972). *A. Rev. Biochem.* **41**, 673–702.

MERZ, W. and ROELCKE, D. (1971). *Eur. J. Biochem.* **23**, 30–35.

MEYER, K. and CHAFFEE, E. (1941). *J. biol. Chem.* **138**, 491–499.

MORA, P. T., BRADY, R. O., BRADLEY, R. M. and MCFARLAND, V. W. (1969). *Proc. natn. Acad. Sci. U.S.A.* **63**, 1290–1296.

MORAWIECKI, A. (1964). *Biochim. biophys. Acta* **83**, 339–347.

MORGAN, W. T. J. (1967). *In* "Methods in Immunology and Immunochemistry", (C. A. Williams and M. W. Chase, eds), Vol. 1, pp. 75–81, Academic Press, New York and London.

MORGAN, W. T. J. and WATKINS, W. M. (1969). *Br. med. Bull.* **25**, 30–34.

MULLINGER, R. N. and MANLEY, G. (1968). *Biochim. biophys. Acta* **170**, 282–288.

NE'EMAN, Z., KAHANE, I., KOVARTOVSKY, J. and RAZIN, S. (1972). *Biochim. biophys. Acta* **266**, 255–268.

NELSON, C. A. (1971). *J. biol. Chem.* **244**, 4406–4412.

NESKOVIC, N. M., NUSSBAUM, J. L. and MANDEL, P. (1970). *J. Chromat.* **49**, 255–261.

162   SURFACE CARBOHYDRATES OF THE EUKARYOTIC CELL

NORTHCOTE, D. H. (1969). *In* "Essays in Biochemistry", (P. N. Campbell and G. D. Greville, eds), Vol. 5, pp. 89–137, published for the Biocheimcal Society by Academic Press, London.

NORTHCOTE, D. H. and HORNE, R. W. (1952). *Biochem. J.* **51**, 232–236.

OKADA, S. and O'BRIEN, J. S. (1969). *Science, N.Y.* **165**, 698–700.

OLSSON, I. and GARDELL, S. (1967). *Biochim. biophys. Acta* **141**, 348–357.

PARDOE, G. I. (1971). *Nouv. Rev. Franc. d'Haemat.* **11**, 863–877.

PARDOE, G. I., UHLENBRUCK, G. and REIFENBERG, U. (1971). *Med. Lab. Technology* **28**, 255–283.

PEARSON, F. G., MARCHESSAULT, R. H. and LIANG, C. Y. (1960). *J. Polymer Sci.* **43**, 101–116.

PEPPER, D. S. and JAMIESON, G. A. (1968). *Nature, Lond.* **219**, 1252–1253.

PEPPER, D. S. and JAMIESON, G. A. (1970). *Biochemistry, N.Y.* **9**, 3706–3713.

PITESKY, D. and TERRY, T. M. (1972). *Biochim. biophys. Acta* **274**, 95–104.

PRESANT, C. A. and KORNFELD, S. (1972). *J. biol. Chem.* **247**, 6937–6945.

RACE, R. R. and SANGER, R. (1962). "Blood Groups in Man", 4th edition, Blackwell Scientific Publications, Oxford.

RAMCHANDRAN, G. N. (ed.) (1962). "Aspects of Protein Structure", Academic Press, London.

RAMCHANDRAN, G. N. (ed.) (1967). "Treatise on Collagen", Academic Press, London.

RANDALL, R. F., STODDART, R. W., METCALFE, S. M. and METCALFE, J. C. (1972). *Biochim. biophys. Acta* **255**, 888–899.

RAY, A., REYNOLDS, J. A., POLET, H. and STEINHARDT, J. (1966). *Biochemistry, N.Y.* **5**, 2606–2616.

REES, D. A. (1969). *J. chem. Soc.* (B), pp. 217–230.

REVEL, J. P. (1970). *In* "Chemistry and Molecular Biology of the Intercellular Matrix", (E. A. Balazs, ed.), Vol. 3, p. 1485–1502, Academic Press, London.

REYNOLDS, J. A. and TANFORD, C. (1970). *Proc. natn. Acad. Sci. U.S.A.* **66**, 1002–1007.

RICH, A. and CRICK, F. H. C. (1961). *J. molec. Biol.* **3**, 483–506.

ROBBINS, P. W. and MACPHERSON, I. A. (1971). *Nature, Lond.* **229**, 569–570.

RODÉN, L. and SMITH, R. (1966). *J. biol. Chem.* **241**, 5949–5954.

ROELCKE, D., EBERT, W., METZ, J. and WEICKER, H. (1971). *Vox Sang.* **21**, 352–361.

ROSENBERG, S. A. and GUIDOTTI, G. (1968). *J. biol. Chem.* **243**, 1985–1992.

ROSENBERG, S. A. and GUIDOTTI, G. (1969). *J. biol. Chem.* **244**, 5118–5124.

ROTHFUS, J. A. and SMITH, E. L. (1963). *J. biol. Chem.* **238**, 1402–1410.

SEGREST, J. P., JACKSON, R. L. and MARCHESI, V. T. (1972). *Biochem. biophys. Res. Commun.* **49**, 964–969.

SENTENDREU, R. and NORTHCOTE, D. H. (1968). *Biochem. J.* **109**, 419–432.

SHAPIRA, J. (1969). *Nature, Lond.* **222**, 792–793.

SHAPIRO, A. L., VIÑUELA, E. and MAIZEL, J. V. (1967). *Biochem. biophys. Res. Commun.* **28**, 815–820.

SHAW, D. H. and MOSS, G. W. (1969). *J. Chromat.* **41**, 350–357.

SHIMIZU, S. and FUNAKOSHI, I. (1970). *Biochim. biophys. Acta* **203**, 167–169.

SIMON, F. R., BLUMENFELD, O. O. and ARIAS, I. M. (1970). *Biochim. biophys. Acta* **219**, 349–360.

SMITH, D. F. and WALBORG, E. F. (1972). *Cancer Res.* **32**, 543–549.

SPIK, G., STRECKER, G. and MONTREUIL, J. (1969). *Bull. Soc. Chim. biol.* **51**, 1287–1295.

SPIRO, R. G. (1970a). *In* "Chemistry and Molecular Biology of the Intercellular Matrix", (E. A. Balazs, ed.), Vol. 1, pp. 195–216, Academic Press, London.

SPIRO, R. G. (1970b). *In* "Chemistry and Molecular Biology of the Intercellular Matrix", (E. A. Balazs, ed.), Vol. 1, pp. 511–534, Academic Press, London.

SPIRO, R. G. (1972). *In* "Glycoproteins: their composition, structure and function", (A. Gottschalk, ed.), part B, pp. 964–999, Elsevier, Amsterdam, London and New York.

SPOONER, R. L. and MADDY, A. H. (1971). *Immunology* **21**, 809–816.

SPRINGER, G. F. and ANSELL, N. J. (1958). *Proc. natn. Acad. Sci. U.S.A.* **44**, 182–189.

STODDART, R. W. (1970). *Biochem. J.* **120**, 30P–31P.

STODDART, R. W. and KIERNAN, J. A. (1973a). *Histochemie* **33**, 87–94.

STODDART, R. W. and KIERNAN, J. A. (1973b). *Histochemie,* **34**, 275–280.

SWEELEY, C. C. and WALKER, B. (1964). *Analyt. Chem.* **36**, 1461–1466.

SWEELEY, C. C., BENTLEY, R., MAKITA, M. and WELLS, W. W. (1963). *J. Am. chem. Soc.* **85**, 2497–2507.

TAKAYAMA, K., MacLENNAN, D. H., TZAGOLOFF, A. and STONER, C. D. (1966). *Archs. Biochem. Biophys.* **114**, 223–230.

TANNER, M. J. A. and BOXER, D. H. (1972). *Biochem. J.* **129**, 333–347.

THIERRAULT, D. G. (1963). *J. Am. Oil Chem. Soc.* **40**, 395–399.

THOMAS, D. B. and WINZLER, R. J. (1969a). *Biochem. biophys. Res. Commun.* **35**, 811–818.

THOMAS, D. B. and WINZLER, R. J. (1969b). *J. biol. Chem.* **244**, 5943–5946.

TRIPLETT, R. B., SUMMERS, J., ELLIS, D. E. and CARRAWAY, K. L. (1972). *Biochim. biophys. Acta* **266**, 484–493.

TURNER, L. P. (1969). *Analyt. Biochem.* **28**, 288–294.

UHLENBRUCK, G. (1969). *Vox Sang.* **16**, 200–210.

UHLENBRUCK, G. (1973). *In* "Membrane Mediated Information: Function and Biosynthesis of Membrane Lipids and Glycoproteins", (P. W. Kent, ed.), Oxford International Symposium 1972, Deutsche Akademischer Austausch-dienst (in press).

UHLENBRUCK, G., WINTZER, G., VOIGTMANN, R., SALFNER, B. and COHEN, E. (1972). *In* "Glycoproteins of Blood Cells and Plasma", (G. A. Jamieson and T. J. Greenwalt, eds), pp. 74–93, Lippincott, Philadelphia and Toronto.

VICKER, M. G. and EDWARDS, J. G. (1972). *J. Cell Sci.* **10**, 759–768.

WARREN, L., CRITCHLEY, D. and MACPHERSON, I. A. (1972). *Nature, Lond.* **235**, 275–278.

WATKINS, W. M. (1958). *In* "Proceedings of the 7th International Congress on Blood Transfusion", Rome, p. 206.

WATKINS, W. M. (1964). *In* "The Red Blood Cell: a comprehensive treatise", (C. Bishop and D. M. Surgenor, eds), pp. 363–364, Academic Press, New York.

WATKINS, W. M. (1966). *In* "Glycoproteins: their composition, structure and function", (A. Gotschalk, ed.), 1st edition, B.B.A. Library, Vol. 5, pp. 462–515. Elsevier, Amsterdam, London and New York.

WATKINS, W. M. (1972). *In* "Glycoproteins: their composition, structure and function", (A. Gottschalk, ed.), part B, pp. 830–891, Elsevier, Amsterdam, London and New York.

WEBER, P. and WINZLER, R. J. (1969). *Archs. Biochem. Biophys.* **129**, 534–538.

WEICKER, H. (1968). *Z. klin. Chem. klin. Biochem.* **6**, 398–406.

WEIGEL, H. (1963). *Adv. Carbohydrate Chem.* **18**, 61–97.

WEISS, J. B., LOTE, C. J. and BOBINSKI, H. (1971). *Nature New Biology* **234**, 25–26.

WHISTLER, R. L. and WOLFROM, M. L. (1963–5). "Methods in Carbohydrate Chemistry", 5 Vols, Academic Press, New York.

WHITTEMORE, N. B., TRABOLD, N. C., REED, C. F. and WEED, R. J. (1969). *Vox Sang.* **17**, 289–299.

WINZLER, R. J. (1969a). *In* "Red Cell Membranes, Structure and Function", (G. A. Jamieson and T. J. Greenwalt, eds), pp. 157–171, Lippincott, Philadelphia and Toronto.

WINZLER, R. J. (1969b). *In* "Cellular Recognition", (R. J. Smith and R. A. Good, eds), pp. 11–18, Appleton Century Crofts, New York.

WINZLER, R. J. (1970). *In* "Blood and Tissue Antigens", (D. Aminoff, ed.), p. 117–126, Academic Press, New York.

WINZLER, R. J. (1972). *In* "Glycoproteins: their composition, structure and function", (A. Gottschalk, ed.), Part B, pp. 1268–1293, Elsevier, Amsterdam, London and New York.

WRAY, V. P. and WALBORG, E. F. (1971). *Cancer Res.* **31**, 2072–2079.

YOKOYAMA, M. and TRAMS, E. G. (1962). *Nature, Lond.* **194**, 1048–1049.

YOSIZAWA, Z. (1972). *In* "Glycoproteins: their composition, structure and function", (A. Gottschalk, ed.), Part B, pp. 1000–1018, Elsevier, Amsterdam, London and New York.

ZAHLER, P. H. (1968). *In* "Membrane Models and the Formation of Biological Membranes", (L. Bolis and B. A. Pethica, eds), pp. 181–189, North-Holland Publishing Company, Amsterdam.

ZINBO, M. and SHERMAN, W. R. (1970). *J. Am. chem. Soc.* **92**, 2105–2114.

ZVILICHOVSKY, B., GALLOP, P. M. and BLUMENFELD, O. O. (1971). *Biochem. biophys. Res. Commun.* **44**, 1234–1243.

4

Surface Carbohydrates of Plant Cells

I. Basic Analytical Problems

In the case of plant cells, the carbohydrates of the cellular surface are largely concentrated in the cell wall, which is a more or less distinct structure lying external to the plasmalemma. In general the wall is composed predominantly of carbohydrate (mostly as polysaccharide), though proteins, lipids and polyphenols are other important components.

Although the cell wall appears in the optical and electron microscopes as an organelle distinct from the bimolecular leaflet of the plasmalemma, and can be substantially separated from the plasmalemma by plasmolysis in solutions of high osmotic strength, it is intimately associated with the metabolic activities of this underlying membrane. For, despite the cell wall's being separated from the cytoplasm by the major permeability barrier of the cell, it is not unchanging, but undergoes continued modification as the cell grows and specializes. New components are added to the wall, either by the action of complex enzyme systems of the cell membrane, or by the passage of materials through the plasmalemma after their synthesis deeper within the cell. The cell wall can also be degraded and eroded as the cell differentiates, so as to achieve a structural specialization appropriate to its function. It should not be thought that the cell wall is subject only to passive modification by the cellular membranes, for it can contain some enzymatic functions of its own and considerable metabolism of carbohydrate may occur *in situ*. Thus, though the plasmalemma and cell wall are histologically—and to some extent mechanically—separable, they are a metabolic unity.

Cell walls of plants vary greatly in their chemistry between species and between cell types, but some generalizations are possible. In the lower plants there is great variation among the orders, though most cell walls contain some acidic polysaccharides, and in most cases the biological functions of the polysaccharides are not well defined. In the higher algae and in terrestrial plants, there does appear to be a basic structure common to almost all cell walls. In this there is a highly ordered fibrillar component, consisting largely of cellulose, arranged in a less ordered or amorphous matrix of hemicelluloses, pectins, protein and other substances. It is this type of cell wall which has attracted the greatest degree of attention with regard to its chemistry, biosynthesis and physical properties and which will, consequently, be discussed in detail in this chapter.

A major problem in any consideration of the polysaccharides of plant cell walls is the lack of any means of assaying them for their biological activity. As a consequence, they cannot be defined biologically, as can an enzyme, and are usually defined in terms of ideal chemical compositions, though these compositions may represent only a part of the biologically active macromolecule. Thus cellulose is defined as a β-1,4-glucan on the

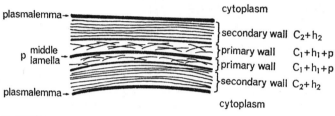

p = pectins
C_1 = cellulose of primary wall
C_2 = cellulose of secondary wall
h_1 = hemicellulose of primary wall
h_2 = hemicellulose of secondary wall

Fig. 1. The arrangement of two adjacent plant cell walls. The figure shows the relationship of the various layers of cell walls to each other and the plasmalemmata of the cells that produced them. Note that pectins tend to be associated with the oldest part of the wall, the middle lamella and primary wall, while cellulose is deposited in primary wall and, very extensively, in secondary wall (where it is highly ordered). Many complex specializations of this structure are possible.

basis of a substance isolated from the cotton boll under specified conditions, though the polysaccharide isolated from most living plant tissues under similar conditions contains small amounts of mannose and xylose. These, it can be argued, are mere contaminants, albeit strongly adsorbed, and can be removed by suitably harsh procedures to leave an ideal cellulose. This, however, is an unsatisfactory argument, since these same conditions will cleave some glycosidic bonds, so that it could be objected that these sugars are really an integral part of the macromolecule. Since the sugars are present only in small quantities, both arguments are quite persuasive and can be resolved absolutely only by the demonstration of the presence in partial hydrolysates of small and characterizable oligosaccharides containing all the sugars in question in covalent linkage. Even then, care must be taken to avoid the generation of artefacts by reversion of free sugars to oligosaccharides by forming glycosidic bonds during hydrolysis. Because of the enormous disparity in lability among different glycosidic linkages it is not always possible to achieve suitable sites of cleavage to display such oligosaccharides, and so proof positive is often lacking. On the other hand, suggestive evidence may be available from measurements of molecular weights of polysaccharides. Though polysaccharides in plant cell walls are generally polydisperse and show some microheterogeneity, they generally tend towards an upper limit of molecular weight which is often very high. Some of the observed heterogeneity and polydispersity may arise from chemical degradation during isolation, in which case the distribution of molecular weights will be distorted (usually broadened, somewhat skewed and shifted to a lower mean, causing disparity between number-average and weight-average values), and the measured molecular weight (by whatever means) is lowered relative to undegraded polymer. A broad spread of values of molecular weight and a low average value is strongly suggestive of degradation, as is well illustrated by the pectic substances (which are considered later in this chapter), where the literature-values of molecular weight have risen by at least two orders of magnitude over twenty years. Another possible source of degradation during isolation is from the action of endogenous glycosidases, and other enzymes. It is seldom easy to guard against this and the possibility of their having acted is often ignored.

In any discussion of the biosynthesis and function of polysaccharides in the cell wall caution must be exercised in making generalizations, since relatively few species have received close attention and it is unusual for metabolic studies to be made on chemically well defined walls. With the exception of the coleoptile of the oat (*Avena*) it is the dicotyledons that have received most biochemical and mechanical study, and the gymnosperms, ferns, liverworts and mosses have largely been ignored. Chemically, the dicotyledonous angiosperms and the gymnosperms are fairly widely explored, but the liverworts and mosses are almost unknown. Moreover, it is often the case that biochemical and mechanical studies relate to histologically heterogeneous material, such as whole stems, and some polysaccharides (such as pectic substances) are best known in chemical terms in a particular type of organ (such as a fruit) where they may be structurally typical only of those parts of the plant used in biochemical and other investigations.

Clearly, if cell wall polysaccharides are often heterosaccharides, there are enormous possibilities for variations in their structure by changes in sequences, branching or configuration of residues, many of which may occur not only between species, but also between adjacent cell types in a stem, for example, or even between different regions in the wall of a single cell, such as a sieve tube of phloem. This will, again, lead to heterogeneity in isolated polysaccharides. Accordingly the definitions of polysaccharides, applied here and in subsequent chapters, should be understood to allow for some deviation from an ideal chemical definition, and in those cases where a heterogeneous tissue has been used, or degradative conditions of isolation applied, the polydispersity of the product may be in part an artifact.

The cell walls of higher plants are essentially biphasic structures in which an ordered lattice of fibrillar elements is arranged within a matrix, which may also show some striated structure, but is much less ordered than the other component. Considerable morphological specialization may occur and the whole wall may show many layers and zones within it, in addition to its fundamental structure. The fibrillar lattice-work is composed of interwoven mats of microfibrils of about 50–150 Å in diameter, which

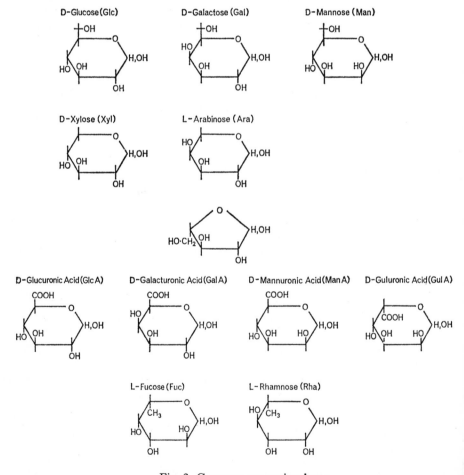

Fig. 2. Common sugars in plants.

are so regularly packed that certain regions form quasi-crystalline structures from which X-ray diffraction patterns can be obtained. Between these regions of high order are less ordered zones, through which the microfibrils appear to be substantially continuous.

II. Selective Extraction of Polysaccharides

It is possible to extract the non-fibrillar polysaccharides of plant cell walls by the sequential application of suitable solvents, and a variety of techniques have been used. Polyphenols, especially lignin, are first removed by converting them to soluble derivatives. Alkaline hypochlorite solutions are frequently used (Thornber and Northcote, 1961a and b), and convert the polyphenols to halogenated derivatives which are soluble in organic solvents such as ethanol/ethanolamine. A similar result is achieved by treating the cell walls with acidified bisulphite solutions at temperatures of 120°–150°, under pressure, or by using alkaline sulphite at 150°–170° (Aspinall, 1970). In these latter two processes, a substantial part of the non-fibrillar polysaccharide of the wall is also removed. The lignin-free residue is termed "holocellulose".

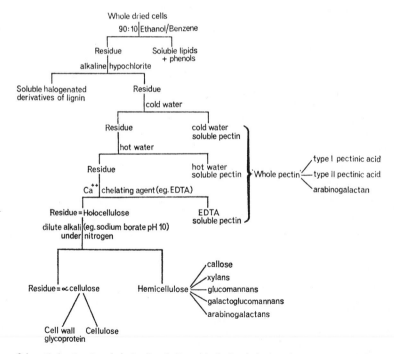

Schematic fractionation of plant cell walls to explain the terminology used

Fig. 3. Scheme showing a typical sequence of extraction procedures for the several major components of a plant cell wall.

Pectic substances may be extracted from whole walls (or some lignin-free residues) by means of cold water, hot water and hot solutions of chelating agents for calcium, applied in sequence. Terms such as "cold-water soluble pectin" do not imply chemical homogeneity in the extract, and the various pectinic acids occur in differing proportions in each type of crude pectin fraction. A variety of chelating agents have been used, including disodium ethylenediamine tetra-acetic acid, ammonium oxalate and sodium hexametaphosphate. This last is to be preferred, since it has been shown to cause minimal degradation of pectinic acids (Stoddart *et al.*, 1967).

After extraction of lignins, hemicelluloses are extracted from holo-cellulose with dilute alkaline solutions (e.g. sodium borate-sodium hydroxide mixture, above pH 9·5, Heath and Northcote, 1971) under nitrogen, to minimise degradation. The hemicellulose fraction is normally very heterogeneous and the residue is termed α-cellulose. This fraction is also a mixture.

The contents of each of these crude fractions varies greatly from tissue to tissue and species to species, and varies especially with the state of differentiation of the cells. In cotton fibres cellulose accounts for about 98% of the dry tissue be weight, while in typical woods the content is about 45%.

Comparative studies of different cell types can be made by using standardized extraction procedures of this general type, and these have proved very valuable (Thornber and Northcote, 1961a and b; Jensen and Ashton, 1960), but the accurate definition of structural differences between corresponding polysaccharides in two different cell types usually depends on the application of special techniques for the isolation of these polysaccharides alone in fairly undegraded form.

In this chapter emphasis will be placed on the structures of cellulose and the pectins, since the former illustrates well the complexities of even an apparently simple polymer, while the latter shows how essential careful and critical analysis is to the description of a complex polysaccharide. Moreover, these two types of polymer are ubiquitous among the terrestrial plant cell walls.

III. Cellulose

A. Distribution

As Timell (1965) has pointed out, the occurrence of cellulose is remarkably general in terrestrial plants, and it appears always to have the same structure. With the exception of pectins, no other polysaccharide is known to be so widespread in distribution in plant cell walls, for the structures of polysaccharides of hemicellulose are very variable between species.

Cellulose is not restricted to plants, for essentially similar polysaccharides are known from several species of bacteria and from the tunicates (see Chapter 3). It may well be that the ability to produce polymers of this sort is an ancient property of the plasmalemma which has been lost in some lower plants and more generally in animals.

B. Isolation and Composition

Cellulose accounts for at least 98 %, weight for weight, of the dry fibre of the cotton boll, and this has been the source most generally used in chemical studies. In most woody tissues cellulose accounts for rather less than half the dry weight of the cell wall and in young and actively dividing cells its content may be lower still, but it is in proportion to the amount of fibrillar material in the walls. Differential extraction of the cell wall with alkali removes the hemicelluloses and pectins, leaving a residue that is defined as being α-cellulose. Though this residue contains some protein it is relatively rich in cellulose, for it contains all the cellulose initially present in the wall and retains a fibrillar structure. If the cellulose is disorganized by treatment with cellulase, or alkaline cuprammonium salts, the fibrillar structure is lost, and so there is good evidence that the microfibrils of the cell wall are at least substantially composed of cellulose.

The cellulose obtained from the cotton boll after removal of other materials is termed cellulose I, and this can be converted to another form, cellulose II, by solubilization in alkaline cuprammonium salts, followed by ethanolic precipitation, or by regeneration of cellulose from derivatives of cellulose I, such as the acetate. Cellulose II has the thermodynamically preferred configuration and it has often been prepared in the process of purifying cellulose, so that many physical-chemical studies upon cellulose have used this modified form.

Celluloses I and II differ from α-cellulose in their very low content of sugars other than glucose, for α-cellulose contains substantial quantities of mannose, xylose, galactose and arabinose and traces of other sugars. The galactose and arabinose are largely attributable to a wall glycoprotein (q.v.) that is separable from the cellulose of the α-cellulose fraction, but the nature of the other sugars is more obscure. Attempts to isolate oligosaccharides from α-cellulose in which glucose of cellulose is covalently linked to other sugars have failed thus far, but if such linkages do exist they might be more labile than the β-1,4 links of cellulose and they are likely to be present only in very small quantities, perhaps too small for ready isolation. Against this, it is known that xylans can adsorb to celluloses irreversibly, and this could easily happen with a variety of polysaccharides during preparation of α-cellulose. Some celluloses have been claimed to possess acid labile linkages in very low concentration. Though these might represent glycosidic links other than β-1,4-links, they could also represent strained linkages that have resulted from abnormal hydrogen bonding between glucosyl residues and they cannot be taken as firm evidence for the presence of sugars other than glucose. Thus, though it is not possible to assert that sugars other than glucose do occur in cellulose, and though their presence in α-cellulose can be explained away, it is conceivable that very small quantities of mannose and xylose, for example, are present in cellulose chains at levels well below the minimum that X-ray crystallography can detect. In proving the β-1,4-glucan structure of cellulose it was necessary to show the type and configuration of the glycosidic bond and presence of only one type of glycosidic bond in the molecule. The studies which achieved this are a good example of the classical analytical procedures of carbohydrate chemistry.

C. Structure

1. SEQUENCE AND LINKAGE OF RESIDUES

Irvine and Hirst (1923) made the first important step in the elucidation of the structure of cellulose, by showing that exhaustive methylation, followed by hydrolysis, gave 2,3,6-tri-O-methyl-D-glucose in a yield almost theoretical for a long 1,4-linked polymer of D-glucopyranose units, or a 1,5-linked polymer of D-glucofuranose residues. Haworth and his collaborators (Haworth et al., 1927) distinguished between these alternatives by analysis of the structure of cellobiose, the disaccharide obtained, together with higher saccharides, by partial acid hydrolysis of cotton cellulose. They (Haworth et al., 1927) showed that the disaccharide could not arise in the quantities observed by reversion from glucose, and also showed that acetolysis of methylated cellulose gave rise to the completely methylated glycosides of cellobiose and higher oligosaccharides. Hence, both the possibilities of reversion and rearrangement during hydrolysis were excluded, and cellobiose must have been derived directly from cellulose. By oxidation of the reducing end of cellobiose they obtained cellobionic acid, subjected it to an exhaustive methylation and hydrolysed the product. This yielded 2,3,4,6-tetra-O-methyl-D-glucose and 2,3,5,6-tetra-O-methyl-gluconolactone, the latter of which could only arise in a 1,4-linked polymer.

That the linkage is β-1,4 was proved by showing that emulsin will degrade cellobiose, whereas maltase will not. Since emulsin is specifically a β-glucosidase, the linkage must have been β. This has been confirmed by measurements of optical rotation, by crystallography and by nuclear magnetic resonance.

The demonstration by Freudenberg and Blomqvist (1935) that the kinetics of hydrolysis of cellulose in 51%, weight for weight, sulphuric acid are compatible with the hydrolysis of a uniform, linear polymer, showed that the proportion of glycosidic bonds of a lability different from

the β-1,4 type must be less than 1%, and no other type of glycoside has
been proved to exist in cellulose.

2. THREE-DIMENSIONAL STRUCTURE

In 1937, Meyer and Misch proposed a crystal structure for cellulose,
on the basis of data obtained from X-ray diffraction patterns of cellulose II.
They proposed that the fibre repeat distance (10·3 Å) was that of a cellobiose unit and that the chains in the cellulose were straight (i.e. that all
the glucose residues of any chain were coplanar). This model was
modified by Hermans in 1943, since it would lead to steric hindrance

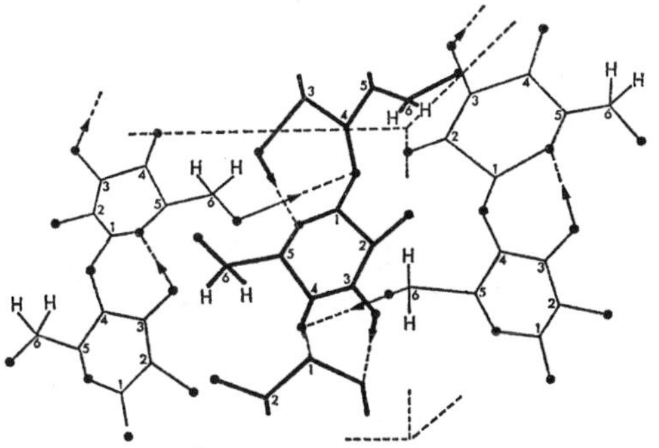

● Oxygen
Numbers refer to carbon atoms of glucosyl residues

Fig. 4. Model of the unit cell of cellulose (after
Liang and Marchessault, 1959). One cellulose
chain is drawn more heavily than the other two.
Note the antiparallel arrangement of the
chains.

between $O_{(2)}$ and $C_{(6)}$ and strain between certain hydrogens, and he
proposed a bent configuration in which alternate glucose residues lie in
the same plane (Fig. 4). This model accommodates extensive hydrogen
bonding, and Liang and Marchessault (1959) have shown that it is in
agreement with relaxation data using polarized infra-red irradiation of
cellulose II. The main difficulty with this model is that, like that of Meyer
and Misch (1937), it proposes an antiparallel arrangement of the adjacent
β-1,4-glucan chains. This poses great problems for the biosynthesis of
cellulose in that it could place considerable and complex restrictions
upon the structure of the site in the plasmalemma at which the microfibril
originates, or upon the subsequent folding of the β-1,4-glucan chains
in the microfibril.

The evidence that the β-1,4-glucan chains of native cellulose are
arranged in an antiparallel manner is by no means clear cut. Meyer and
Misch (1937) used the argument that cellulose II is essentially similar to

cellulose I, and that it forms from the latter without general rearrangement of the main chains. Marchessault and Sarko (1967) have argued against this, and in the absence of clear-cut evidence the question remains open. Chitin, a closely related linear polymer of N-acetylglucopyranosylamine residues, does possess an antiparallel structure with a repeat of 10·3 Å, and to this extent such a structure is feasible for cellulose.

Measurements of the molecular weight of cellulose have been made by means of a variety of physical properties, and by estimation of the amount

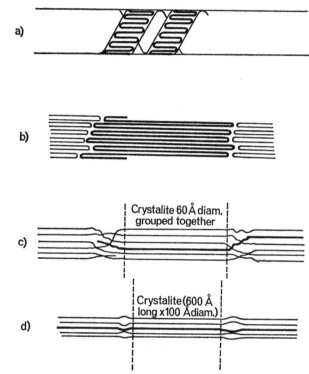

Fig. 5. Various models for the structure of the microfibril of cellulose. In cases (a) to (d) one cellulose chain is drawn heavily, to clarify its arrangement.

(a) Model of Manley (after Manley, 1964). Each molecule is folded upon itself in a plane and is then twisted.

(b) Model of Marx-Figini and Schulze (after Marx-Figini and Schulze, 1966). The molecule is again folded, but parallel to the long axis of the fibre.

(c) Model of Hess, Mahl and Gütter (after Hess et al., 1957). Fibres run from crystallite to crystallite and start and end in intervening regions of lesser order.

(d) Model of Rånby (after Rånby, 1958). This is similar to (c) above, but the fibres start and end in the crystallite.

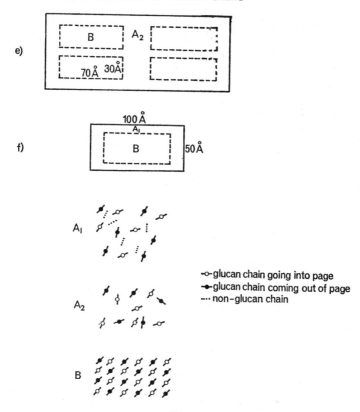

Fig. 5

(e) Model of Frey-Wyffling.
(f) Model of Preston and Cronshaw (after Preston and Cronshaw, 1958). The fibril is shown in transverse section.
Models (e) and (f) are compatible with (b), (c) and (d). Mark *et al.* (1969) have argued strongly against the folded-chain models of cellulose.

of 2,3,4,6-tetra-*O*-methyl-D-glucose liberated upon acid hydrolysis of exhaustively methylated cellulose (which gives a measure of the amount of non-reducing terminal residues). In a series of measurements based upon light scattering, Goring and Timell (1962) showed that cellulose from several sources, all of then plant, had a maximum molecular weight of $1\cdot3 - 1\cdot6 \times 10^6$ (corresponding to a degree of polymerization[1] of $8 - 10 \times 10^3$), which is in excess of the old physical and chemical values by an order of magnitude. Marx-Figini and Schulze (1963) obtained a similar result with cotton cellulose, and reported a fraction of a degree of

---

[1] Degree of polymerization here means the number of monosaccharide residues per molecule of polysaccharide, and is thus a measure of molecular size.

polymerization of $11 \cdot 5 \times 10^3$, thought to be in secondary wall, and a smaller molecular weight fraction, of degree of polymerization $1 \cdot 5 \times 10^3$, from primary wall. A fraction of intermediate molecular weight (degree of polymerization $= 5 \cdot 5 \times 10^3$) was considered to be a degradation product.

X-ray diffraction studies upon whole microfibrils show an ordered structure, similar to that of isolated cellulose, but with a lesser degree of regularity than purified cellulose. Any attempt to accommodate the unit cell structure into a model of a microfibril must not only allow for some degree of irregularity, but must also take account of the fact that the length of the microfibril is much greater than that of an individual cellulose molecule. It must also take account of the possibility of an antiparallel arrangement of the unit cell, even though individual glucan chains may all arise from the plasmalemma in the same direction and sense. A number of such models have been proposed.

Manley (1964) suggested that the β-1,4-glucan chains of cellulose are folded back upon themselves, many times, so as to form narrow flat ribbons, with the mean axis of the glycosidic links at an angle (less than a right angle) to the long axis of the ribbon. It is suggested that this angle is such as to allow the ribbon to be folded into a close helix, in which the axis of the helix is parallel to the mean axis of the glycosidic links (Fig. 5a). Such a model incorporates an antiparallel arrangement of the chains as a consequence of folding after synthesis, but it does not explain in detail the linking together of glucan chains in the whole microfibril, or explain the lack of order in microfibrils relative to cellulose, for no specifically disordered regions are included in the model. Likewise, it is not clear how non-cellulosic components can be incorporated.

In their recent model, (Fig. 5), Marx-Figini and Schulze (1966) have also achieved an antiparallel arrangement of chains by repeatedly folding them back and forth. It is supposed that there are long, linear portions of the chains, which end with sharp bends, so that the chains are turned back, and by repetition of this, each glucan chain forms a short, wide unit (relative to that of Manley, 1964). The axis of the glucan chains is parallel to the axis of the fibril, and all regions of folding occur within a fairly narrow zone of the fibril and are the regions where adjacent cellulose molecules will interact, and where non-cellulosic components might be concentrated. This model thus accommodates distinct zones of lessened regularity, as well as antiparallel chains.

Both Hess *et al.* (1957) and Rånby (1958) have put forward models of other types in which a highly ordered array of linear chains is interspersed with disordered zones (Fig. 5c and d), in which cellulose molecules can begin and end and across which many cellulose molecules are continuous. Rånby (1958) suggests that the cellulose molecules lie across such less ordered zones and may contain monomers other than glucose in these regions, but such postulates may not be necessary and hemicelluloses could be present as well. Such models account for many of the observed features of microfibrils as seen in the electron microscope, but they do not necessarily allow for antiparallel glucan chains.

A further class of model of microfibrillar structure is that in which a regular quasi-crystalline array of glucan chains lies at the core of the structure, and less regular groups of cellulose and hemicellulose molecules are packed around them. In the model of Preston and Cronshaw (1958) the microfibril is supposed to have a single rectangular crystalline core about 30 Å by 80 Å in section, with a region of lesser order about 10 Å wide around it. Frey-Wyssling (1954) proposes that the microfibril is four-cored and that the dimensions of the core are similar to those of Preston and Cronshaw (1958). Both these models are to some extent compatible with the foregoing.

Clearly, any detailed description of the structure of microfibrils is necessarily dependent upon a proper understanding of the structure of cellulose, and here the major problems to be resolved are the questions of the covalent linkage of sugars other than glucose in the molecule. and the possible antiparallel arrangement of the glucan chains. From the point of view of the biosynthesis of cellulose, it is probably easier to envisage an ordered synthesis of chains of the same polarity, rather than an equal mixture of antiparallel chains, though the possibility of an intricate folding of newly synthesized chains would make it necessary to postulate a rather sophisticated organizational system. Likewise, the synthesis of a uniform homopolymer is more easily envisaged than that of a heteropolymer with a few unusual residues at long intervals (though this occurs in pectins, q.v.). A particularly difficult point is raised by the molecular weight values of cellulose in primary and secondary walls, for something must stop cellulose synthesis at a certain degree of polymerization, and this seems to differ with the state of development of the cell. The biosynthetic problems of cellulose are discussed in detail in Chapter 5.

IV. Pectins

A. Distribution

Acidic polysaccharides of a wide variety of different kinds are found among plants; in particular among the algae (Smith and Montgomery, 1959). Pectic substances, though found in some algae (see below), are typical of terrestrial vascular plants, and their chemistry has been studied in detail in very few species of these. The presence of pectin has been claimed in many algae, usually upon the basis of staining with ruthenium red, but the results are equivocal (Lewin, 1962). Farr (1963) reported the presence of pectin in *Chorella pyrenoidosa* and Desikachary and Dweltz (1961) in diatoms: in neither case is the evidence good. Pectin does occur in the *Characeae*. It has been clearly demonstrated in *Chara* spp. (Anderson and King, 1961b) and in *Nitella* spp. (Anderson and King, 1961a and c) where it resembles the pectins of higher plants in its content of galactose, arabinose, xylose and rhamnose. This is of great interest in view of the supposed descent of higher plants from algae similar to the *Characeae*. The *Zosteraceae* also contain pectins (Ovodova and Ovodov, 1969).

Very little is known of pectins of mosses and liverworts, though there is evidence of polysaccharide complexes of this type in *Funaria hygrometrica* and *Marchantia polymorpha* (Stoddart, unpublished results). The former contained uronic acid, galactose, arabinose, rhamnose, xylose, glucose and mannose, the latter uronic acid, galactose, arabinose, rhamnose and xylose only. Timell (1962a, b) has reported the presence of pectin in fourteen different species of fern. He found that they contained 2–9%, weight for weight, of pectin in their cell walls, which is a figure similar to those obtained with gymnosperms and angiosperms. No data were given to indicate the neutral sugar contents of these pectins, but the walls as a whole contained less arabinose than those of gymnosperms. Timell (1964) has also examined three species of *Equisetum* and three species of *Lycopodium*. A very high content of pectin was claimed in *Equisetum*, though Stoddart (unpublished) did not find this to be true of *Equisetum arvense*, which had a normal content of pectin. The *Lycopodium* species contained rather little pectin. In each case Timell gave no data upon the contents of neutral sugars, though Stoddart indentified galactose, arabinose and rhamnose in pectin of *Equisetum arvense*.

Pectins have been reported from several gymnosperms, though interest in these plants has emphasized their hemicelluloses. Among the angiosperms, the dicotyledons—and more especially their fruits—have attracted more attention than the monocotyledons, though Aspinall and Cañas-Rodriguez (1958) studied sisal pectin in some detail. In all these different pectins, the neutral sugars present are galactose, arabinose, rhamnose and only traces of others. Thus, it appears that pectic substances generally have the same fundamental structure, whatever their source. This is of great importance, since Timell (1965) has emphasized that the cellulose and glucomannans have very constant structures throughout the terrestrial vascular plants, whereas hemicelluloses vary widely. This may possibly be related to a terrestrial habitat, but very little is known of how such complex molecules evolved or of their functions in cell walls.

B. Histology

Evidence for the location of pectic substances in the cell wall is of two types, derived from histochemistry and from enzymatic degradation. There is circumstantial evidence for the presence of pectin in the cell plate. Staining with ruthenium red (Mangin, 1893) and ferric chloride/ hydroxylamine (Gee et al., 1959) has shown the presence of a partially esterified polycarboxylic acid which probably represents a pectin (Carré and Horne, 1927; Kerr and Bailey, 1934; Albersheim et al., 1960a; Albersheim and Killias, 1963). Similarly the primary wall and middle lamella stain. Enzymatic degradation with polygalacturonase has also shown that the middle lamella is rich in pectin (Zaitlin and Coltrin, 1964), but there is no direct evidence that it is composed in a manner similar to the cell plate, even though it is in an equivalent site in the wall. Pectino-lytic enzymes are known to "macerate" plant tissues, to lessen the adhesion of the cells, and so pectins are presumed to be associated with regions of cellular contact. These results are in general agreement with the known phases of polyuronide deposition in plant cell growth (Thornber and Northcote, 1961a and b), that is with cell division and extension, primary growth, but not with secondary thickening. Thus pectins have long been regarded as probable agents and regulators of cell extension. This is considered in detail in Chapter 6.

C. Structure

1. HISTORICAL

Though the solubilization of pectic polysaccharides by water and their selective precipitation with alcohol was described by Vauquelin as early as 1790, the elucidation of their structure has proved particularly difficult, and is still by no means completed. Early attempts to hydrolyse crude preparations of the whole pectins of many plant tissues led to the detection of galactose and arabinose in the hydrolysates, and a variety of structures were proposed to accommodate them, including some cyclic structures (see Kertesz, 1951). Hirst and Jones (1939a and b) used selective acid precipitation to purify a crude preparation of apple pectin and showed that there were two major components present, one rich in galactose and arabi-nose and lacking uronic acid, and the other rich in uronic acid. They showed that the polyuronide component was composed of 1,4-linked galacturonosyl residues, with some methyl esterification (Hirst and Jones, 1939a), and suggested that there were separate arabinan and galactan components in the neutral fraction (Hirst and Jones, 1939b). From these results the view developed that the acidic component of crude pectin was free of neutral sugars, and that the associated arabinose and galactose were really a hemicellulose. A "pectic triad" was proposed, of "protopectin" (insoluble, native pectin), "pectic acid" (an ideal galac-turonan) and "pectinic acid" (its ideal methyl ester). These ideas were embodied into the official definitions of pectinic acids (Kertesz et al., 1944), which still stand. In his massive review of the pectic substances of 1951, Kertesz states this attitude badly, and it is still widely held (for example,

Roudier and Galzin, 1966). Because of this view, most of the physical-chemical studies of pectic substances take no account of neutral sugars, or use preparations that lack them (which are now known to be degraded), and have concentrated attention on the effects of methyl ester groups and of acetylation of hydroxyl groups in the polymers. Though this has yielded some valuable information, especially with regard to gelling in pectins of industrial usefulness, it does leave an extensive gap in knowledge of these molecules.

Some workers (for example Speiser, 1947) continued to maintain that neutral sugars were present in pectinic acids, because they could only be fully removed by clearly degradative procedures, but they were generally unheeded until the work of Aspinall and his collaborators.

2. MODERN CHEMICAL ANALYSIS

a. *Presence of Neutral Sugars*

In 1958 Aspinall and Cañas-Rodriguez isolated a "pectic" acid from sisal flesh and subjected it to methylation analysis. From the methylated sugars obtained upon hydrolysis, they showed that L-rhamnose, L-arabinose, D-galactose, as well as D-galacturonic acid, were covalently linked in the starting material. Aspinall and Fanshawe (1961) followed this with a study of pectic substances in lucerne (*Medicago sativa*) and showed the presence of "aldobiouronic" acids (disaccharides containing neutral sugar attached to galacturonic acid), thereby establishing for the first time a covalent linkage of neutral sugars to polyuronide chains in a pectin. Application of methylation studies also showed that arabinose was present in a branched polymer. This, taken with the fact that long galacturonan chains were well known in pectins, indicated that the molecule must contain substantial neutral, as well as acidic, regions, rather than a mixture of monomers and small oligomers of each.

b. *Stability of Pectins*

Another important consequence of this work was that many of the earlier results, obtained by hydrolysing pectins, now became explicable upon the basis of the great range of stabilities towards various kinds of hydrolysis that might be expected from the different types of glycosidic bond present in these polymers. Any use of acidic precipitation, such as that used by Hirst and Jones (1939b), would lead to a degradation of the polysaccharide, releasing neutral sugars and oligosaccharides similar in composition to the neutral arabinan-galactans usually extracted together with pectinic acids (Hirst and Jones, 1939a and b). Such a degradation at pH2 has been described with apple fruit pectinic acid (Barrett and Northcote, 1965). Though the free galacturonan is unusually stable to acid hydrolysis below the pK value of the uronide carboxyl group, and the pyranose rings are usually extensively degraded upon hydrolysis, it is rather more labile than a similar galactan above the pK, because of participation of the ionised carboxyl group in the hydrolysis (Smidsrød et al., 1966). At alkaline pH pectins are highly unstable, even at low temperature (Vollmert, 1950), and both saponification of methyl ester groups and cleavage of the polyuronide chain may occur. If oxygen is present, oxidation of the reducing terminals will also take place. The ester groups are also liable to slow saponification at neutrality (Doesburg, 1965), while Albersheim (1959) and Albersheim et al. (1960b) found that the polyuronide chain would break (by a β or transelimination) at methyl esterified residues if heated at neutral pH. Since this transelimination reaction has a larger temperature coefficient than saponification it tends to predominate upon heating pectins. Consequently, pectins are stable only in the range pH 3–4·5 on heating, and within rather wider limits in the cold.

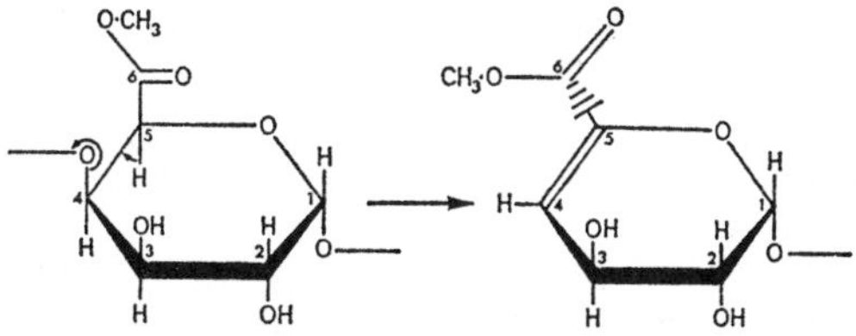

Fig. 6. The mechanism of β-elimination (or trans-elimination) of a pectin. The reaction occurs under neutral or alkaline conditions on heating or, more slowly, in the cold. The greater the alkalinity the more rapid the reaction. It is necessary for the carboxyl groups to be esterified and saponification and β-elimination compete in alkaline solution. The latter has the greater temperature coefficient and so predominates upon heating (Albersheim et al., 1960). The unsaturated product left after the breakage of the polyuronide chain fragments to yield formyl pyruvic acid, which gives a red product with thiobarbituric acid.

This lability of the pectinic acid molecule has not been widely appreciated, with the result that most extraction procedures used in the preparation of pectin, even now, are such as to cause considerable degradation, especially by transelimination. EDTA solutions are effective calcium-binding agents only at neutral or alkaline pH (Letham, 1962) under which conditions transelimination will occur (Joslyn and Deuel, 1959—unpublished data cited in Joslyn, 1962; Albersheim et al., 1960b), while their use at acid pH (Thornber and Northcote, 1962) is largely superfluous. Ammonium oxalate, a widely used extractant (Aspinall and Fanshawe, 1961), has been reported to catalyse decarboxylation of polygalacturonic acid (Anderson et al., 1961). A safe extractant appears to be sodium hexametaphosphate (Letham, 1962), which chelates efficiently at pH 4 and does not cause degradation in sycamore callus and apple pectins

(Stoddart *et al.*, 1967). This point is of especial importance where pectin must be extracted for metabolic studies.

c. *Arrangement o Neutral Sugars*

Since the experiments of Aspinall and Fanshawe, a considerable amount of further information has been gained about the chemistry of pectins and related gums (q.v.) from several sources, and a general picture of the structures of these molecules is emerging. Barrett and Northcote (1965) reported a new study of a pectin from apple fruit, which they obtained by

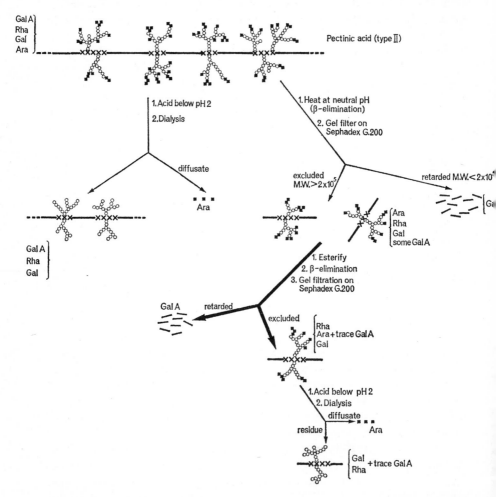

Fig. 7. General scheme for the structure and degradation of a pectinic acid of type II, illustrating the principles of the experiments of Barrett and Northcote (1965). For details see the text.

hot water extraction followed by selective precipitation with cetyl-pyridinium chloride. They confirmed the observation of Hirst and Jones (1939a and b) that a neutral polymer, or polymers containing arabinose or galactose, was present, but they also found that these sugars, as well as rhamnose and traces of other sugars, were associated with the polyuronide precipitable with cetylpyridinium ions. The pectinic acid appeared to be reasonably homogenous by zone electrophoresis upon glass-fibre paper, but after transelimination resolved into two fractions. These were separated into an almost pure galacturonan (molecular weight below 10^4) and a high molecular weight (above 2×10^5) component, which contained almost all the neutral sugar present initially, plus some uronic acid. A further cycle of esterification and transelimination of this second component further reduced its uronic acid content, without loss of neutral sugar. A feature of particular interest is that the rhamnose, which other studies have shown to be in the polyuronide chains, remained associated with the arabinose and galactose. An acid treatment of this transelimination product led to a loss of neutral sugar, especially arabinose. Barrett and Northcote (1965) also identified aldobiouronic acids in partial hydrolysates from this pectin, including galacturonosyl 1,2-rhamnose. They concluded that the great bulk of the neutral sugar was present as neutral "blocks" attached to, or interspersed in, the polyuronide chains, and that all the galactose and arabinose could be so accounted.

This study firmly established that pectins can contain neutral sugars as large and distinct portions of the whole molecule, and that they are not distributed at random as small branches upon the polyuronide chains. It also served to emphasize the importance of controlling the procedures for isolation of these polysaccharides most carefully. At low pH (below about 2·5) there is an appreciable loss of arabinosyl residues by a "peeling" of the sugar from the outer parts of the neutral blocks, while at any pH above 6·5 the pectin is liable to undergo transelimination at esterified uronosyl residues. Hence, there seemed every reason to suspect a great many of the claimed preparations of "pectin" as being highly degraded, and to reconsider the possible derivation of many "hemicelluloses" from pectins.

Gould et al. (1965) showed that the pectic arabinan of white mustard was a highly branched polymer containing 1,3- and 1,5-links, and that the pectinic acid from the same source contained neutral polymers of arabinose, which seemed to resemble the neutral arabinan. Both arabinose polymers and single residues of xylose occurred as branches, via C3 of galacturonic acid or C4 of rhamnose. Thus, this pectin has rather simple neutral "blocks".

In 1967, Stoddart et al. reported a study of the pectins of several actively growing tissues, which further demonstrated the complexity of the chemistry of these molecules. They found that cambial tissues of apple (*Pyrus malus*, var. Bramley's Seedling) and of sycamore (*Acer pseudoplatanus*), and callus tissues derived from these in culture (as well as callus tissues of bean, *Phaseolus vulgaris*, and virginian creeper, *Parthenocissus tricuspidatus*), contained two distinct types of pectinic acid, plus a neutral arabinan-galactan. All of these could be separated from each other in small quantities by zone electrophoresis at pH 6·5 upon glass-fibre paper. The three polysaccharides of sycamore callus tissue were also separated in highly pure form by differential precipitation procedures, with calcium ions, ethanol, acetone and cetylpyridinium chloride. The electrophoretic procedure enabled esterifications, transeliminations and saponifications to be followed on the very

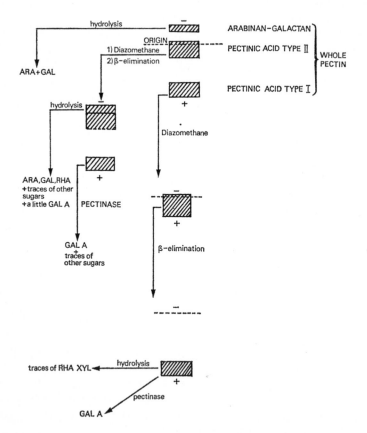

Fig. 8. General scheme for the chemical degradation of a whole pectin from a callus tissue as followed by electrophoresis, to indicate the approach in the experiments of Stoddart, Barrett and Northcote (1967). For details see the text.

small scale needed for this study, and made it possible to define the major chemical features of each polysaccharide.

The most mobile of the three sycamore carbohydrates at pH 6·5 was a pectinic acid which contained very little neutral sugar and was especially lacking in galactose and arabinose. It was unaffected by pectin pectyl hydrolase (pectin methylesterase, E.C.3.1.1.11), while esterification and transelimination gave no evidence for neutral blocks. This appeared, therefore, to be essentially similar to the classical type of pectinic acid save that it had an appreciable content of rhamnose and xylose. It was suggested that it lacked any neutral "blocks", but still contained short branches containing xylosyl residues and had rhamnosyl residues in the polyuronide chains. Pectinic acids of this sort which lack neutral blocks, will be referred to in this book as of "type I".

The other pectinic acid present in sycamore callus was similar to that of the apple fruit (see above), in that it contained neutral blocks, though of a different composition from that of the apple. Pectinic acids containing such neutral blocks will be referred to as of "type II". Comparative studies of the whole pectins of the various callus and cambial tissues, showed the same two classes of pectinic acids in each, though with substantial variation in their detailed neutral sugar contents. It seems likely, therefore, that the pectinic acids of type I might be characteristic of actively growing and dividing tissues.

The neutral polysaccharides present in the sycamore pectin did not resolve into two components upon zone electrophoresis in sodium tetraborate (0·05 M, pH 9·2), and no evidence was obtained to suggest that it was not a copolymer of arabinose and galactose, as in the apple fruit, which were the only two sugars present in it. However, no attempt was made to isolate the heterodisaccharide from it, so its composition as an arabinan-galactan is not proven.

Studies by Stoddart and Northcote (1967) showed metabolic interrelations between these three components of pectin in sycamore callus, and these are discussed in Chapter 5.

A considerable amount of new chemical information about the structure of pectinic acids has come from a variety of workers, and the total literature of gross pectin compositions is now large, but usually uncritical and often unsound. Consequently, ideas about details of structure still mostly depend upon a small number of analyses. Of these probably those of Aspinall and his colleagues on Soybean pectin (Aspinall *et al.*, 1967a, b and d), and on the pectic gum tragacanth (Aspinall *et al.*, 1967c), are the most important.

The analysis of partial hydrolysates of pectins, isolated by quite violent procedures, from soybean hull and meal enabled a large number of di, tri and tetra saccharides to be characterized. The results are summarized in

OLIGOSACCHARIDES of PECTINS, CONTAINING NEUTRAL SUGARS

SUGAR	LINKAGE
D-Galactose	β-D-Gal p1, 4-D-Gal p
	β-D-Gal p1, 2-D-Xyl p
	β-D-Glc pA1,6-D-Gal
	β-D-Glc pA1,4-D-Gal p
L-Rhamnose	D-Gal pA1, 2-L-Rha
	D-Gal pA1, 2-L-Rha 1,4-D-Gal pA1, 2-LRha
	D-Gal pA1, 2-L-Rha 1, 2-L-Rha
	D-Gal pA1, 4-D-Gal pA1,2-L-Rha p1, 2-L-Rha
	D-Gal pA1,4-D-Gal pA1, 2-L-Rha
L-Fucose	∝-L-Fuc p1, 2-D-Xyl
	β-D-Glc pA1, 4-Fuc p
D-Xylose	∝-L-Fuc p1, 2-D-Xyl
	β-D-Gal p1, 2-D-Xyl
	β-D-Xyl p1, 3-D-Gal A
L-Arabinose 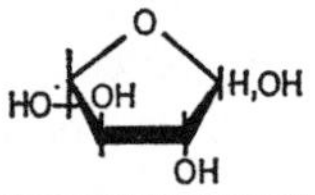	L-Araf 1,5-L-Araf
D-Glucuronic Acid	β-D-Glc p A1,6-D-Gal
	β-D-Glc p A1,4-D-Gal p
	β-D-Glc p A1,4-L-Fuc p
	β-D-Glc pA1,2-D-Man
D-Mannose 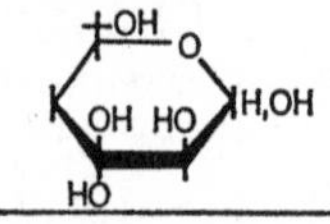	β-D-Glc p A1,2-D-Man

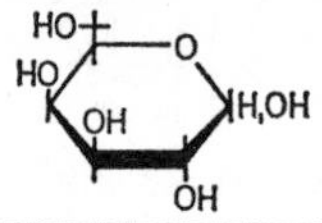

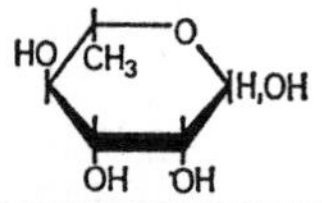

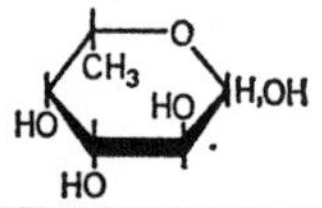

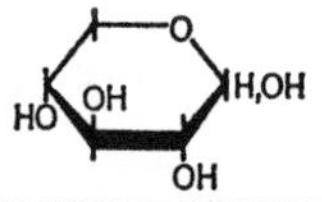

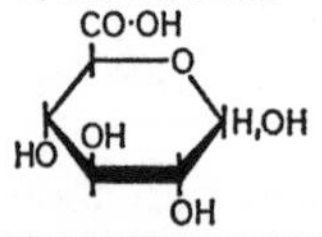

Fig. 9. Various oligosaccharides now identified in partial hydrolysates of pectins. Chiefly based upon the analysis of soybean pectins by Aspinall *et al.* (1967a, b and d). Many of these saccharides are known from several other pectins. They are incorporated into the structure shown in the next figure (Fig. 10).

Fig. 9. Because of the conditions used in the extraction of the pectin, arabinose was largely lost and no conclusions could be drawn about its site within the molecule. It appeared to be present as a branched polymer of the furanoside form, and was probably in the outer part of the molecule. Hence it was very labile to acid.

The polyuronide portion of the molecules contained both galacturonoside and rhamnoside residues. The latter could be both contiguous with each other, and alternate with galacturonic acid. In view of the results of Barrett and Northcote (1965), who showed the association of rhamnosyl-rich regions of pectin with neutral blocks, and the virtual absence of rhamnose from the acidic product of transelimination, it follows that there can be long polyuronide sequences without any rhamnose in them, and that rhamnose occurs near the point of attachment of neutral blocks. By methylation studies Aspinall et al. (1967a) showed that some rhamnosyl residues were substituted at C4, a result similar to that previously obtained by Aspinall and Fanshawe (1961) for lucerne (*Medicago sativa*) pectin.

It is not yet possible to distinguish how many types of branching from the polyuronide chain can occur in pectinic acids. At least two types of linkage can occur, to C3 of galacturonic acid and C4 of rhamnose, while there are at least two types of branch—the single xylosyl residue, or small xylosyl oligosaccharide, attched to galacturonic acid, and the larger, neutral block. By analogy with tragacanthic acid (see below), disaccharide branches (-2-Fuc *p*. 1—Xyl.*p*.1 and β-L-Gal.*p*—2β—Xyl *p* 1—) might be expected (Aspinall and Baillie, 1963). This site of attachment of the large, neutral blocks is unknown, but it is tempting to suggest that they may be linked through C4 of rhamnose, in contrast to xylose and the xylose-containing oligosaccharides linked to C3 of galacturonic acid. The resolution of this question is one of the most important outstanding problems of pectin chemistry, since the rhamnose-rich regions may be recognition sites for the attachment and assembly of neutral blocks (see Chapter 5).

The identification of glucuronic acid in both the soybean hull and meal pectins was of great interest in view of the widespread occurrence of this sugar in gums and hemicelluloses (q.v.). Stoddart et al. (1967) found no glucuronic acid in callus or cambial whole pectins, nor in their purified fractions of sycamore callus pectin, though small traces could easily have escaped detection. Aspinall et al. (1967c) have pointed out that the aldobiouronic acid intially identified as galacturonosyl galactose is, in fact, glucuronosyl galactose and hence there could be glucuronic acid in both apple (Barrett and Northcote, 1965) and sycamore (Stoddart et al., 1967) pectins, which has not been identified as such. Glucuronic acid is a trace component of citrus and lucerne pectins (Aspinall et al., 1967d citing Aspinall, Craig and Uddin, unpublished) and -O-β-D-glucopyranosyl uronic acid -L-fucose has been identified in these. The linkage of mannose [see opposite page

to glucuronic acid is interesting, in that it is common in gums (Smith and Montgomery, 1959). Mannose is known in sycamore callus pectins and cambial pectins (Stoddart *et al.*, 1967). It was not found in other callus tissues, though it occurred in all three polysaccharides of sycamore, as separated by electrophoresis. The release of 3-*O*-β-D-xylopyranosyl-D-galacturonic acid upon the enzymatic degradation of soybean hull pectin (Aspinall *et al.*, 1967d) proved the linkage of xylose to galacturonic acid, which corresponded to the single stub branches demonstrated in the pectic tragacanthic acid (Aspinall and Baillie, 1963; Aspinall *et al.*, 1967c). Xylose has also been found linked to galactose and to fucose through the 2-position, again like tragacanthic acid. Galacturonosyl-xylose was demonstrated in sycamore whole pectin (Stoddart *et al.*, 1967). Barrett and Northcote (1965), and Stoddart *et al.* (1967) found 3-*O*-methyl xylose as a trace sugar in apple fruit and sycamore pectins. Stoddart *et al.* (1967) did not find fucose in callus or cambial pectins, though 2-*O*-methyl fucose was present in sycamore. The unequivocal demonstration of covalently linked fucose in soybean pectin is, therefore, of very great interest, especially since this sugar is associated with the cell membrane of animal cells (q.v.) and is quite common in gums (Smith and Montgomery, 1959). Fucose and 2-*O*-methyl fucose were found in apple fruit pectinic acid by Barrett and Northcote (1965).

This information makes it possible to advance probable general structures for the two types of pectinic acids. It is clear that the galacturonan chains of pectins are interspersed with rhamnosyl residues, which tend to be clustered together with long intervals between clusters. These rhamnosyl residues lie close to, or at, the sites of attachment of the neutral blocks in pectinic acids of type II, and are present in pectinic acids of type I, where they are not associated with neutral blocks. Shorter branches are also present and contain xylose, fucose, glucuronic acid and, probably, other sugars. These branches occur in both types of pectinic acid and are

---

Fig. 10. Detailed structure of a pectin. A schematic drawing is shown in (a) of a short section of a galacturonorhamman chain, to which are attached various known types of branches (i.e. a pectinic acid of type I). Most pectins would be less densely substituted than this *in vivo*. Some methyl esterification is shown and the glucuronosyl groups are also probably esterified. In native pectins some acetylation and methyl esterification (mostly at C4 of xylosyl groups) is found. Some rhamnosyl residues are substituted at C4 by an unknown group (R). Quite possibly this group represents a neutral block, since the site of attachment of these is unknown, but close to the rhamnosyl groups (see text). A small part of the proposed structure of a typical neutral block of a pectinic acid of type II is shown in (c). The grouping shown in (b) is part of the structure of the type I and II pectinic acids, but cannot yet be located. Many variations of these structures occur.

probably attached to galacturonosyl residues. In pectinic acids of type II neutral "blocks" are present, of a high degree of polymerization. Their major components are galactose and arabinose, and the latter appears to be largely in the furanoside configuration and is more external than the galactose. A few residues of other sugars may be present.

This model is far from complete, but it does allow room for great structural variation between pectins. More recent studies have made it clear that such variation is to be found, but have added little to the basic model. With the improved separations of specific glycosidases made possible by techniques such as electrofocusing, and the increasing availability of lectins, a new attack upon the chemistry of pectins might now prove feasible and valuable.

d. Pectins and Differentiation

An important consequence of the new complexity of pectin chemistry is that the possible changes in the structure of the molecule, that might attend differentiation, are far more numerous than formerly seemed to be the case. Hence, a great area of plant physiology needs re-appraisal (see Chapter 6) in the light of possible changes in neutral sugars during differentiation. The results of Stoddart et al. (1967) showed that the whole pectins of apple callus and sycamore callus were very similar to each other, but differed from apple fruit pectin in their lower arabinose content. Hence, not only were there differences in the contents of pectinic acids of types I and II in these pectins, but the structures of the type II pectinic acids seemd to vary with growth conditions rather than species. The study of Nevins et al. (1967) further supported this. They found that the cell walls of sycamore suspension callus tissue, grown upon a variety of carbon sources, showed changes in the levels of wall polysaccharides which depended on the substrate. Notably, the largest effects were in the levels of arabinose and galactose, and to some extent this is now explicable in metabolic terms (see Chapter 5).

e. Pectic Gums

Gums are, by definition, polysaccharides which are not part of the cell wall, but which form exudates and slimes. However, some classes of gums resemble normal wall polysaccharides so closely that it is difficult to believe that they have not evolved from them. The pectic gums are a particularly good example. Some study has been made of them on a basis of taxonomy (Shaw and Stephen, 1965; Stephen and Schelpe, 1964), and this has considerable metabolic implications.

(i) *Galacturonans*: Gums, such as gum tragacanth, contain polysaccharides which are essentially identical with pectinic acids of type I.

Tragacanthic acid has a 1,4-linked chain of D-galacturonopyranosyl residues, punctuated by 1,2-linked rhamnopyranosyl residues, which may be substituted at C4. The galacturonosyl residues can be substituted at C3. Several types of side-chain are known. Some simple xylosyl residues occur, linked in the β-configuration, some are further 1,2 substituted at the L-fucopyranosyl residues and others by a β-1,2-O-galactopyranosyl residue. Possibly oligomers of xylose also occur, simple and branched. Glucuronopyranosyl residues can be β-1,6 linked to galactose, or β-1,4. linked to fucose, and some arabinopyranosyl residues are present. There is both esterification and acetylation *in vivo*.

Several other gums closely resemble gum tragacanth and some, like it, contain a separate, neutral polymer of arabinose and galactose. Some contain more complex arabinogalactans.

(ii) *Arabinogalactans:* These consist of branched polymers of β-1,3-linked-galactopyranosyl residues, branching via β-1,6-links. Some residues are further substituted by L-arabinofuranoside groups (linked 1,3 and1,5) or by glucuronosyl (or 4-O-methyl-D-glucuronosyl) groups joined via β-1,6-links. Some L-arabinopyranoside groups are linked β-1,3 to the arabinofuranosides, and L-rhamnopyranosyl residues occur attached 1,4 to glucuronosyl groups.

Sometimes more than one type of polymer is present, and neutral arabinogalactans of the pectic type are quite well known.

V. Hemicelluloses

Hemicelluloses are a very varied group of polysaccharides, defined empirically in terms of the methods of their extraction, rather than by their sharing common structural features. They appear to occur in all terrestrial plants, but are too varied from group to group to show anything like the generality of structure of cellulose or pectins. It is far beyond the scope of this chapter to survey all the partial structures now demonstrated for hemicelluloses and what follows is, of necessity, a catalogue of the basic features of the major types of hemicellulose.

A. Glucomannans

Glucomannans generally consist of β-1,4-linked-glucopyranosyl and mannopyranosyl residues, arranged apparently at random. Branching can occur, and small amounts of galactose are often present, linked as 1,6-residues, in the gymnosperms, but not the angiosperms. Glucomannans are a major component of the hemicellulose of gymnosperms, but a relatively minor component of that of angiosperms It should be noted that the "cellulose" fraction often contains mannose upon hydrolysis, and this may represent very insoluble glucomannan of a cellulose-like nature. Those glucomannans which occur in the hemicellulose fraction generally contain about three times as much mannose as glucose, in gymnosperms, while galactose accounts for no more than about 2% by weight of the molecule and seems to be in terminal positions.

The gymnosperms have larger glucomannans than the angiosperms, with molecular weights of about 30–35, $\times 10^3$ daltons against 10–15, $\times 10^3$ daltons. Possibly this is partially an artefact of isolation, since there is considerable variation in physical properties, especially solubility, among glucomannans, depending upon size, degree of acetylation (at C2 or C3 of the mannosyl residues) and degree of substitution by galactose.

B. Galactoglucomannans

Galactoglucomannans are found in gymnosperms and are basically like the glucomannans of these plants, but differ in having a greater degree of substitution by galactosyl residues, and in having molecular weights of 15–18 $\times 10^3$ daltons. They are much more soluble than glucomannans, and are normally extracted with aqueous alkali and then precipitated. As much as 20% of the weight of the molecule may be galactose, though the ratio of glucose to mannose is similar to that of glucomannans.

It seems likely that in gymnosperms these two classes of polysaccharides from hemicellulose are closely related. If the growth of chains is terminated by the addition of galactosyl residues, then the greater size of glucomannans than galactoglucomannans is to be expected. It is notable that the galacto-glucomannans in particular can occur as soluble gums and mucilages, for example the slime of *Lilium henryi*. In this respect they may be analoguous to the pectic gums, and represent a specialized evolution.

C. Xylans

A wide variety of different types of xylan are known, both from higher plants and from algae, though too few of the latter are known to permit of safe structural generalization. Among the higher plants, the xylans share a common basic structure, of β-1,4-linked chains of D-xylopyranosyl residues, to which a variety of side chains may be attached.

The gymnosperms are relatively poor in xylans, though they do contain them. Their main xylan chains are substituted by L-arabinopyranosyl residues, linked 1,3 and by 4-O-methyl-D-glucuronosyl residues linked 1,2. There is usually about twice as much uronic acid present as arabinose, and the xylose to uronic acid ratio is about 5·5 to 1. It is probable that part of the arabinose, which is very labile, has been lost in some preparations.

In the angiosperms the same general structure prevails, though poly-saccharides are known with arabinosyl, arabinosyl plus glucuronosyl, and glucuronosyl substitutions. The first group are uncommon, but are found in endosperms, and it is quite likely that they are really storage and

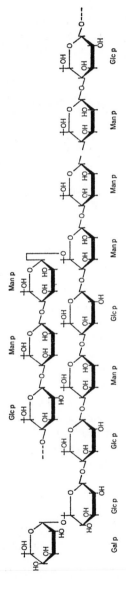

Fig. 11. A galactoglucomannan. This figure shows a section of the chain of a galactoglucomannan, with a galactosyl group at its non-reducing terminal and the likely type of branch. Most galactoglucomannans are more heavily galactosylated than that shown, and must be more branched. There is generally more mannose than glucose in the chains, which appear to be random. Glucomannans have related structure, but without galactose, and are larger than the galactoglucomannans in gymnosperms (but not angiosperms).

not wall polysaccharides. The second group are very common in the *Graminae* and are closely similar to the polysaccharides of the gymnosperms, though they sometimes show a substitution of all the arabinosyl side chains by xylosyl terminals, linked 1,2 and probably xylofuranoside in form. The third group are also widespread and have only one uronosyl group to approximately every ten xylosyl residues, and those residues are

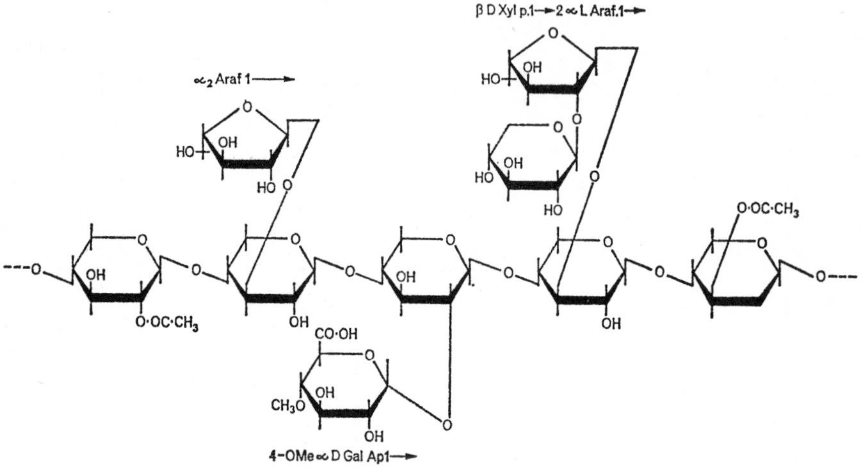

Fig. 12. A xylan. Basic structural features of a typical xylan are shown.

often esterified. Thus these xylans probably do not contribute greatly to the ion-exchange properties of the cell wall. Careful extraction has shown that there is extensive acetylation of these xylans *in vivo*, mostly via C3 of xylosyl residues. A fourth type of xylan is known in apple and cherry, both members of the *Rosaceae*, where the xylan chain is branched; otherwise these polymers are like those of the third group above.

Again gums are known which are very similar to all these xylans, suggesting that they have evolved from a common origin.

D. Arabinogalactans

The pectic arabinogalactans have already been discussed (see above) and the arabinans of, for example, white mustard belong to the same sort of polysaccharides. Possibly no arabinogalactan should be regarded as a hemicellulose, though those of conifers conventionally are so classed. The best known are arabinogalactans A and B of larch, which consist of β-1,3 and β-1,6 linked branched chains of D-galactopyranoside residues. The branches are substituted by L-arabinopyranoside groups, linked 1,3 or D-L-arabinopyranosyl 1,3-L-arabinofuranoside groups, similarly linked. The two polymers differ very slightly in structure, and in proportions.

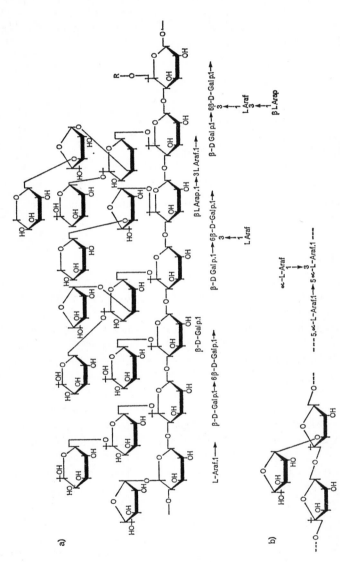

Fig. 13. An arabinogalactan. Basic structural features of a typical arabinogalactan, such as those of gymnosperms, are shown. Many are probably more highly branched and arabinosylated than that illustrated, particularly when related to pectins. Arabinogalactans of gums contain traces of other sugars, especially glucuronic acid and rhamnose. The galactan core can be 1,3-linked or 1,4-linked (as shown).

E. Callose

Callose is a distinctive β-1,3-linked, linear-glucan of wounds and sieve tubes in vascular plants. It is the polysaccharide deposited around sieve plates and on the sides of the sieve-tube pores to the companion cells. A little uronic acid, about 2%, is also present. The peculiarities of the synthesis of this polysaccharide are discussed in Chapter 5.

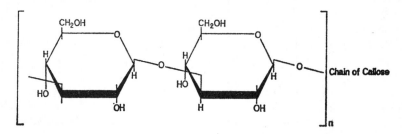

Fig. 14. Callose, the repeating unit. The native molecule contains traces of glucuronic acid also.

VI. Cell Wall Glycoproteins

In 1960 Dougall and Shimbayashi, and Lamport and Northcote, separately reported the existence of hydroxyproline in a protein of the plant cell wall.

In two papers Steward et al. (Steward et al., 1967, and Israel et al., 1968) attacked this view, holding that the hydroxyproline-rich protein was not really a cell wall component, but was adsorbed. However, there is now a substantial body of work from several sources (Clark and Ellinger, 1967; Jennings et al., 1968; Pusztai and Watt, 1969) which leaves little doubt that such proteins are present in the cell wall, and that there are also hydroxyproline-containing proteins in the cytoplasm. A certain amount of work has been done on their biosynthesis (see Chapter 5). This protein was termed "extensin" by Lamport, who has reviewed the field (Lamport, 1970).

A series of studies by Lamport (1965, 1967b, 1969) have established that a carbohydrate, rich in arabinose, is attached via the 4-hydroxyl group of hydroxyproline, and a recent, and very elegant, study by Heath and Northcote (1971) has greatly amplified our knowledge of the nature of this linkage.

These workers used sycamore suspension callus tissue as their source of cell walls, and took meticulous care to ensure that cytoplasm was removed from the walls, by washing them exhaustively after sonic disruption. They found that 18·1 mol. % of the whole wall protein was hydroxyproline. Upon a simple wall fractionation, by successive extraction with 0·5% EDTA (pH 6·8) and alkaline sodium borate, the protein remained in the α-cellulose residue and 28·6 mol. % of hydroxyproline were present in this fraction. Hydrazinolysis was used to liberate soluble glycopeptides from this material, under conditions where glycosidic bonds were not split, and high and low molecular weight fragments were separated on Sephadex G25. This accounted for 22% by weight of this fraction. Glyco-peptides were identified by feeding ³H-proline and ¹⁴C-arabinose to the cells in culture, and then isolating material as above and fractionating it on Sephadex G25. One glycopeptide was further purified and found to contain 10% hydroxyproline, 35% arabinose and 55% galactose by weight. All the hydroxyproline residues were substituted and the average degree of polymerization of the oligosaccharides was 9. The peptide-carbohydrate links were probably furanosidic. No evidence for cross-linking was found.

An important consequence of this study is the implication that hydroxy-proline residues may occur in sequence, or nearly so. This is in agreement with the work of Lamport *et al.* (1971) who found sequences of four such residues in the tomato cell wall. Heath and Northcote (1971) discuss this, and point out that the carbohydrate would be on the outside of a poly-hydroxyproline helix and would not perturb it very much. The whole structure would be very rigid.

Glycoproteins could play an important part in the organization of the deposition of plant cell walls, and the elucidation of their structure is one of the most important challenges to the plant biochemist and the carbo-hydrate chemist. It is essential that they are now extracted in an undegraded form, and the possibility of localized regions of polyhydroxyproline properly explored in the context of the structure of the whole polypeptide. Many of the techniques used in the isolation of membrane glycoproteins of animals may well prove valuable here—and it is a safe prediction that they will soon be applied to good effect.

VII. Carbohydrates of the Plant Plasmalemma

The plasmalemmata of plant cells appear by electron microscopy, both in their sections and by freeze-etching, to be fundamentally similar to those of animals (Cook, 1971). In the few cases studied, it appears that their protein: lipid ratio is comparable to that typical of animal cells. In general, plant lipids contain fatty acids of greater chain length than their animal counterparts, but there is insufficient evidence to be sure how far they are present in the plasmalemma.

The question of the presence of sugars in the plasmalemma, as opposed to the cell wall, is very difficult. Plant cell membranes can only be obtained after loosening of the cell wall by degradation with glycosidases; usually those of the gut of the snail (*Helix pomatia*) are used. Inevitably these cause some destruction of oligosaccharides in the plasmalemma. If the free plant protoplasts are allowed to regenerate their membrane-bound saccharides in culture they will also start to resynthesize their walls, and no means exists to inhibit one without the other. Accordingly there is little direct evidence for the presence of specifically membrane-bound saccharides, unless cellulose is considered to be one, and certainly there is no evidence against them.

Sialic acids have not yet been conclusively demonstrated in plants, since sucrose, which is almost ubiquitous in plant cells, will give false positive colour reactions for sialic acid in the presence of amines or amino acids. This could represent a major difference between higher plants and animals, and it would be interesting to know if sialic acids are absent from all groups of algae.

The ability of plant cells to synthesize glycoproteins is unquestioned, and seems to be akin to that of animal cells (Salton, 1965; Roberts, 1970; Roberts *et al.*, 1971), though there is no evidence that the materials synthesized are located in the plasmalemma. Thus, the question of saccharides in plant cell membranes, as distinct from walls, is not settled, but there is no good reason to doubt their presence.

References

ALBERSHEIM, P. (1959). *Biochem. biophys. Res. Commun.* **1**, 253–256.
ALBERSHEIM, P. and KILLIAS, U. (1963). *Am. J. Bot.* **50**, 732–745.
ALBERSHEIM, P., MÜHLETHALER, K. and FREY-WYSSLING, A. (1960a). *J. biophys. biochem. Cytol.* **8**, 501–506.
ALBERSHEIM, P., NEUKOM, H. and DEUEL, H. (1960b). *Archs. Biochem. Biophys.* **90**, 46–51.
ANDERSON, D. M. W., BEWS, A. M., GARBUTT, S. and KING, N. J. (1961). *J. chem. Soc.* pp. 5230–5234.
ANDERSON, D. M. W. and KING, H. J. (1961a). *Biochim. biophys. Acta* **52**, 441–449.
ANDERSON, D. M. W. and KING, N. J. (1961b). *Biochim. biophys. Acta* **52**, 449–454.
ANDERSON, D. M. W. and KING, N. J. (1961c). *J. chem. Soc.* pp. 5333–5338.
ASPINALL, G. O. (1970). "Polysaccharides", p. 44, Pergamon, Oxford.
ASPINALL, G. O. and BAILLIE, J. (1963). *J. chem. Soc.* pp. 1702–1714.
ASPINALL, G. O. and FANSHAWE, R. S. (1961). *J. chem. Soc.* pp. 4215–4225.
ASPINALL, G. O. and CAÑAS-RODRIGUEZ, A. (1958). *J. chem. Soc.* pp. 4020–4027.
ASPINALL, G. O., BEGBIE, R., HAMILTON, A. and WHYTE, J. N. C. (1967a). *J. chem. Soc.* C. pp. 1065–1070.
ASPINALL, G. O., COTTRELL, I. W., EGAN, S. V., MORRISON, I. M. and WHYTE, J. N. C. (1967b). *J. chem. Soc.* C. pp. 1071–1080.
ASPINALL, G. O., DAVIES, D. B. and FRAZER, R. N. (1967c). *J. chem. Soc.* C. pp. 1086–1088.
ASPINALL, G. O., HUNT, K. and MORRISON, I. M. (1967d). *J. chem. Soc.* pp. 1080–1086.

BARRETT, A. J. and NORTHCOTE, D. H. (1965). *Biochem. J.* **94**, 617–627.

CARRÉ, M. H. and HORNE, A. S. (1927). *Ann. Bot.* **41**, 193–235.

CLARKE, E. M. W. and ELLINGER, G. M. (1967). *J. Sci. Fd. Agric.* **18**, 536–540.

COOK, G. M. W. (1971). *A. Rev. Pl. Physiol.* **22**, 97–120.

DESIKACHARY, T. V. and DWELTZ, N. E. (1961). *Proc. Indian Acad. Sci.* B. **53**, 157.

DOESBURG, J. J. (1965). "Pectic Substances in Fresh and Preserved Fruits and Vegetables", Institute for Res. on Storage and Processing of Horticultural Produce, Wageningen, Netherlands.

DOUGALL, D. K. and SHIMBAYASHI, K. (1960). *Pl. Physiol., Lancaster* **35**, 396–404.

FARR, W. K. (1963). "Research on the Cytochemistry of Micro-organisms", U.S. Govt. Res. Rept. **39**, Suppl. B. 142 (*Chem. Abstr.* (1965) **62**, 10867c).

FREUDENBERG, K. and BLOMQVIST, G. (1935). *Ber. dt. Chem. Ges.* **68**, 2070–2082.

FREY-WYSSLING, A. (1954). *Science, N.Y.* **119**, 80–82.

GEE, M., REEVE, R. M. and MCCREADY, R. M. (1959). *J. agric. Fd. Chem.* **7**, 34–38.

GORING, D. A. I. and TIMELL, T. E. (1962). *Trans. Am. Pulp and Paper Inst.*, tech. Sect. **45**, 454–460.

GOULD, S. E. B., REES, D. A., RICHARDSON, N. G. and STEELE, I. W. (1965). *Nature, Lond.* **208**, 876–878.

HAWORTH, W. N., LONG, C. W. and PLANT, J. H. G. (1927). *J. chem. Soc.* pp. 2809–2814.

HEATH, M. F. and NORTHCOTE, D. H. (1971). *Biochem. J.* **125**, 953–961.

HESS, K., MAHL, H. and GÜTTER, E. (1957). *Kolloidzeitschrift* **155**, 1–19.

HIRST, E. L. and JONES, J. K. N. (1939a). *J. chem. Soc.* pp. 452–453.

HIRST, E. L. and JONES, J. K. N. (1939b). *J. chem. Soc.* pp. 454–460.

IRVINE, J. C. and HIRST, E. L. (1923). *J. chem. Soc.* **123**, 518–532.

ISRAEL, H. W., SALPETER, M. M. and STEWARD, F. C. (1968). *J. Cell Biol.* **39**, 698–715.

JENNINGS, A. C., PUSZTAI, A., SYNGE, R. L. M. and WATT, W. B. (1968). *J. Sci. Fd Agric.* **19**, 203–213.

JENSEN, W. A. and ASHTON, M. (1960). *Pl. Physiol., Lancaster* **35**, 313–323.

JOSLYN, M. A. (1962). *Adv. Fd Res.* **11**, 1–107.

KERR, T. and BAILEY, I. W. (1934). *J. Arnold Arbor.* **15**, 327–349.

KERTESZ, Z. I. (1951). "The Pectic Substances", Interscience Publishers Inc., New York and London.

KERTESZ, Z. I., BAKER, G. L., JOSEPH, G. H., MOTTERN, W. H. and OLSEN, A. G. (1944). *Ind. and Eng. Chem.* (*News*) **22**, 105–106.

LAMPORT, D. T. A. (1965). *Adv. bot. Res.* **2**, 151–218.

LAMPORT, D. T. A. (1967a). *Fedn Proc. Fedn Am. Socs exp. Biol.* **26**, 608.

LAMPORT, D. T. A. (1967b). *Nature, Lond.* **216**, 1322.

LAMPORT, D. T. A. (1969). *Biochemistry, N.Y.* **8**, 1155–1163.

LAMPORT, D. T. A. (1970). *A. Rev. Pl. Physiol.* **21**, 235–270.

LAMPORT, D. T. A. and NORTHCOTE, D. H. (1960). *Nature, Lond.* **188**, 665.

LAMPORT, D. T. A., KATONA, L. and ROERIG, S. (1971). *Fedn Proc. Fedn Am. Socs exp. Biol.* **30**, 1279.

LETHAM, D. S. (1962). *Exp Cell Res.* **27**, 352–355.

LEWIN, R. A. (1962). "Physiology and Biochemistry of Algae", Academic Press, New York.

LIANG, C. Y. and MARCHESSAULT, R. H. (1959). *J. Polymer Sci.* **37**, 385–395.

MANGIN, L. (1893). *C. r. Lebd. Séanc. Acad. Sci., Paris* **116**, 653–656.

MANLEY, R. ST. J. (1964). *Nature, Lond.* **204**, 1155–1157.

MARCHESSAULT, R. H. and SARKO, A. (1967). *Adv. Carbohydrate Chem.* **22**, 421–482.

MARK, R. E., KALONI, P. N., TANG, R. C. and GILLIS, P. P. (1969). *Science N.Y.* **164**, 72–73.

MARX-FIGINI, M. and SCHULTZE, G. V. (1963). *Makromolek. Chem.* **62**, 49–65.

MARX-FIGINI, M. and SCHULTZE, G. V. (1966). *Biochim. biophys. Acta* **112**, 81–101.

MEYER, K. H. and MISCH, L. (1937). *Helv. chim. Acta* **20**, 232–244.

NEVINS, D. J., ENGLISH, P. D. and ALBERSHEIM, P. (1967). *Pl. Physiol., Lancaster* **42**, 900–906.

OVODOVA, R. G. and OVODOV, YU. S. (1969). *Carbohydrate Res.* **10**, 387–390.

PRESTON, R. D. and CRONSHAW, J. (1958). *Nature, Lond.* **181**, 248–250.

PUSZTAI, A. and WATT, W. B. (1969). *Eur. J. Biochem.* **10**, 523–532.

RÅNBY, B. G. (1958). *In* "Handbuch der Pflanzenphysiologie", (Ruhland, W. ed.), Vol. 6, pp. 268–304, Springer-Verlag, Berlin.

ROBERTS, R. M. (1970), *Pl. Physiol., Lancaster* **45**, 263–267.

ROBERTS, R. M., CONNOR, A. B. and CETORELLI, J. J. (1971). *Biochem. J.* **125**, 999–1008.

ROUDIER, A. J. and GALZIN, J. (1966). *Bull. Soc. chim. Fr.*, pp. 2480–2485.

SALTON, M. R. J. (1965). *A. Rev. Biochem.* **34**, 143–174.

SHAW, D. H. and STEPHEN, A. M. (1965). *S. Afr. ind. Chem.* **19**, 146–147.

SMIDSRØD, O., HAUG, A. and LARSEN, B. (1966). *Acta Chem. Scand.* **20**, 1026–1034.

SMITH, F. and MONTGOMERY, R. (1959). "The Chemistry of Plant Gums and Mucilages", Reinhold Publishing Corporation, New York; Chapman and Hall, Ltd., London.

SPEISER, R. (1947). *J. Polymer Sci.* **2**, 281–289.

STEPHEN, A. M. and SCHELPE, E. A. C. L. E. (1964). *S. Afr. ind. Chem.* **18**, 12–15.

STEWARD, F. C., ISRAEL, H. W. and SALPETER, M. M. (1967). *Proc. natl. Acad. Sci. U.S.A.* **58**, 541.

STODDART, R. W., BARRETT, A. J. and NORTHCOTE, D. H. (1967). *Biochem. J.* **102**, 194–204.

STODDART, R. W. and NORTHCOTE, D. H. (1967). *Biochem. J.* **105**, 45–59.

THORNBER, J. P. and NORTHCOTE, D. H. (1961a). *Biochem. J.* **81**, 449–455.

THORNBER, J. P. and NORTHCOTE, D. H. (1961b). *Biochem. J.* **81**, 455–464.

THORNBER, J. P. and NORTHCOTE, D. H. (1962). *Biochem. J.* **82**, 340–346.

TIMELL, T. E. (1962a). *Svensk Papperstidn.* **65**, 122–125.

TIMELL, T. E. (1962b). *Svensk Papperstidn.* **65**, 266–272.

TIMELL, T. E. (1964). *Svensk Papperstidn.* **67**, 356–363.

TIMELL, T. E. (1965). *Adv. Carbohydrate Chem.* **20**, 409–483.

VAUQUELIN, M. (1790). *Ann. chim. et phys.* **5**, 92–106.

VOLLMERT, B. (1950). *Makromol. Chem.* **5**, 110–138.

ZAITLIN, M. and COLTRIN, D. (1964). *Pl. Physiol., Lancaster* **39**, 91–95.

5

Biosynthesis of Surface Heterosaccharides

I. Animal Cells

The biosynthesis of glycosylated membrane materials poses several important questions; such as the mechanism by which the cell synthesizes a particular macromolecule, its site of synthesis and the means by which the cell controls this synthesis. Answers to such questions may not only be of general biochemical interest, but may throw light on the biological rôle of glycosubstances at the cellular periphery. Membrane heterosaccharides, or complex carbohydrates,[1] are general and rather loose terms which can encompass glycoproteins, glycolipids, polysaccharides and glycosaminoglycans (mucopolysaccharides). These last two classes of materials are likely to occur as major components of the plasma membrane in only relatively few cases, where the cell, such as a connective tissue cell, is actively producing such molecules for export. These may not be structural components of the membrane but could have a transient association only with the cellular periphery. In view of the occurrence of glycolipids within the glycosylated macromolecule of the cell surface, and because of the widespread interest in their biological function, some mention will be made of them where appropriate, though it is to the glycoproteins that this section is chiefly devoted.

A. Intact-Cell Studies with Radioactive Precursors

A number of studies on the biosynthesis of membrane glycoproteins have been performed during the last decade. In all of these much attention has been turned to the process of glycosylation. The mechanism of protein synthesis has been outlined in considerable detail, and it is assumed that a common mechanism exists for the biosynthesis of the polypeptide portion of glycoproteins and non-glycosylated proteins.

The glycoproteins of erythrocytes have been the subject of several investigations and here the work of Eylar and Matioli (1965) and Laico

---

[1] The term "complex carbohydrates" has been used by Neutra and Leblond (1966a) in histochemical studies to cover glycoproteins, mucopolysaccharides, glycogen and glycolipids. However, in the latter case these authors point out that in view of the solubility of glycolipids in organic solvents, it is possible that when the term is used in histochemical studies it need not necessarily include glycolipids.

and Eylar (1966) on the biosynthesis of membrane glycoproteins by reticulocytes is of special importance. The former paper is particularly useful, as the study of the biosynthesis of glycoprotein was performed with intact cells. Eylar and Matioli (1965) studied the reticulocytes of thalassaemia, a condition in which the erythropoietic cells have a low haemoglobin level; in particular haemoglobin A is low, though haemoglobin F and A2 are usually elevated. The erythroblasts and erythrocytes of thalassaemic patients show an accumulation of a material showing a positive periodic acid-Schiff reaction, likely to be glycoprotein, and Eylar and Matioli (1965) were able to equate the uptake of ^{14}C labelled glucosamine into reticulocyte stroma with this material, as demonstrated histochemically. Their evidence rules out the possibility that the results are due to production of, and incorporation of radioactivity into, glycogen. Indeed they were able to measure a sizeable (50–80%) elevation in the sialic acid content of stroma isolated from thalassaemic erythrocytes. Eylar and Matioli (1965) suggests that it is possible that this illustrates a defective synthesis of plasma membrane, and that there is a block or shunt at the point where glycoprotein is associated with lipid and protein. They argue that the regulating mechanism would be strongly directed towards synthesis and that this would explain the increased incorporation of glucosamine. Unfortunately these authors did not report further on the nature of their radioactive product. None the less, this study focused attention on the need to take account of the question of surface heterosaccharides, at a time when little work was being performed on the biosynthesis of these materials in biological membranes. Justification for the use of this amino sugar for studying glycoprotein biosynthesis in erythroid cells may be found in the work of Kornfeld and Noll (1968), who showed, using principally rabbit material, that immature erythroid cells of bone marrow can utilize both fructose and glucosamine for the synthesis of UDP-N-acetylglucosamine (see Fig. 1), which is the "activated" form of the amino sugar required for glycoprotein synthesis. At the reticulocyte stage, cells can still utilize glucosamine (which accords with the work of Eylar and his colleagues on reticulocytes, described above), but they begin to lose the enzyme that converts fructose-6-phosphate to glucosamine-6-phosphate (see Fig. 1). On the other hand, the mature erythrocyte which is no longer synthesizing membrane glycoprotein, cannot utilize either fructose or glucosamine for the synthesis of UDP-N-acetyl hexosamine, since it has lost both L-glutamine-D-fructose-6-phosphate-amido-transferase, and also the enzyme(s) which converts glucosamine-6-phosphate to the sugar nucleotide.

Apart from an early interest in glycoproteins of erythrocytes, considerable attention has been paid to the presence of complex carbohydrates in the surfaces of tumour cells. This interest in tumour cell surfaces stems from

BIOSYNTHESIS OF MEMBRANE GLYCOPROTEIN:SCHEMATIC DIAGRAM SHOWING THE POSSIBLE PATHWAYS

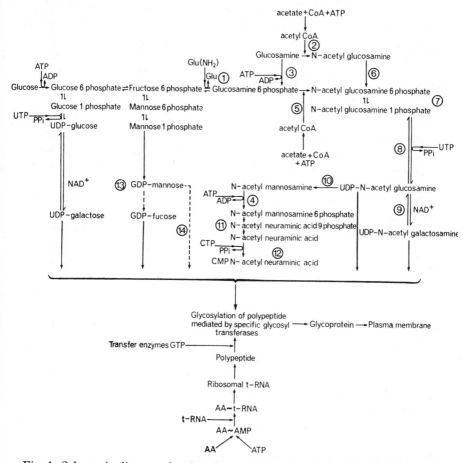

Fig. 1. Schematic diagram showing the possible pathways for the biosynthesis of membrane glycoproteins. A comprehensive review on the synthesis of amino sugars and sialic acids has been presented by Warren (1966) and gives extensive details concerning the amination of sugars (1), the synthesis of the acetamido derivatives (2) (3) (5) (6) and the various amino sugar nucleotides (8) (9). This review also provides a detailed treatment of the biosynthesis of sialic acids (4) (11) and the activation of sialic acids (12). The metabolism of sialic acids (11) (12) and D-mannosamine (4) (10) were also reviewed earlier by Roseman (1962) and covers much of the initial work in this field, though the speculation that nucleotide trisaccharides are involved in the glycosylation of protein now seems unlikely (Roseman, 1970). The epimerization of mannose to fucose occurs at the sugar nucleotide level (13), and this and the details of oligosaccharide biosynthesis has been reviewed by Leloir (1964). Addition of mannose to protein may occur via a lipid intermediate (see text) (14). A glucosamine-phospholipid intermediate with assumed participation in the glycosylation of protein has been described (Ikehara et al., 1971).

a desire to relate the invasive properties of malignant cells to a change in the molecular composition of the cellular periphery, and both electro-kinetic and histochemical studies have demonstrated clearly the presence of glycosylated molecules in the plasma membranes of a number of malignant cells. Furthermore, the availability of gram quantities of tumour cells, especially ascites cells, whose surfaces have been well characterized physicochemically, has prompted their use in biosynthetic studies. Though a number of the cell types studied suffer from the disadvantage of the lack of a suitable normal counterpart for comparative studies, work on such cells has, none the less, provided much of our fundamental knowledge of the sequence and sites of reactions involved in the metabolism of membrane glycoprotein.

It has been seen (Chapter 2) that the electrokinetic properties of the Ehrlich ascites carcinoma cell is largely caused by the presence of sialo-glycoproteins at the periphery of the plasma membrane. In 1965 Cook et al. studied the incorporation in vitro into the membranes of the Ehrlich ascites cell of glucosamine-^{14}C together with serine-^{14}C and leucine-^{14}C, as a measure of glycoprotein and protein synthesis respectively, in the presence and absence of various inhibitors. Using the microsomal fraction prepared after cellular homogenization in the French pressure cell as a source of plasma membrane fragments, Cook et al. (1965) were able to demonstrate that the intact cell rapidly incorporated amino sugar and amino acids into the cell membrane. When the incorporation was followed in the presence of the inhibitors of protein synthesis, puromycin and tenuazonic acid, it was found that although the uptake of labelled amino acid was almost completely abolished, as expected, the incorporation of glucosamine over a two-hour period was inhibited to a lesser extent. After a four-hour period of incubation with puromycin, the inhibition of hexosamine incorporation was trebled. Cook et al. (1965) suggested that this result indicates that a sizeable amount of membrane protein exists in the cell, and that the glycosylation takes place after the protein has left the polysomal complex. To investigate the latter suggestion, Cook et al. (1965) used a sucrose density-gradient analysis of rough-surfaced mem-branes dissociated with deoxycholate. They showed that while appreciable quantities of labelled leucine were associated with polysomal and ribosomal RNA, as would be expected in normal protein synthesis, it was not possible to demonstrate any association of labelled amino sugar with ribosomes or polysomes. Molnar et al. (1965a) studied the biosynthesis of glycoprotein in Ehrlich ascites carcinoma cells, and also found that puromycin had a marked effect on protein synthesis, but had little effect on glucosamine incorporation. By preincubating the intact cells for periods of up to four hours with puromycin, Molnar et al. (1965a) were able to increase the inhibition of glucosamine incorporation from 14% to about 50%.

These results are substantially the same as those found by Cook *et al.* (1965) and, as pointed out by Molnar *et al.* (1965a), could be explained by the slow formation of large oligosaccharides as well as the exchange of the oligosaccharide portion of glycoproteins. This latter interpretation had also been considered by Cook *et al.* (1965). These studies would indicate that attachment of carbohydrate to polypeptide is unlikely to be regulated at the RNA level by means of a specific RNA-amino acid-hexosamine intermediate. This is particularly interesting, since Sinohara and Sky-Peck (1965) unsuccessfully examined murine liver for such an intermediate and considered "that the carbohydrate moiety of glyco-protein is attached after the completion of the polypeptide chain but before it is released into cytoplasm".

While the above studies could be considered to describe the synthesis of membrane glycoproteins, this may not be entirely correct. Molnar *et al.* (1965a) have presented data which indicates that Ehrlich ascites cells produce many of the glycoproteins of the ascitic fluid, and that these materials are probably not solely derived from the process of cell destruction. This is important, because as Eylar (1965) has pointed out the mechanism by which membrane glycoproteins are synthesized may differ from that by which secretory glycoproteins are produced. The fact that puromycin has been shown (Molnar *et al.*, 1964) to inhibit incorporation of glucosamine-^{14}C into liver and plasma proteins, may well be indicative of the operation of such a different mechanism for glycoproteins destined for export to the body fluids, as opposed to membrane glycoproteins, and may explain why such results on liver differ from those observed with Ehrlich ascites carcinoma cells. Indeed, there are a number of studies with liver which indicate that some hexosamine is incorporated into protein while the polypeptide is still attached to the ribosome, and further sugar residues, including hexosamine, are added to the protein as it traverses the channels of the rough and smooth endoplasmic reticulum. Working with radioactive glucosamine and rat liver, Helgeland (1965) found small, but significant, amounts of radioactivity associated with the ribosomes after deoxycholate extraction. This author was very cautious in his interpretation of the result, pointing out that although insufficient extraction of the ribosomes is unlikely, adsorption of labelled material cannot be excluded. Helgeland (1965) considered that because of the high radioactivity of the extractable proteins of the membrane it was tempting to consider that these were the actual site of glycosylation of protein, but urged caution with such an interpretation in view of the evidence of association of label with the ribosomes. He suggested that detailed time-course studies of glucosamine incorporation might help to resolve the problem. The kinetics of incorporation *in vivo* of ^{14}C-gluco-samine into rat liver glycoprotein has been studied by Lawford and

Schachter (1966). They indicated that glucosamine residues are often the carbohydrate by which the sugar moiety is attached to protein, and hence the use of this sugar is particularly relevant for studying the early stages of the glycosylation of protein. Evidence was presented to support the view that the membrane-bound ribosome is the site where hexosamine is first incorporated into protein. Puromycin *in vitro* released appreciable quantities of acid-insoluble glucosamine, incorporated in the rat liver ribosomes *in vivo*, which supports this latter contention. Hallinan *et al.* (1968), also working with rat liver, found a five to twelve-fold greater activity for glycoprotein prosthetic group initiation with bound, as opposed to free, ribosomes, and in the latter case the activity present could be ascribed to contamination with membranes. There is much evidence in the case of rat liver to indicate that glycosylation of protein is initiated while the polypeptide is still associated with the ribosome, and thus differs from the mechanism proposed by Cook *et al.* (1965) who suggested that sugar was added after the polypeptide had left the polysome. Noting this apparent discrepancy between studies using liver and Ehrlich ascites cells, Lawford and Schachter (1966) admit that "the detailed pattern of glycoprotein biosynthesis may differ from one tissue to another". The synthesis of glycoprotein on membrane-bound ribosomes, as suggested by Eylar (1965), could be of fundamental importance in the secretion of proteins.

In order to examine further the question of whether the biosynthesis of membrane glycoprotein involves addition of sugar to polypeptide still attached to polysomes, one may turn to the work of Bosmann *et al.* (1969) on the biosynthesis of glycoprotein and glycolipid in HeLa cell membranes. These authors regard HeLa cells as being particularly useful for the study of the biosynthesis of membrane components, as nearly all the glycosubstances synthesized are membrane-bound or intracellular, extracellular material represents only 1–4% of the cellular glycoprotein activity. At first sight this point might appear to be at variance with the earlier work on the metabolism of glucosamine in HeLa cells by Kornfeld and Ginsburg (1966). They showed that cells grown in the presence of glucosamine-^{14}C for 36 hours, to label extensively hexosamine and N-acetylneuraminic acid residues of the cellular components, eventually lose two thirds of the label by secretion of high molecular weight material into the medium. The other third of the label remained with the cells and was gradually metabolized. Bosmann *et al.* (1969) studied the incorporation of labelled glucosamine, fucose and leucine for up to one hour into cellular and secreted material, as well as pulse chase experiments which were concluded within a total of four hours. They were careful to use HeLa cells which were in the log phase of growth, when it is likely that material incorporated into the plasma membrane will be retained within this organelle rather than "secreted" into the medium (as is more likely with cells that are non-dividing). The time scales used by the two groups of authors might well explain the apparently differing results with regard to "secreted" material.

In their pulse chase experiments, Bosmann et al. (1969) were able to show that soluble and smooth membranes were labelled first and that the labelled material then migrates progressively to the plasma membrane, which suggests that the smooth internal membranes of the cell represent the site where the addition of sugar to protein takes place. The authors envisage that a naked polypeptide chain, synthesized at the polysomes, migrates to the smooth internal membranes (probably the Golgi apparatus) where the glycosyl transferases are located. Although these workers were able to show labelling of microsomes with leucine-^{3}H, presumably as a result of labelling of protein remaining on the ribosomes, when the incorporation of glucosamine-^{14}C in microsomes was studied, no significant incorporation of labelled sugar was found. In microsomes which had been labelled with glucosamine, less than 2% of the activity was located in membranes associated with RNA, while 96% of the label was found to be in membranes lacking ribosomes, on careful density gradient purification. This result, the authors contended, shows that glucosamine is not incorporated to a significant extent into microsomes or ribosomes. These results with intact HeLa cells are in agreement with the previous studies on Ehrlich ascites cells, but contrast, in some respects, with the mechanisms apparently operating in the liver.

So far, it has been suggested that the addition of the carbohydrate prosthetic groups takes place as a post-ribosomal event, predominantly, if not exclusively, in the case of membrane glycoproteins in the smooth membranes of the cell. It is now widely accepted that the synthesis of the carbohydrate moieties of glycoprotein, in common with other complex carbohydrates, proceeds by means of the operation of a multienzyme group of highly specific glycosyl transferases, which are responsible for transferring, in stepwise fashion, glycosyl residues from the appropriate sugar nucleotide, which is the "activated" form of the monosaccharide (see Leloir, 1964), to the acceptor molecule. The acceptor is an aglycone which is either the appropriate amino acid residue[2] in the polypeptide chain, or a sugar residue in the growing oligosaccharide chain. No attempt will be made to detail all the various pieces of work which have gone into

---

[2] The presence of a particular sequence of amino acids around a potentially functional amino acid residue (e.g. asparagine) may determine whether carbohydrate attachment takes place. The probable importance of the sequence asparagine-X-serine (or threonine) has been discussed (Hunt and Dayhoff, 1970). This may not, however, be a sufficient condition for the formation of a carbohydrate protein linkage, and it is possible that there are restrictions as to the nature of the amino acid X. The question of the significance of the primary structure of the protein of glycoproteins, with respect to the formation of carbohydrate linkages, has recently been reviewed by Marshall (1972).

establishing the essential reactions involved in the synthesis of the various sugars from glucose and their conversion to sugar nucleotides. These reactions, which are in the main common to the metabolism of both secreted and membrane glycoproteins of a variety of tissues, have been widely reviewed elsewhere and will therefore not be described here. However, a diagram illustrating the major pathways of membrane glycoprotein metabolism is given (see Fig. 1), together with appropriate literature references to serve as a guide to those readers less familiar with this area of biochemistry.

It will be obvious from an examination of Fig. 1 why the use of glucosamine as an intermediate in studies on membrane glycoprotein biosynthesis is so widespread. Labelled glucosamine has been shown to be taken up first into the cell's uridine diphosphate-N-acetyl hexosamine pool. In the studies of Molnar et al. (1965a) on Ehrlich ascites carcinoma cells, evidence that some glucosamine ^{14}C was converted into UDP-N-acetyl hexosamines was obtained from a chromatographic examination of Dowex-1-Cl of acid-soluble radioactive components. About two-thirds of the UDP-N-acetyl hexosamine was UDP-N-acetylglucosamine, and the remaining third UDP-N-acetylgalactosamine. Both radioactive glucosamine and galactosamine were identified as present in the proteins of the Ehrlich ascites cells, suggesting that these sugars are incorporated via their respective UDP precursors. These activated sugars were detected in the acid soluble fraction, which contrasts with the situation in the rat liver where no measurable galactosamine is found in liver or plasma proteins, although radioactivity appears in the UDP-N-acetylgalactosamine pool. Cook et al. (1965) also found that appreciable quantities of glucosamine ^{14}C were converted into galactosamine by Ehrlich ascites cells and that was found to be associated with glycoprotein of the smooth membranes of the cell. In support, Langley and Ambrose (1967) also report galactosamine to be present in a glycopeptide released from the plasma membrane of Ehrlich ascites cells by trypsin. Both these results are in accord with the finding of a significant pool of UDP-N-acetylgalactosamine in this cell (Molnar et al., 1965a), and it is perhaps rather surprising that further studies by Molnar (1967) indicated that the sole hexosamine of smooth and rough microsomal membrane fractions of Ehrlich ascites carcinoma was glucosamine. Molnar (1967) offered no explanation for this discrepancy. However, this group (Molnar et al., 1965a) had demonstrated that the "secreted fraction" of these cells contained equal amounts of radioactivity in galactosamine and glucosamine, and only very little activity in galactosamine in material derived from cellular elements, so it could be argued that the detected pool of UDP-N-acetylgalactosamine is largely present as an intermediate for glycoproteins destined for "secretion". This is another indication for the need to distinguish between membrane and serum glycoprotein metabolism in biosynthetic studies. The galactosamine detected by Cook *et al.* (1965) may represent material to be exported from the cell, which is associated with membrane fractions at the time of isolation. The possibility that the number of galactosamine residues is small in relation to the total hexosamine content of the cellular membranes cannot be discounted, and is in accord with the results of Molnar *et al.* (1965a) from acid hydrolysates of sub-cellular fractions. The enzyme treatment of Langley and Ambrose (1967) might be selective for that part of the membrane glycoprotein containing such galactosamine residues, and their method might thus tend preferentially to concentrate such a moiety. Alternatively, it has been pointed out (Kraemer, 1967a) that Langley and Ambrose (1967) did not provide evidence that during the incubation of cells with enzyme the cell count and viability remained constant. Thus it is possible that the glycopeptide detected was a degradation product of a damaged cell, and as such it would be difficult to determine whether it is derived from a macromolecule destined to be "secreted" or from a cellular component, and not necessarily from the plasma membrane.

Labelled glucosamine provides a sensitive measure for studying those components of the cell that contain amino sugar. Kornfeld and Ginsburg (1966) showed, for instance, that over 90% of ^{14}C-1-glucosamine is initially taken up by HeLa cells into the uridine diphosphate-N-acetyl hexosamine pool, and that over a twelve-hour period this pool loses its activity to the amino sugar residues of the cell's components. Blood cells, especially erythrocytes, are an ideal model for studying the structural aspects of membrane glycoproteins (Chapter 3). White cells have also been used for examining the biosynthesis of membrane glycoproteins. Indeed, in view of the large amount of structural knowledge of the glycoproteins of blood cells, relative to other cells, biosynthetic studies with these cells are particularly informative and desirable. While labelled glucosamine may be an ideal marker for studying the synthesis of complex carbohydrates in tumour cells, Kornfeld and Gregory (1968) provided evidence that a considerable amount of glucosamine is converted to glucose in human white blood cells. Further, while phytohaemagglutinin caused a three to five fold increase in incorporation of label, they found it had little effect on the portion of glucosamine converted to glucose. In contrast to these results Hayden *et al.* (1970), also studying the biosynthesis of the membrane glycoproteins of the human lymphocyte, with glucosamine-1-^{14}C, found that some 84% of the label was present as uridine diphosphate-N-acetyl-glucosamine. Glucosamine was subsequently incorporated into glycoprotein of the plasma membrane, from which it could be released as glycopeptides by digestion of the surface with proteolytic enzyme. They (Hayden *et al.*, 1970) explain that the difference between their results and those reported earlier by Kornfeld and Gregory (1968) are almost certainly caused by the presence of platelets in the latter authors' preparations. Apparently platelets deplete the medium of glucose and then take up glucosamine vigorously. A major fraction of this is deaminated and introduced into the glycolytic pathway. Two points therefore emerge from this work. Firstly, a knowledge of the cell types present in the material under study is essential to any metabolic study. Secondly, as Kornfeld and Gregory (1968) have demonstrated, when examining the labelling properties of glucosamine in a hitherto unstudied tissue it is wise to establish that it is in fact specific for glycoprotein metabolism in that tissue, and not diverted to other pathways. The studies with intact cells used together with various inhibitors of protein synthesis have indicated that the glycosylation of membrane protein occurs as a post ribosomal event primarily, if not exclusively, in the smooth endoplasmic reticulum. Additional evidence for this mechanism has come from studies with cell-free systems which were only developed after experiments with intact cells had been initiated. This aspect of the synthesis of membrane glycoproteins will be covered in the following section, together with a more detailed treatment of the glycosyl transferases—enzymes which have been characterized by means of such systems.

B. Development of Cell-Free Systems

Chronologically the development of the first cell-free systems for studying glycoprotein biosynthesis occurred in the latter part of 1965 (Eylar and Cook, 1965; Sarcione and Carmody, 1966) after the initial studies on intact cells had been reported; though the majority of these latter studies had not been confined to the question of the biosynthesis of membrane glycoproteins. In view of their earlier studies on intact Ehrlich ascites carcinoma cells (Cook et al., 1965), Eylar and Cook (1965) used the same material to develop a cell-free system for the study of the synthesis of membrane glycoprotein. This represented an attempt to amplify the knowledge gained from studies in whole cells, especially with regard to the site of synthesis of the oligosaccharide units of membrane glycoproteins, as well as the means by which the cell regulates the construction of the final membrane. Eylar and Cook (1965) found that enzymes which incorporated glucosamine and galactose (the latter was shown to be incorporated from UDP galactose) into soluble and membrane glycoproteins were present in a post-microsomal particulate fraction. In contrast to rough surfaced membranes, this particulate fraction, which consisted of smooth membranes and was suggested at the time to represent components of the Golgi apparatus, was inactive in protein synthesis. The membranes bearing ribosomes were completely inactive in the synthesis of oligosaccharides though, as expected, they actively supported protein synthesis. Hence, it was suggested that oligosaccharides were synthesized in the smooth membranes of the cell, while the polypeptide acceptor was assembled on the ribosomes of the rough membranes. At the time it was thought that oligosaccharides might be attached to protein at membranes which bear ribosomes, though the present evidence (see the section upon the Golgi apparatus) is that the protein is glycosylated successively as it moves through the channels of the endoplasmic reticulum. Nevertheless, the postulate that the smooth membranes of the cell are the major site of intracellular glycosylation, as suggested by Eylar and Cook (1965), is entirely in accord with the more recent ideas upon the synthesis of glycoproteins. Caccam and Eylar (1970) have re-examined the Ehrlich ascites system with particular reference to the glycoprotein: galactosyl transferase activity, which they were able to purify some 46 fold. They showed that the enzyme was intimately bound to the smooth internal membranes of the cell, confirming the earlier suggestion of Eylar and Cook (1965) on its localization within the Golgi apparatus. The enzyme could, however, be solubilized with non-ionic detergents such as Triton X-100, or, enzymatically, by phospholipase A.

A second cell-free system was described by Sarcione and Carmody (1966) in which the incorporation of D-galactose was studied in rat liver. Since deoxycholate is known to disrupt the vesicles of the endoplasmic reticulum that are found in the microsomal fraction, it was reasoned that by solubilizing their limiting membranes, while leaving the ribosomes appreciably intact, it would be feasible to look for the enzyme system (glycoprotein:galactosyl transferase) which catalyses the incorporation of galactose into protein, in the membranous fraction of liver microsomes. Sarcione and Carmody (1966) demonstrated that the deoxycholate-soluble fraction of the liver microsomes contained glycoprotein:galactosyl transferase activity, and that UDP-galactose is the galactosyl donor in the reaction. It was also shown that the addition of increasing amounts of ribosomes to the solubilized fraction markedly stimulated galactosyl incorporation into protein; the authors emphasized that this observation does not imply that ribosomes are the site of the incorporation of galactose *in vivo*, the ribosomes purely provide a source of a suitable protein acceptor. This work is particularly relevant to the controversial question of when and how the initial glycosylation of the polypeptide chain takes place. Evidence from studies upon intact cells indicates that the addition of the first sugar to the polypeptide chain does not occur via a tRNA-glucosamine-amino acid complex. The point at dispute is whether the glycosylation of a growing polypeptide chain is still associated with the polysomes, or whether the glycosylation only takes place after the completed polypeptide is released. In the case of membrane glycoproteins the latter mechanism appears to be operative, while with rat liver glycoproteins where an

N-glycosidic linkage occurs, the first sugar, *N*-acetylglucosamine, was attached while the growing peptide was ribosomally bound. This should not be taken to imply that *N*-glycosidic bonds are absent from membrane glycoproteins. Admittedly the system described by Sarcione and Carmody (1966) does not immediately answer this problem because the sugar studied was galactose, which is likely to be present in those parts of the complete oligosaccharides which are more distal to the polypeptide sugar bond. Nevertheless, this system presents an opportunity to attempt to resolve this controversy. In reviewing this problem, Winzler 1970) points out that McGuire and Roseman (1967) have used a ribosome-free system to show that a UDP-*N*-acetylgalactosamine transferase can transfer the acetyl hexosamine to ovine sub-maxillary mucin which has been stripped of its sugar residues. In addition, Hagopian and Eylar (1968), working with bovine submaxillary glands, describe a similar polypeptidyl-*N*-acetyl hexosamine transferase, and point out that the analogous enzyme from HeLa and Ehrlich ascites carcinoma cells (Hagopian *et al.*, 1968) is located in the smooth membranes of the cell. This provides further evidence that formation of *N*-acetylgalactosaminyl-hydroxyamino acid linkage is a post-ribosomal event. All these results might point to a variety of mechanisms of assembly for different glycoproteins, depending on the type of peptide-carbohydrate linkage involved and the destination, either cellular or extracellular, of the completed macromolecule. The study of Caccam *et al.* (1969) on the biosynthesis of mannose-containing glyco-proteins is important here. These authors addressed themselves to the question of whether the sequential action of glycosyl transferases represented a general mechanism, and whether those glycoproteins which are secreted into the blood plasma or egg white are assembled like those in the HeLa cell. In their study, Caccam *et al.* (1969) used mannose, a monosaccharide found in soluble, secreted glycoproteins but rarely in mucins. Working with a number of tissues and a cell-free system derived from rabbit liver, they found that mannose was incorporated from GDP mannose into glycoprotein and lipid. The mannolipid closely resembled a mannosyl-1-phosphoryl-polyisoprenol compound, and it was suggested that it represented an intermediate in the biosynthesis of mannose-containing glycoproteins, since mannose is not found in glycolipids. Furthermore, it appears that two enzymes would be involved, one enzyme being responsible for the synthesis of mannolipid and the other for the transfer of the sugar from the lipid intermediate to the glycoprotein. This represents a more complicated mechanism for the synthesis of mannose-containing secretory glycoproteins than the "one-linkage-one enzyme" mechanism which seems to operate in the case of membrane glycoproteins. The site of mannose incorporation is probably the smooth membranes of the cell where the enzyme activities are most concentrated.

Studies with cell-free systems support the contention that the smooth membranes of the cell, especially those of Golgi apparatus, represent a major site of protein glycosylation, and in the particular case of membrane glycoproteins the addition of sugar residues takes place post-ribosomally. It can be seen from the "one-enzyme–one-linkage" concept that the ultimate structure of the carbohydrate moiety is dictated by the specificity of each of the transferases for the acceptor molecule, genetic control over the structure of the final oligosaccharide unit is exerted by the presence or absence of the particular glycosyl transferase. This is of special importance when one is considering genetic control over the production of molecules with blood-group specificity. Cellular antigens may be considered as secondary gene products, while the primary gene products are the glycosyl transferases. It is the specificity of these enzymes that determine which specific structures are produced. (See Watkins, 1970 and Kobata et al., 1970 for a discussion on the rôle of glycosyl transferases and blood-group types; see also Chapter 3.)

C. Isolation of The Golgi Apparatus and Glycosyl Transferase Assays

The Golgi apparatus is a major site of protein glycosylation (see above) and this section of the chapter will examine, in some detail, the work which has gone into isolation and characterization of the Golgi apparatus.

The nature and existence, to say nothing of the rôle, of the Golgi apparatus has been a source of some doubt and controversy ever since its discovery by Golgi (1898a and b) at the end of the last century. The historical aspects of the subject have been dealt with very thoroughly in previous reviews (Beams and Kessel, 1968; Dalton, 1961; Mollenhauer and Morré, 1966; Northcote, 1971). Here it is the rôle that the Golgi apparatus plays as a site of synthesis of those complex carbohydrates that are destined to form integral parts of the animal cell surface, which is to be considered. Its rôle in the synthesis of plant cell walls is discussed later in this chapter. Two recent reviews (Whaley et al., 1972; Cook, 1972) have dealt specifically with the Golgi apparatus in relation to the cell surface and complex carbohydrates, but nevertheless, because of the importance of this organelle, it is intended here to detail the major evidences which implicate the Golgi apparatus as an agent for membrane synthesis.

1. ISOLATION AND CHARACTERIZATION OF GOLGI APPARATUS-ENRICHED FRACTIONS

In order to determine the rôle of the Golgi apparatus in the biosynthesis of the glycoproteins, it was necessary for biochemists to isolate this organelle in a relatively purified state. In 1965, when Eylar and Cook first suggested that their post-microsomal particulate (PMP) fraction represented fragments of Golgi apparatus, methods for the isolation of the organelle from animal tissues were not available, and this hampered the characterization of the PMP fraction. An added complication was that although Eylar and Cook (1965) were able to determine that their PMP fraction contained smooth membranes, the stacking of parallel cisternae, seen in thin sections and characteristic of the dictyosomes of the Golgi apparatus, was destroyed by the homogenization procedure. Indeed, until 1968, there was no general method available for the isolation of the Golgi apparatus from animal cells other than those of the epididymis. However, considerable progress had been made (Morré and Mollenhauer, 1964) on the isolation of Golgi apparatus-enriched fractions from plant cells, by using conditions of minimum shear. The presence of compounds such as dextran in the homogenization medium, ensured that structural integrity of the organelle could be sufficiently maintained to enable the isolation of a fraction containing structures resembling the dictyosomes of the intact cell. Additional stabilization of the Golgi apparatus can be obtained by the use of glutaraldehyde (Morré et al., 1965), though a disadvantage is that although morphological features may be retained (Cook, 1972) enzymatic activity may be destroyed, depending on the concentration of stabilizing agent used. None the less it is possible, with murine leukaemic tissue, for example, to use a concentration of glutaraldehyde where glycoprotein:galactosyl transferase activity is retained with maximal preservation of morphological features (A. Warley, personal communication). By the latter part of the 1960s the isolation of morphologically identifiable fractions enriched in the Golgi apparatus was well documented for plant tissues, though little biochemical data was available on this material.

At the beginning of this decade a number of papers on the preparation of such fractions from animal tissues, principally liver, were published. As with the plant tissues, a major contribution to the field has been made by Morré and his colleagues (Morré et al., 1969; Morré et al., 1970; Cheetham et al., 1970) together with the work of Leelavathi et al. (1970) and Fleischer et al. (1969). Basically these techniques favour the use of a gentle method of homogenization. In the case of the method described by Morré et al. (1970), the Golgi apparatus recovered in a "nuclear" pellet (2,000 x g for 20 minutes) was resuspended in cell sap material and then purified by centrifugation in a sucrose density gradient. The appropriate fraction could be identified as being enriched in the Golgi apparatus by means of the electron microscope, using both negative-staining and thin-section techniques. As Morré et al. (1970) point out, the morphology of the Golgi apparatus is "so characteristic that it serves as a reliable marker". In negatively stained preparations the tubular nature of this organelle is obvious, while the characteristic pattern of dictyosomes, secretory vesicles and numerous small vesicular profiles is seen in thin sections of the material. All the published methods make use of extensive electron microscopical evidence for the characterization of the isolated material. Both negative staining of the preparation and an examination of thin sections of the material should ideally be performed, though some published methods (Schachter *et al.*, 1970) confine the microscopical evaluation to negative staining only. In addition, the property of taking up a great deal of osmium after incubation with osmium tetroxide at elevated temperatures for long periods of time, which is considered peculiar to the Golgi apparatus, has been used by Fleischer *et al.* (1969) in their characterization of fractions enriched in the organelle.

Early work on the isolation and characterization of dictyosome-enriched material relied heavily on morphological criteria for the identification of the fractions. This was inevitable since, at the time, no marker enzyme was known which could be shown to be located exclusively in the Golgi apparatus. Morré *et al.* (1969) found that fractions morphologically rich in dictyosomes derived from rat liver were low in 5′-nucleotidase and glucose-6-phosphatase activity when compared with fractions enriched in plasma membrane and endoplasmic reticulum respectively. These two activities have been regarded as markers for plasma membrane and endoplasmic reticulum. A number of nucleotide di- and tri-phosphatase activities have been demonstrated in Golgi apparatus-enriched fractions, none of which can really be regarded as unique to this material. The activities which have been discussed so far may all be regarded as of a "negative type", in the sense that they are all markers for other organelles and have been used to demonstrate that the fraction is depleted in those contaminating components. On the other hand, the glycosyl transferases "may be used as excellent markers for the Golgi apparatus" (Roseman, 1970).

Much of the work on the isolation and characterization of Golgi apparatus-rich material has been with liver, and consequently most of the data on glycosyl transferases in the Golgi apparatus have been derived from this material. At the end of the 1960s Wagner and Cynkin (1969a) described the incorporation of glucosamine into endogenous acceptor by a particulate enzyme preparation, which transfers the amino sugar from UDP-*N*-acetyl-^{14}C-glucosamine to the acceptor. These authors noted that the transferase activity was greatest in smooth microsomal membranes, especially those present in their post-microsomal fraction. By the use of pronase digestion, which converts the radioactive product into a form soluble in 5% trichloroacetic acid, Wagner and Cynkin (1969a and b) were able to demonstrate that the acceptor present in the system was protein or glycoprotein. They considered that their post-microsomal fraction contained fragments of the plasma membrane and Golgi apparatus, and noted that work published by Eylar and Cook (1965) (see section on cell-free systems) some three years previously had also demonstrated that the enzymes responsible for the synthesis of the carbohydrate moieties of glycoproteins of the Ehrlich ascites tumour cell plasma were also found in a post-microsomal particulate fraction. This fraction, Eylar and Cook (1965) suggested, represented fragments of the plasma membrane or Golgi apparatus. Being thus aware of the likely importance of the Golgi apparatus in the biosynthesis of glycoproteins, Wagner and Cynkin (1969b) prepared a fraction enriched in the Golgi apparatus from rat liver, using the techniques of Morré et al. (1970) which had been communicated to them prior to publication. They were able to confirm that the Golgi apparatus contained the enzymes necessary to mediate the transfer of amino sugar from UDP-N-acetylglucosamine to endogenous protein acceptors, while plasma membrane and endoplasmic reticulum fractions were devoid of perceptible activity, figures of 0 and 5% of the activity of the total homogenate being obtained with plasma membranes and endoplasmic reticulum respectively. In the fraction rich in the Golgi apparatus the transferase activity was enriched 150-fold, and accounted for 48% of the total activity of the crude homogenate. Similar results were obtained by Morré et al. (1970) who again found that over half the total glycosyl transferase activity of the original homogenates was recovered in the Golgi apparatus fraction; further purified endoplasmic reticulum fractions had very much lower activity, while the plasma membranes and cell sap fractions contained little or no enzyme. In addition to examining their fractions for enzymatic transfer of glucosamine to endogenous protein acceptors, Morré et al. (1970) examined their fractions for N-acetylglucosamine: galactosyl transferase activity (N-acetyl-lactosamine synthetase) with the formation of N-acetyl-lactosamine. This activity, in which a defined low molecular weight acceptor is used, was also found by Morré et al. (1970) to provide a useful marker enzyme for the Golgi apparatus, at least in rat liver. The post-microsomal fraction showed an absence of this transferase, while plasma membrane and endoplasmic reticulum showed much lower specific activities, though the evaluation of the activity in the plasma membrane was complicated by high levels of non-specific hydrolysis of the nucleotide sugar donor. This latter work should not be taken as being in contradiction to the finding in other systems (Eylar and Cook, 1965) of transferase activity in the post-microsomal fraction, because the methods of cell breakage used in the two cases were completely different. In the latter experiments, the Golgi apparatus is reduced to small vesicular fragments which are recoverable in the post-microsomal fraction, while in the method used for the specific isolation of the Golgi apparatus relatively intact structures sediment in the "nuclear" fractions on differential centrifugation. In addition to these studies, Leelavathi et al. (1970) also found considerable amounts of N-acetylglucosamine: galactosyl transferase activity in their Golgi apparatus preparation, and concluded that this enzyme "can indeed be used as a specific marker for Golgi membranes". Evidence that the glycosyl transferases are suitable marker enzymes for monitoring for the presence of the Golgi apparatus in cell fractionation studies is not confined to rat liver, as a similar conclusion can be made from the studies of Fleischer *et al.* (1969) on bovine liver. In the case of other cell types, Bosmann *et al.* (1968) have described the preparation from HeLa cells of a "smooth internal membrane" fraction which they suggest represents Golgi apparatus. In the isolation procedure used by these workers the characteristic morphology of this organelle was lost so the identification of their fractions rests heavily on enzyme markers. Of particular interest, however, is the finding of Hagopian *et al.* (1968) that a polypeptidyl:N-acetylgalactosamine transferase activity, in 56% yield with a 47-fold purification, is located in their Golgi apparatus fraction. This enzyme is important in that it is responsible for the formation of the protein-carbohydrate linkage between seryl and threonyl residues of the polypeptide receptor and N-acetylgalactosamine. Since only a small percentage of the glycoproteins of the HeLa cell are secreted, while the majority of glycoproteins produced are incorporated into membranes, the finding of polypeptidyl:N-acetylgalactosaminosyl transferase in a Golgi-derived fraction would suggest that the addition of the first sugar to the polypeptide may not, for membrane glycoproteins, take place on the rough endoplasmic reticulum (as may be the case for secreted glycoproteins). In addition to this activity, Hagopian *et al.* (1968) also recovered 33% of the glycoprotein:galactosyl transferase of the homogenate with a 28-fold purification; in their assays the authors were careful to use a well characterized exogenous acceptor, as well as examining for endogenous transferase activity.

That glycosyl transferases should be suitable markers for the Golgi apparatus draws attention to the importance of this organelle as a major site of glycoprotein biosynthesis, a conclusion which correlates with a number of autoradiographic studies made on the synthesis of complex carbohydrates.

2. MEASUREMENT OF GLYCOSYL TRANSFERASE ACTIVITY

Before discussing this evidence, it is appropriate to mention here some of the techniques that have been used to measure glycosyl transferase activity in the various cellular fractions. All the assays used monitor transferase activity radio-chemically. The majority of sugar nucleotides suitably radioactively labelled, and necessary as donors in the assay, are now obtainable commercially. Schachter *et al.* (1970) make use of high voltage electrophoresis to separate the products. The reaction mixture is applied to Whatman 3MM paper and subjected to high voltage electrophoresis, so that free sugar and excess sugar nucleotide migrate away from the required product—either protein bound radioactivity or labelled *N*-acetyl-lactosamine—which remains at or near the origin. On completion of electrophoresis, this portion of the paper may be dried and the activity present measured by liquid scintillation counting. In the case of *N*-acetyl-glucosamine:galactosyl transferase assay (Fleischer *et al.*, 1969; Morré *et al.*, 1969) the products of the reaction have also been separated by means of resin chromatography. Using a suitable anion exchange resin, in the Cl⁻ form, the excess UDP-galactose remains bound to the column allowing the labelled reaction product, *N*-acetyl-lactosamine, and any free galactose, to pass through the column. In the studies conducted by Fleischer *et al.* (1969) the products of the galactosyl transferase assay were characterized by high voltage electrophoresis in borate buffer, as well as by chromatography on sheets of DEAE paper. As an alternative to the methods described above where protein acceptors are being studied, use may be made of a combined precipitation and washing technique. Such a technique has been used for a number of glycoprotein:glycosyl transferases by Wagner and Cynkin (1969b) and others (Eylar and Cook, 1965; Bosmann *et al.*, 1968; Grimes, 1970). In discussing the various assay techniques reference has been made to the use of endogenous, as well as exogenous, acceptors. In the latter case these are either small molecular weight acceptors, for example *N*-acetylglucosamine, or larger molecules such as fetuin, the carbohydrate moieties of which have been specifically modified by treatment with glycosidases. Determinations of endogenous activity are documented for a number of systems, though as Schachter *et al.* (1970) have pointed out, the use of endogenous acceptors can give misleading results since tissues often contain many glycosyl transferases, some of which may use the same sugar nucleotide but different acceptors, with the resulting difficulty in defining the measured enzyme. In addition, the endogenous acceptor may be the rate-limiting component in the assay rather than the enzyme.

3. RADIOAUTOGRAPHIC STUDIES

Although biochemical studies implicate the Golgi apparatus as the major site of protein glycosylation, and even though one is able to isolate the organelle in a relatively intact state and demonstrate that glycosyl transferases are an integral constituent of it, the radioautographic technique presents an unique opportunity to study the relationship of the Golgi complex to other cellular constituents involved in the synthesis of these macromolecules. It should be remembered that autoradiographic studies were initially involved in the study of cells which were synthesizing large amounts of glycoprotein for secretion and, although such studies are of obvious relevance to the general problem of glycoprotein biosynthesis, there may be features whereby membrane and soluble glycoproteins differ in the mechanism of their assembly. More recently, the problem of the biosynthesis of membrane glycoproteins has received special attention from workers using the autoradiographic technique.

The elegant studies of Leblond and his colleagues (Neutra and Leblond, 1966a, b; Whur *et al.*, 1969) on the formation of the products of secretory cells of rat tissues have unequivocally demonstrated the rôle of the Golgi apparatus in the anabolism of these glycoproteins. Neutra and Leblond (1966a) showed by autoradiography using the light and electron microscopes that if glucose-^{3}H was administered to the intact animal, it was incorporated initially into substances present in the flattened saccules of the Golgi apparatus of goblet cells. With increasing periods of time (from five minutes to four hours), labelling of Golgi saccules and nearby mucigen granules was seen and then the label, in mucigen granules, was found to migrate to the apical membrane where the intact mucigen granules were released. It is unlikely that these results can be explained in terms of diffusion, and Neutra and Leblond (1966a) consider it more reasonable that the carbohydrate is added to the protein to form the mucus in the Golgi saccule; the saccule then becomes distended by the products of the reaction to become a mucigen granule. Apparently the dictyosomal stacks must be continuously renewed to accord with this mechanism. In an accompanying paper, Neutra and Leblond (1966b) used galactose-^{3}H to examine not only the production of mucus in goblet cells, but also extended their observations to a whole series of secretory cells of the rat. Galactose is a particularly useful label for investigating glycoprotein biosynthesis, since it is a constituent of the carbohydrate moiety of several glycoproteins and is not so readily broken down as glucose, which is subject to degradation by several metabolic paths. From their study, Neutra and Leblond (1966b) found that in contrast to their results with glucose, only certain mucus cells incorporated the galactose label. However, when the label was first located it was always in the Golgi region and only later in the secretion product. The results of both these studies indicate that the addition of the carbohydrate moieties of these glycoproteins takes place in the Golgi apparatus. The term glycoprotein is used advisedly; Neutra and Leblond (1966b) use the term complex carbohydrates to cover glycoproteins and mucopolysaccharides (glycosaminoglycans) as well as glycogen and glycolipids, though as these authors point out the last class of molecules are largely removed from the tissues by the organic solvents used in the histological processing. Further, Neutra and Leblond (1966b) assessed glycogen formation by treating specimens with alpha-amylase, and the radioactive material in the Golgi apparatus was found not to be glycogen, though when galactose was supplied to the liver, intestinal epithelium and muscle all the scattered cytoplasmic label was found to be glycogen. With glucose, much of the scattered label was not glycogen. Hyaluronidase had no effect on the radioactivity of epithelia, though in tooth pulp, cartilage, aortic wall and other connective tissues some label became labile with time. A sequence of treatment with peracetic acid and β-glucuronidase, indicated that material synthesized in the Golgi region was glycoprotein.

More recently, Leblond and his colleagues (Whur et al., 1969) have turned their attention to the incorporation of galactose-^{3}H, mannose-^{3}H and leucine-^{3}H by rat thyroids in vitro in a study of the stages of thyroglobulin synthesis. This work is particularly important since it set out to examine whether different regions of the carbohydrate component were synthesized in distinct and separate locations within the cell. Whur et al. (1969) were able to show that with leucine and mannose (which sugar is located in the inner core of glycoproteins) incorporation took place in the rough endoplasmic reticulum. The sugar was added as soon as the polypeptides were released into the cisternae of the rough endoplasmic reticulum. Puromycin almost completely inhibited the incorporation of leucine-^{3}H and mannose-^{3}H, but had no effect, during the first hour, on galactose-^{3}H incorporation. This latter sugar, which usually occupies a penultimate position to the terminal sialosyl and fucosyl residues in completed glycoproteins, was found to be added to the growing molecule in the Golgi apparatus. Subsequently Haddad et al. (1971) investigated the incorporation of fucose ^{3}H in this system, and showed that some of the oligosaccharide side chains are completed within the Golgi apparatus. So far investigations employing the autoradiographic technique have not been able to define the site of incorporation of sialic acids into glycoproteins.

These experiments have examined, in the main, secreted glycoproteins and attention will now be turned to a consideration of the part that autoradiography has played in the study of membrane glycoproteins.

The results of Ito (1969) and Bennett (1970) are important here, if the synthesis of the "fuzzy coat" associated with the intestinal microvilli is regarded as a general model for membrane glycoprotein biosynthesis. Ito (1969) has shown, by autoradiography, that galactose or mannose administered to cat intestinal cells accumulates first of all in the Golgi apparatus, before migrating to the cell surface. A similar sequence of events was found to take place by Bennett (1970) when galactose-^{3}H was administered to young rats and the duodenal columnar cells examined at various time intervals after the injection of the label. It is likely that labelled material is transported from the Golgi apparatus to the cell surface as smooth-surfaced vesicles, for a similar picture has been seen in other systems (Rambourg et al., 1969). Thus, it is clear that membrane proteins are not glycosylated while they are associated with the rough endoplasmic reticulum. Admittedly, the question of how and where the first sugar of an oligosaccharide sequence is attached to the polypeptide chain is still a matter of some controversy; it may well be that the initiation of glycosylation occurs in different sites in the cell, depending on the type of glycoprotein or cell involved. Is glycosylation confined to the Golgi apparatus for membrane glycoproteins, while for secreted glycoproteins the initial sugar is transferred while the polypeptide is still associated with the rough endoplasmic reticulum, though not necessarily while still attached to the polysomes? Are O-glycosidic linkages produced at one site and N-glycosidic at another? The finding of polypetidyl:N-acetyl-galactosaminyl transferase activity in the Golgi apparatus of the cell may be a reflection of a mechanism peculiar to the O-glycosidic linkage, and for this the results of Reith et al. (1970) are particularly relevant. These workers examined the incorporation of glucosamine-^{3}H in HeLa cells by light and electron-microscope autoradiography. Besides occupying more exterior positions in the carbohydrate moieties of glycoproteins, N-acetylglucosamine can provide the linkage point between protein and carbohydrate in the N-glycosidic form of linkage. In their studies Reith et al. (1970) showed that the Golgi apparatus was the site of incorporation of this sugar. If N-acetylglucosamine is subsequently shown to be involved in N-glycosidic linkages in HeLa cell membrane glycoproteins then this study is of particular significance.

D. Glycolipids

The previous sections have covered the evidence which indicates that the biosynthesis of the oligosaccharide components of glycoprotein involves three stages, namely the synthesis of the various monosaccharides from glucose, followed by their "activation" by conversion to the corresponding sugar-nucleotide (in some cases the monosaccharides are formed at the sugar nucleotide level, see Fig. 1), and finally the polymerization of the monosaccharides which takes place in a totally different manner to that employed in the synthesis of nucleic acids and proteins. In the latter cases a template mechanism exists, while in the case of the glycoproteins the oligosaccharide moieties are synthesized by the means of sequential action of glycosyl transferases which transfer the sugar from the sugar nucleotide to the appropriate acceptor. The carbohydrates of the glycolipids are synthesized in an identical manner, and some of the features of this group of specific glycosyl transferases, called a multiglycosyltransferase system, has been reviewed by Roseman (1970).

Changes which take place in the composition of glycoproteins and glyco-lipids at the cell surface have been correlated with various alterations in biological functions, and will be discussed in Chapter 6. As an illustration of the rôle of glycosyl transferases in glycolipid biosynthesis, and the means by which they may be studied, reference may be made to the work of Fishman et al. (1972). These investigators studied the activities of four glycosyl transferases involved in ganglioside (a sialoglycolipid; some authors prefer to restrict the term to sialoglycolipids of the central nervous system) biosynthesis in normal mouse fibroblast cells, and SV40 and polyoma virus transformed variants of these lines. The enzymes were assayed by procedures similar to those used with glycoproteins, save that the glycolipid acceptors were added to the reaction tubes in organic solvent and then reduced to dryness. The sugar donor, buffer, appropriate detergent and cell fraction were then added, followed by vigorous swirling. The reaction was stopped by adding chloroform-methanol and chloroform-methanol-water mixtures. Isolation of radioactive glycolipid product from precursor radioactive sugar-nucleotide was achieved by passage through small columns of G-25 Sephadex. Using these techniques, Fishman et al. (1972) found that the only consistent change in all of the transformed cells was a greatly reduced level (15–20% of normal cells) of haematoside[3]: N-acetylgalactosaminyl transferase. No significant differences were found with any of the glycosyl transferases with increased cell density, and cell to cell contact.

E. Membrane Biosynthesis

When dealing with the biosynthesis of a glycoprotein one is faced with questions of the location in the cell of the various components of the macromolecule assembled, and the controlling mechanisms of this process. Such questions are also relevant to the wider topic of membrane biosynthesis which is undoubtedly closely related to the subject of membrane structure and function. In any study aimed at elucidating the nature of membrane biosynthesis, it is essential to distinguish between those systems in which there is a net synthesis of new membrane and those in which one is examining membrane turnover alone.

1. MEMBRANE TURNOVER

The phenomenon of membrane turnover has been widely studied by Warren and Glick (1968) in the case of plasma membranes isolated from animal cells, as described in detail in the first chapter. In their studies, Warren and Glick (1968) used L-cells (mouse subcutaneous fibroblasts) as a model and employed ^{14}C labelled precursors of membrane constituents such as D-glucose, L-leucine, L-valine and D-glucosamine. When the incorporation of a general precursor, such as D-glucose, into surface membrane was examined in both dividing and non-dividing cells,

---

[3] Haematoside; N-acetylneuraminyl galactosyl glucosylceramide. In the particular transferase studied the donor is UDP-N-acetylgalactosamine.

Warren and Glick (1968) found that the rate of increase in specific activity was approximately the same in both cases. A rise in specific activity might be expected in the dividing cell, though turnover cannot be totally excluded in this case. However, for the non-dividing cell the experimental finding is compatible with the turnover of membrane constituents, since the mass of membrane in the cell would be expected to remain constant. Similar results were also reported (Warren and Glick, 1968) with the precursors of protein and carbohydrate. In a series of experiments designed to test the extent of turnover in the dividing cell, these authors followed the removal of ^{14}C-labelled material from cells grown in cold medium following heavy labelling. The rate of fall in specific activity was slightly lower than that predicted by taking into consideration dilution caused by an increase in cell number. It would, therefore, appear that growing and non-growing cells synthesize similar amounts of membrane which is incorporated into the dividing cell, where it remains at the surface, while in the non-dividing cell turnover takes place with rejection of some of the membrane constituents and incorporation of fresh components. This pattern is not limited to the surface membrane, for when working with a particulate fraction, which was considered to consist largely of the internal membranes of the cell, Warren and Glick (1968) found a similar pattern of labelling. If L-cells can be taken as a representative model of the non-dividing cell, turnover takes place with a constant replacement of membrane components, while in the dividing cell the turnover is greatly reduced. A note of caution is to be sounded in accepting the L-cell as a representative model, because an examination of a number of L-cell cultures has shown that the cells contain a number of virus-like particles (Kindig and Kirsten, 1967). The particles are associated with intracellular vesicles and also bud from the surface of the cell. On being released from the cell, the virus-like particles would acquire a membrane, derived from the surface membrane of the host cell, and therefore an apparently high rate of membrane turnover in the non-dividing cell could in effect be a measure of the synthesis and release of virus-like particles from the cells.

A feature of membrane turnover is that it provides a mechanism whereby the cell can maintain the integrity of its surface in the absence of any mechanism for responding to specific sublethal damage of the plasma membrane. Warren (1969) cites evidence to show that when L-cells have been treated with trypsin there is no specific repair mechanism to deal with the damage, but rather the surface membrane is replaced in the normal cause of turnover. In this respect, the studies of Hughes et al. (1972) on the regeneration of the surface glycoproteins of TA3 cells[4] after treatment with neuraminidase is particularly relevant. They found

[4] TA3 cells: a transplantable spontaneous mammary adenocarcinoma of an A strain mouse.

that the pattern of D-(^{3}H) glucosamine incorporation into the glycoprotein of non-growing TA3 cells was very similar whether or not they had been treated with neuraminidase. This result is not compatible with a specific repair effected by the transferance of sialic acid to these oligosaccharide chains which had been enzymically depleted of their sialosyl residues, but rather indicates that there is a *de novo* synthesis of complete carbohydrate chains. These results with TA3 are entirely in accordance with the suggestion of Warren (1969) that the cell does not have a specific repair mechanism, but relies on membrane turnover to maintain the integrity of the cell surface. In addition, the TA3 cells were found to have a rate of turnover comparable to that found for L-cells by Warren and Glick (1968). Furthermore, the results of Hughes *et al.* (1972) substantiate the view that, in the dividing cell, new membrane material is incorporated into the cell surface, where it remains, but in the absence of cell multiplication the overproduction of membrane components is compensated for by a relatively high rate of turnover.

Considering the turnover and rate of synthesis of plasma membrane components during the cell cycle, Warren (1969) cites work in which KB cells grown in synchronous culture were "pulsed" for one hour periods with L-leucine and D-glucosamine-^{14}C; from the labelling patterns obtained, it appears that these cells are able to make surface membrane at all stages of the cell cycle, but that there is a burst of synthesis in the G1 period just after the cell divides. There is negligible elevation in the synthetic rate during M (mitotic) phase. As an explanation, Warren (1969) suggests that the increase in rate of synthesis at G1 phase is designed to bring the two relatively small daughter cells to normal size. In another system, Bosmann and Winston (1970) examined the synthesis of glycoproteins and glycolipids in synchronous cultures of L5178Y cells, a murine lymphoma cell, and found that cellular lipid and glycolipid synthesis was confined almost exclusively to the G2 and M periods. However, cellular protein and glycoprotein was found to be synthesized throughout the cell cycle, being highest in the S period. These results are quite different from those of Warren and his colleagues (Warren and Glick, 1968; Warren, 1969) and certainly no extra burst of synthesis was observed in the G1 phase by Bosmann and Winston (1970). Instead they (Bosmann and Winston, 1970) found that lipid and glycolipid are synthesized late in the cell cycle, suggesting that these are the last molecules to be incorporated into membranes. Obviously this result may be pertinent to the major problem of membrane biosynthesis, however, Bosmann and Winston (1970) unlike Warren and his co-workers did not obtain their data using isolated membrane fractions and, although glycolipids and glycoproteins may be regarded as giving a measure of membrane biogenesis, it is preferable when making deductions about the cell surface and plasma membrane to isolate the various classes of membranes in a known state of purity.

As explained earlier, at various stages in the cell cycle the cell surface reacts differently to lectins and, with the existing evidence, it would be difficult to consider whether such results correlate with the studies above. The results observed with the lectins could well be a reflection of reorientation of membrane components, rather than a measure of net synthesis. The earlier, electrokinetic studies of Kraemer (1967) are particularly pertinent here. Mayhew (1966) had shown a transient elevation of mean cell electrophoretic mobility which is related to neuraminidase-susceptible groups at the surface of the cell immediately before and during the division wave of parasynchronous RPMI No. 41 cells. Kraemer (1967) investigated the question of whether this result was caused by transient elevation of the density of surface sialic acid, or a reorientation of the same macro-molecule with respect to the electrokinetic plane of shear. Kraemer (1967) computed the mean cell surface area of chinese hamster ovary cells at various times after release from thymidine blockade of division, and made a chemical estimation of surface sialic acid. He was able to show that, when surface sialic acid is expressed as molecules per μ^2 surface area, no sudden shifts in this parameter occurred, but rather a gradual increase in surface density of sialic acid is seen. Kraemer (1967) is of the view that the transient neuraminidase-susceptible shift of mobility observed by Mayhew (1966) could be explained by any one, or all, of the following mechanisms. Firstly, the surface density of terminal sialic acid molecules could change; secondly, the effective (*in situ*) pK of the carboxyl groups of the terminal sialic acid residues might alter; or thirdly, a change in the topology of surface sialic acid residues in relation to the electrokinetic surface might occur. Certainly these results of Kraemer (1967), if generally applicable, would tend to exclude the first possible mechanism, and he favours a conformational change as the most likely explanation, considering that the electrokinetic surface and the surface that bears sialic acid residues accessible to neuraminidase, are not identical. However, experiments by Rosenburg and Einstein (1972) with synchronized Roji lymphoid cells in culture suggest that cell membrane sialic acid may be synthesized primarily in late G2 (interval in cycle following DNA synthesis and prior to mitosis) phase of the cell cycle, and as such would provide a molecular basis for Mayhew's (1966) observations. Alternatively, the natural rate of destruction of sialic acid is decreased.

Recently Brown (1972) has examined the spectrum of glycopeptides released by trypsin treatment from a range of cells including mouse LS-cells. The glycopeptides were eluted from DEAE cellulose with a linear sodium chloride gradient. In general, Brown (1972) found that they fell into four groups, eluting at approximately 0·03, 0·04, 0·06 and

0·07 M Na Cl respectively, which he classifies as glycopeptides 1, 2, 3 and 4. Brown (1972) makes it clear that he does not make any claims as to the homogeneity of these fractions, and points out that "all could be mixtures of similar glycopeptides". Using this method of classification, Brown (1972) found that classes 1, 2 and 4 were present in the cell, but, for experiments performed with synchronously dividing populations of cells, he showed that one class of glycopeptides, class 4, was found to accummulate at the cell surface preferentially at the time of cell division. Apparently once this glycopeptide fraction has appeared on the cell surface it does not later become masked, modified or unavailable to the action of trysin. It would be interesting to know whether the burst of synthesis following mitosis observed by Warren (1969) might be related to the synthesis of class 4 glycopeptides, especially as Brown (1972) finds that the other glycopeptides, present in classes 1 and 2, appear at a relatively constant rate throughout the cell cycle. The finding that the class 4 glycopeptides do not become unavailable to trypsin, once present at the cell surface, would suggest that if there is some reorientation of the sialoglycoproteins within the membrane at division, as suggested by Kraemer (1967), then it is too slight to prevent the action of trypsin on the surface of the LS-cells. Though these studies only demonstrate the assembly of glycoproteins in the cell surface, and not necessarily glycoprotein biosynthesis *per se*, they demonstrate that the biosynthesis of the surface membrane of the cell is very complex and certainly illustrate the dynamic nature of this structure.

2. MECHANISMS OF MEMBRANE BIOGENESIS

Turning to a consideration of membrane biogenesis, as opposed to the biosynthesis of component macromolecules, it can be argued that a membrane could be synthesized by either a single-step, or a multiple-step, process. In the case of a single-step process it is implied that all the components of the membrane—carbohydrate, protein and lipid—are assembled simultaneously. Alternatively, a multistep process would involve the synthesis of a relatively stable primary membrane composed of lipid and structural protein to which various components, such as enzymes, are then added. This process would be very much in accord with the bilayer model for membranes, while the former mechanism would be compatible with the subunit structure, assuming that the subunits of assembly necessarily persist in the completed membrane (see Chapter 1).

A system which has proved useful in the study of membrane biogenesis is the developing endoplasmic reticulum of the rat hepatocyte, at the stage three days before, to eight days after, birth (Dallner et al., 1966a, b). Immediately before birth the endoplasmic reticulum, which at that stage is predominantly rough surfaced, increases in volume. After birth, the endoplasmic reticulum continues to increase in volume and this increase is mainly in the smooth-surfaced elements. Dallner *et al.* (1966a) presented evidence from radioactive tracer experiments with ^{14}C-leucine and ^{14}C-glycerol, coupled with membrane fractionation, which suggests that new membrane is synthesized in the rough endoplasmic reticulum, and is then transferred to the smooth membranes. These investigations also showed (Dallner *et al.*, 1966b) that the various constituent enzymes of the micro-somes appeared at different times in the pre- and post-natal periods, being synthesized in the rough membranes and then transferred to the smooth membranes. The synchronous assembly of lipid and protein is in agreement with the single-step process for membrane biogenesis, but the synthesis of the various microsomal enzymes at different rates and times is not compatible with such a mechanism, unless one envisages the mem-brane as being comprised of patches, or tesserae, each bearing a characteristic group of enzymes. Such tesserae could then be assembled into a completed membrane by means of a single-step mechanism in which a number of different "functional subunits" came together. That such membranes are unlikely to be composed of planar aggregates of discrete subunits comes from the observations of Omura *et al.* (1967), who found that the half-lives of the membrane protein and lipid were different. This led them to argue that a multistep assembly, as applied to membrane proteins, is more probable, as membrane lipid turnover appears to be largely independent of that of membrane protein. The finding of Holtzman *et al.* (1970), that ^{32}P is incorporated at widely different rates into the various phospholipids of smooth membranes, would perhaps be better explained in terms of a mosaic type of membrane structure than the classical unit membrane of the Danielli-Davson type. The newer ideas on membrane structure, in which protein is considered to be present in a lipid matrix, would be an appropriate example of such a mosaic type of structure.

The studies above have not been directly concerned with the plasmalem-ma. Although it is possible that different types of membrane may be synthesized in different ways, the finding by Warren and Glick (1968) of a similar labelling pattern in a particulate fraction, as being representa-tive of internal membranes, to that of the surface membranes, would argue against such a suggestion. The finding that the turnover of the various constituents of the surface membrane is similar, would of course be in accord with an assembly process involving subunits. However, it should be remembered that experiments which measure overall turnover are not necessarily sensitive to differences in the turnover of individual components of the membrane. Bosmann *et al.* (1969) have shown that the intracellular site for the assembly of the carbohydrate units of membrane glycoprotein is the smooth membranes in HeLa cells. These authors noted a parallel between the turnover of the glycoprotein and glycolipid of the smooth and plasma membranes, and favour a scheme for the biosynthesis of plasma membranes from preformed subunits. Certainly this is an interesting suggestion, and it would be appropriate to follow components of the plasma membrane, other than glycoprotein and glycolipids, in their system. The studies of Bosmann and Winston (1970), discussed earlier in this section, which indicate that lipid and glycolipid are synthesized late in the cell cycle, serve to emphasize caution in regarding a single-step or subunit assembly mechanism as general. Likewise, the demonstration by Evans and Gurd (1971) that glycolipids are synthesized at a rate different from that of membrane glycoproteins, leads to the suggestion that glycolipids are incorporated into the plasma membranes of mouse liver independently of the newly-synthesized protein components of the membrane.

From all this it is apparent that the glycoproteins are synthesized in a step-wise fashion. Polypeptide synthesized on the polysomes migrates to the cisternae of the rough endoplasmic reticulum where the initial glycosylation may take place, the additional saccharide residues are then added successively to the growing oligosaccharide chains in different intracellular membrane compartments, especially the Golgi apparatus. Presumably membrane glycoproteins are synthesized in this fashion, except that it would appear from the systems studied that the glycosylation of the membrane protein may not take place in the rough membranes of the cell. Nevertheless, even though the bulk of the sugars are added in one cellular compartment, the polypeptide portion of the macromolecule has to undergo a progressive migration from rough endoplasmic reticulum to the smooth internal membrane and thence to the cell surface, and in this sense appears to be following a multistep process of the type first suggested by Palade and his colleagues. The "multistep" or "assembly-line" mechanism as applied to membrane glycoproteins has been used by Hirano et al. (1972) to explain the exclusive localization of carbohydrate on the outer surfaces of the plasma membrane. This exclusive localization of sialosyl residues at the extracellular surface of the plasma membrane was elegantly demonstrated by Benedetti and Emmelot (1967). Subsequently, the use of ferritin-conjugated lectins (Nicolson and Singer, 1971) indicated that this asymmetry is likely to be a general feature for oligosaccharides of the plasma membrane of animal cells. Further electron microscopic studies (Hirano et al., 1972) with fragments of membranes from P3K murine plasmacytoma cells (using ferritin labelled lectins) showed that the asymmetric distribution of oligosaccharide residues in membranes was also a feature of internal cell membranes. The demonstration that concanavalin A binds exclusively on the cisternal side of the membranes of the rough endoplasmic reticulum would be in accord with the addition of mannosyl residues to the polypeptide at this site. The finding that ricin, a lectin specific for galactosyl residues, did not stain rough membranous elements, but only bound to smooth membranes, is consistent with the sequential addition of sugars to the growing oligosaccharide chain. The murine tumour used by Hirano *et al.*(1972), besides synthesizing membrane glycoproteins, also produces an IgG1 myeloma protein, so the identification of a material capable of binding concanavalin A in the fractions enriched in rough endoplasmic reticulum could represent precursors of this protein. The staining in both the rough and smooth endoplasmic reticulum always occurred asymmetrically. This led Hirano *et al.* (1972) to postulate (see Fig. 2) that the biogenesis of plasma membranes involves an "assembly line" mechanism by which membrane components are assembled. They start from the rough endoplasmic reticulum, and then proceed to the Golgi apparatus where the membrane is packaged into vesicles which contain oligosaccharide on the inside surface of the membrane. These vesicles, it is suggested, fuse with the existing plasma membrane to generate new cellular surface. Such possible precursor-vesicles have been seen in the electron microscope (Rambourg and Droz, unpublished results quoted by Rambourg *et al.* 1969) and it was speculated (Rambourg *et al.*, 1969) that they ferry material from the mature face of the Golgi apparatus to the plasmalemma. Certainly, such a mechanism accords with the confinement of carbohydrate to the outer surface of plasma membranes. Further, such a mechanism would seem attractive in the light of the newer concepts of plasma membrane structure, especially in the light of the evidence which suggests that glycoproteins are able to span the cell membrane. The mechanism is in many respects in accord with the multistep theory of membrane biogenesis, albeit that this theory was envisaged in terms of a Danielli-Davson structure for the plasma membrane. On balance, present thought does not appear to favour a subunit structure for plasma membranes, and this must in turn reflect on ideas of membrane biogenesis. However, what is perhaps more important is that biosynthetic studies illustrate the dynamic aspects of membrane chemistry, which are so often overlooked in those studies employing physical techniques where the membrane structure is chemically fixed. The definitive answers as to the exact mechanism applying must await future experiment.

II. Plant Cells

As plant cells grow and specialize the structures of their walls change, and this is paralleled by changes in the chemical composition of the walls and of the polysaccharides composing them. These changes lead to changes in the gross contents of the various monosaccharides comprising the whole wall polysaccharide and, thus, it follows that specialization and differentiation must be reflected by changes both in patterns of macromolecular synthesis, and in the metabolism of polysaccharide precursors of low molecular weight (Harris and Northcote, 1970).

Though knowledge of the details of polysaccharide and glycoprotein assembly in plant cells is still very incomplete, considerable progress

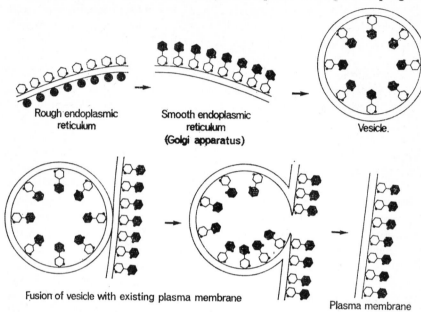

Fig. 2. Schematic mechanism for the biogenesis of plasma membranes of eukaryotic cells after the scheme proposed by Hirano *et al.* (1972). In this scheme an intracellular membrane "assembly line" mechanism is envisaged, leading penultimately to the formation of vesicles, the new surface membrane being generated by the fusion of these vesicles with the previously existing plasma membrane. The large filled circles represent ribosomes. At different stages along the "assembly line" it is proposed that saccharide units (represented by open and filled "pyranose ring" symbols) are added successively to growing oligosaccharide chains on membrane bound glycoproteins and glycolipids. In the scheme proposed by Hirano *et al.* (1972) carbohydrate was envisaged as being added at the rough endoplasmic reticulum, though on present evidence (see text) the glycosylation of membrane protein, as opposed to secreted glyco‑ proteins, takes place predominantly in the Golgi apparatus. Hirano *et al.* (1972) showed diagrammatically in their scheme the carbohydrate as existing on the cell surface as a trisaccharide, and in the vesicle as a disaccharide. In this figure the carbohydrate in the vesicle and on the cell surface is shown diagrammatically as a disaccharide. The portrayal of the same carbohydrate moiety in the vesicle and on the cell surface is done for the purposes of illustrative clarity, and should not be taken as an indication that the synthesis of surface oligo‑ saccharide moieties are necessarily completed at the vesicular stage .The above mechanism explains the observation (see especially Chapter 2) that oligosacchar‑ ides of plasma membranes are exclusively located on the outer surface of the plasma membrane.

has been made towards the elucidation of the metabolism of low-molecular weight precursors of these molecules, and towards the location of their biosynthesis within the cell. Most investigations have been made upon the cells of higher plants, so that little is known of algae generally and of lower terrestrial plants. The metabolism of low-molecular weight precursors of plant polysaccharides will be considered first.

A. Cell Walls: Low Molecular Weight Precursors

Two major routes have been described for the incorporation of glucose into cell wall material. The first pathway is via sugar nucleotides, and its elucidation has been largely the work of Hassid and his collaborators. The second pathway, proposed by Loewus (1965), makes use of cyclitols as intermediates. Each will be described in turn.

Glucose is taken up by plant cells, and it is an effective precursor of the wall polysaccharides as has been shown by many experiments using the radioactive sugar. Sucrose is also a precursor, though it appears to give rise to a rather different pattern of labelling in the wall, suggesting

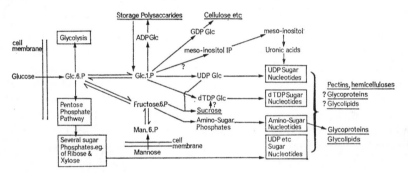

Fig. 3. The fate of glucose and its phosphates in the plant cell, illustrating the complexity of sugar metabolism. Only general pathways, not individual steps, are shown for clarity.

that it is utilized by a route other than that from glucose. Once in the cell, glucose is phosphorylated to glucose-6-phosphate by the action of hexokinase, (E.C.2.7.1.1.), and it is at this stage that the two pathways of its further metabolism diverge. The glucose-6-phosphate can be converted to glucose-1-phosphate in a reversible reaction catalysed by phosphoglucomutase (E.C.2.7.5.1.), in which the equilibrium lies far towards the 6-phosphate. However, the 1-phosphate can undergo further reaction with ATP, GTP, UTP or deoxy TTP to yield the corresponding sugar nucleotides, with elimination of pyrophosphate. The enzymes catalysing these reactions are the nucleotide triphosphate: sugar phosphate nucleotidyl transferases (E.C.2.7.7.9. et. seq.) and the reactions are made effectively irreversible by cleavage of the pyrophosphate under the action of pyrophosphatase (E.C.3.6.1.1.), thereby "pulling" the reaction.

The alternative pathway of glucose-6-phosphate metabolism begins with its oxidative conversion to *meso*-inositol-1-phosphate in the presence of NAD. This pathway will be further discussed below.

Glucose-1-phosphate can serve as a precursor of many different sugar nucleotides in plants, only some of which appear to be on the route of synthesis of wall polysaccharides.

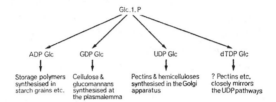

Fig. 4. Sugar nucleotides derived from glucose-1-phosphate, illustrating the general division of function among them.

In general, two methods of approach have been used to investigate the synthesis and metabolism of these intermediates. Their presence in the whole cell has been shown in most cases by isolating them and demonstrating their structure chromatographically. This is no simple task, since they are not present in other than trace amounts and are highly labile. The second approach has been to show their synthesis, interconversions and incorporation into other molecules in cell-free systems, most usually obtained from the mung bean (*Phaseolus aureus*). The great majority of the enzyme activities thus defined are associated with particles in the microsomal fraction of the cell, are unstable to solubilization and are probably associated with the plasmalemma, the smooth endoplasmic reticulum and the dictyosomes. Hassid (1967) has reviewed the field.

A difficulty that arises from studies upon cell-free, particulate systems is that there may be various metabolic intermediates already present. Thus, the apparent incorporation of a labelled sugar nucleotide into a homopolysaccharide might be, in reality, its incorporation into a heteropolysaccharide, or even a glycoprotein, with a concomitant incorporation of an unlabelled, endogenous sugar nucleotide. Similarly interconversions of sugar nucleotides may sometimes be catalysed by enzymes that use endogenous cofactors, and failure to detect some pathway could be the result of endogenous inhibitors. Furthermore, the very complexity of such particulate systems makes it possible that what is described as the action of, for example a sugar nucleotide-polysaccharide glycosyl transferase, is really the summed action of several enzymes and carriers in a multi-enzyme complex. Indeed, the very complexity of structure in many polysaccharides makes such a multi-enzyme action in their synthesis extremely likely.

A further caveat that must be made, is that the demonstration of the metabolism of a given sugar nucleotide by a cell-free system is not proof that this reaction occurs *in vivo*. It is always possible that some other sugar nucleotide is used by the enzyme *in vivo*, and the metabolism of that supplied *in vitro* as a substrate is fortuitous. Even if it is present in the whole cell, this sugar nucleotide may never reach the enzyme because of compartmentation. It is also just possible that some enzyme activities can only be released by cellular breakage. Adenosine diphosphate glucose (ADPG1c) is derived from ATP and α-D-glucose-1-phosphate, and appears to be associated specifically with the synthesis of storage polysaccharides

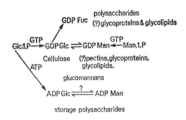

Fig. 5. The GDP glucose and ADP glucose pathways.

in plants, where it is a precursor of amylopectin and phytoglycogen and defects in its metabolism in maize lead to "waxy" mutants. It seems to serve a similar function in bacteria and there is no proof of its being involved in wall synthesis. It may be noted, however, that adenosine diphosphate galactose (ADPGal) and adenosine diphosphate mannose (ADPMan) have been described in corn grain, where their origin and fate is unknown, and the possibility of their being involved in wall synthesis cannot be excluded.

Guanosine diphosphate glucose (GDPGlc), produced from GTP and α-D-glucose-1-phosphate, has been reported as a precursor of cellulose synthesis in the mung bean and the cotton boll. It is known to be able to undergo a 2-epimerization to GDP mannose (GDPMan), and the cell-free system for "cellulose" synthesis from the cotton boll will also incorporate this sugar nucleotide. The GDPMan incorporation greatly enhances the incorporation of GDPGlc, suggesting that the synthesis may be of a glucomannan rather than true cellulose. It is notable that GDPMan can also arise from α-D-mannose-1-phosphate and GTP, and it can act as a glucomannan precursor in the mung bean. In some bacteria GDPGlc is convertible to GDP fucose (GDPFuc), by way of a curious double epimerization, involving a 4-keto-6-deoxy sugar intermediate, and it is known that polymers containing fucose are present in some plant cell walls and probably in cell membranes also. Two GDP uronic acids are known from *Fucus gardneri*, GDP mannuronic acid (GDPManA) and GDP guluronic acid (GDPGulA), which probably act as precursors of alginic acid in that plant.

GDPGlc does not seem to be the only sugar nucleotide which can act as a precursor for cellulose synthesis. In *Lupinus albus* and in oat (*Avena*) coleoptiles, uridine diphosphate glucose (UDPGlc) is a more effective precursor of cellulose than GDPGlc, and in the former case a different set of enzymes are involved from those which use GDPGlc. UDPGlc is also the cellulose precursor in the prokaryotes *Acetobacter xylinum* and *Dictyostelium discoideum*.

UDPGlc was one of the earliest sugar nucleotides to be demonstrated in plant tissues, and has been shown to be a precursor of sucrose, callose, amylose and glucosyl residues in various glycosides. It probably serves as a source of glucosyl groups in hemicelluloses and also pectins.

A large number of sugar nucleotides are known to be derived from UDPGlc, either directly or indirectly, and many of these will act as precursors for synthesis of wall polysaccharides in cell-free systems and for other glycosylation reactions, such as the glycosylation of phenols and steroids. In an irreversible reaction involving oxidation at C6 and catalysed by UDPGlc dehydrogenase and NAD, UDPGlc is converted to UDP glucuronic acid (UDPGlcA). This sugar nucleotide is probably used in both pectin and xylan synthesis and is a glucuronosyl donor to a variety of phenolic, steroidal and other aglycones. It can also be further metabolized to UDP-xylose (UDPXyl) by oxidative decarboxylation at C6, which is an irreversible reaction brought about by UDPGlcA decarboxylase. The UDPXyl produced is again likely to be a donor of xylosyl residues to a variety of wall polysaccharides, and is also probably the xylosyl donor in glycoprotein biosynthesis.

UDPGlc, UDPGlcA and UDPXyl can all be converted to their corresponding 4-epimers by action of the appropriate epimerases. Thus UDPGlc is converted to UDPGal, a widely used precursor of the polysaccharides laid down during primary wall development. Similarly, UDPGlcA is converted to UDPGalA which appears to be a precursor of pectin synthesis (see below). Strominger and Mapson (1957) reported a UDPGal dehydrogenase activity in peas, somewhat analogous to, but distinct from, the UDPGlc dehydrogenase, so another route from UDPGlc to UDPGalA could exist.[5] UDPL-arabinose (UDPAra) is derived from UDPXyl by 4-epimerization, and again serves as a donor for wall polysaccharide synthesis *in vitro*.

[5] However, these authors (Strominger and Mapson, 1957) feel that it does not.

All three epimerases catalyse reversible reactions, with equilibrium constants fairly close to unity. Thus they are unlikely to be used for the regulation of these pathways, though it is likely that their levels will fall during differentiation, as Northcote (1963) has pointed out, as the demands for the galactose series of sugar nucleotides fall relative to those of the glucose series, thereby effecting an economy in protein synthesis. Neufeld

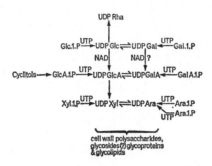

Fig. 6. The UDP glucose and allied pathways.

and Hall (1965). have shown that UDPXyl inhibits UDPGlc dehydrogenases of pea cotyledon, bovine liver and chick cartilage. Low concentrations of UDPXyl showed competitive inhibition of the enzyme, while at higher concentrations the inhibition curve was inflected. UDPAra was also inhibitory, while NAD reversed both inhibitions; an allosteric mechanism was proposed. In a study of UDPGlcA decarboxylase of wheat, Castanera and Hassid (1965) could find no inhibition of the enzyme either by UDPGlcA or UDPXyl. Thus it may be that the levels of uronide and pentoside sugar nucleotides are regulated together, by means of the dehydrogenases.

A further, and separate, reaction of UDPGlc is its conversion to UDPL-rhamnose (UDPRha), by way of a double epimerization and a 4-keto-6-deoxy sugar intermediate. The UDPRha produced by this irreversible reaction is a donor of rhamnosyl residues for the synthesis of phenolic and other glycosides, and is very probably the source of rhamnosyl residues for pectin synthesis.

Deoxy TDP (dTDP) sugars have been known for a number of years in several micro-organisms, but, though they are also well established as present in higher plants, their rôle there is still unclear. dTDP glucose (dTDPGlc) was shown by Milner and Avigad (1965) to be active as a substrate for the sucrose synthetases of wheat endosperm and sugar-beet root. In both cases, UDPGlc completed with dTDPGlc for the enzyme. The enzymes for the synthesis of dTDPGlc and dTDPGal (TDP galactose) from dTTP and the corresponding sugar-1-phosphates have been described, and dTDPGlc-dTDPGal 4-epimerase is known. Both dTDPL-rhamnose (dTDPRha) and dTDPGal have been demonstrated in plant extracts, and the conversion of dTDPGlc to dTDPGalA via dTDPGal has been shown in sugar beet. It is of interest that this oxidation is NADP dependant, unlike the NAD dependant pathway for corresponding oxidation of UDPGlc. Thus it appears that a series of pathways of dTDPGlc metabolism exist which partially, and perhaps wholly, mirror the corresponding pathways of UDPGlc metabolism, but use different enzymes and sometimes different cofactors. The significance of this is not clear, but one possibility is that whereas glucose, and other monosaccharides, are utilized via their UDP derivatives, sucrose may be metabolized via dTDP derivatives. It is also possible that this duplication of pathways indicates differences in intracellular location of the enzymes, and it would be well connected with the control of the synthesis and utilization of sugar nucleotides in the cell.

It should be noted that enzymes are known for the catalysis of the formation of most of these sugar nucleotides, from the nucleotide triphosphate and the appropriate sugar-1-phosphate. In the case of UDPAra two distinct enzymes are known. Since many of these sugar-1-phosphates can, in principle, be derived indirectly from the pentose phosphate pathway, it is possible that this route also contributes towards polysaccharide synthesis.

Figures 3–6 summarize the major known pathways for the interconversion of the sugars, for references to the demonstrations of these pathways, see Hassid (1967).

The alternative type of route from glucose to UDPGalA is that proposed by Loewus (1965). It is suggested that, after phosphorylation of glucose by hexokinase, glucose-1-phosphate is converted to *meso*-inositol-1-phosphate and, thence to *meso*-inositol. This is oxidized, in a reaction which used molecular oxygen, to give glucuronic acid, which is then phosphorylated to glucuronic acid-1-phosphate by glucuronate kinase and ATP. This sugar phosphate can be converted to UDPGlcA, as above, and this can then be 4-epimerized to UDPGalA. UDPXyl and UDPAra would still arise from decarboxylation and epimerization as in "normal" sugar nucleotide metabolism.

There is a considerable amount of evidence from radioisotopic tracing to support this pathway. *Meso*-inositol is a precursor of glucuronosyl and galacturonosyl groups in strawberries and parsley leaves, and the latter are known to be capable of converting glucose to *meso*-inositol (Loewus and Kelly, 1962), without fragmentation of the molecule during cyclization (Loewus, 1965). *Meso*-inositol kinases have been demonstrated in mung bean and sycamore (*Acer pseudoplatanus*) callus tissue, and have been partially purified (Dietz and Albersheim, 1966; English *et al.*, 1966), and such enzymes must be widespread since very many plant tissues can metabolize *meso*-inositol (Albersheim, 1963; Loewus *et al.*, 1962; Roberts and Loewus, 1966; Grubner and Hofman-Ostenhof, 1966; Stoddart and Northcote, 1967a; Harris and Northcote, 1970).

It may be possible that epimerization and substitution of the hydroxyl groups can occur directly from *meso*-inositol, before conversion to sugar nucleotides. Thus galacturonosyl groups might arise directly, without the need for glucuronosyl intermediates, and methylation of the various cyclitols involved could give rise to some of the rare trace sugars found in wall polysaccharides, including pectins. Loewus (1965) has pointed out that any form of blockade at C1 would prevent subsequent decarboxylation and pentose formation, while still allowing uronic acid and uronide ester interconversions; this could form a control mechanism. Harris and Northcote (1970) have shown that, while *meso*-inositol is a good precursor of galacturonic acid, xylose and arabinose, it also gives rise to hexoses in the maize root tip.

It thus appears that the fate of sugar nucleotides in polysaccharide synthesis is to some extent determined by the nucleotide portion of the adduct. ADP sugars are probably destined for the synthesis of storage polysaccharides, and GDP sugars for the synthesis of cellulose and glucomannan. The UDP and dTDP sugars seem to have less specific fates, and can be employed in the synthesis of hemicellulose and pectins, some storage polysaccharides, possibly cellulose and cell-wall glycoprotein, as well as in the glycosylation of steroids, phenols and other classes of aglycone present in plants.

At least two separate routes, and probably more, exist for the biosynthesis of the UDP uronic acids and UDP pentoses, and the dTDP sugar pathways closely resemble those for the metabolism of their UDP sugar analogues. A complex system of control and compartmentation of their metabolism is indicated, since the sugar nucleotides do not seem to accumulate, despite the varying demands made upon them as the cell wall differentiates. Several schemes are possible. For example, when there is a requirement for polyuronide synthesis for type I pectinic acids, during cell plate formation, the cyclitol pathway might be very active in producing galacturonosyl residues. As the requirement shifts towards synthesis of type II pectinic acids, and hemicelluloses, during primary wall deposition, the sugar nucleotide 4-epimerase levels might rise and there could be an increase in pentosyl nucleotide formation. This process could reach its limit with deposition of type II pectinic acids only, for example in fruit. The onset of secondary thickening could lead to a switching-off of the cyclitol pathway and the reduction in epimerase levels, so adjusting to the requirement for a predominance of sugar nucleotides of the glucose series.

B. Cell Walls: Polysaccharide Synthesis

1. SYNTHESIS OF CELLULOSE

The synthesis of cellulose might, at first sight, seem to be a simple process, in which there is a sequential, repetitive addition of glucosyl residues to a pre-existing β-1,4-glucan chain. Two problems have to be faced: the origin of this "primer" must be explained, and the biosynthetic apparatus must produce not just β1,4-chains, but chains in such an array as to account for the observed structural arrangement of native cellulose (see Chapter 4).

Barber and Hassid (1964) showed that a cell-free, particulate preparation from the mung bean would incorporate GDPGlc into a polysaccharide which they considered to be cellulose, using an endogenous acceptor. This appeared to be a general mechanism of cellulose synthesis in mung beans (Elbein et al., 1964; Barber et al., 1964) and applied to the cotton boll also (Barber and Hassid, 1965; Elbein and Hassid, 1966). In the latter case, the synthesis could well have been of a glucomannan, since GDPMan was a strong stimulant of incorporation. It is not clear that these preparations are in any case wholly free of GDPGlc-GDPMan 2-epimerase activity.

Brummond and Gibbons (1964; 1965) have produced evidence that UDPGlc is more effective as a precursor of cellulose than GDPGlc, in *Lupinus albus*, and Ordin and Hall (1967) have reported that the same is true of the oat coleoptile. This does not constitute proof positive that UDPGlc is the functional donor *in vivo*, or indeed, is the more effective donor *in vivo*. The present situation is, thus, rather unclear, but there is evidence in favour of a rôle for GDPGlc and UDPGlc in the synthesis of glucans of the cellulose type.

Colvin (1961) suggested that there is an involvement of an ethanol-soluble intermediate acceptor in cellulose synthesis, which is probably a glucolipid. Glucose would be transferred to this lipid from a sugar nucleotide and then transferred from it to the growing polysaccharide, by a transglycosylation. So far, it has proved impossible to separate this intermediate from the enzyme system which transfers glucose to cellulose, and so direct evidence for it is lacking. However, there is indirect evidence which can be used to support such a proposal. If model acceptors are added to a system in which incorporation of glucose from GDPGlc into polysaccharides is taking place, and there is an excess of GDPGlc present, one might expect a stimulation of incorporation if the transfer is direct. If an intermediate acceptor is involved, and it is already saturated, no promotion of incorporation will occur. The latter was found to be the case, whether cellulose, cellodextrins, cellobiose or glucose were added, and so this evidence is in accord with the idea of some intermediate acceptor. Alternatively, the results can be explained as a saturation of the amount of enzyme (GDPGlc-cellulose glycosyl transferase) available, or as the consequence of a highly ordered site of cellulose synthesis, in which the incomplete chains do not exchange with added acceptors (in any case, the site may not be able to bind them).

There is no evidence that cellulose can be synthesized other than by addition of glucosyl residues to some pre-existing acceptor of fairly high molecular weight (a "primer"). A similar situation obtains with many other polysaccharides, both of cell walls and storage granules, and this poses a problem in postulating a mechanism for the synthesis of a new molecule. A piece of completed molecule must, presumably, be released for use in the priming of the next molecule. It is not known how this takes place in the cellulose molecule, or, for certain, whether it occurs at all.

As was explained in Chapter 4, there are several models proposed for the three-dimensional structure of the cellulose microfibril, all of which require a highly organized system for packing and folding the individual molecules of cellulose. In particular, those models that postulate an anti-parallel arrangement of the chains within the unit cell of the crystallite, pose very severe problems in the devising of a suitable arrangement of the biosynthetic apparatus to achieve such a structure. Only if elaborate folding of the chains is accepted, is it feasible to postulate a system in which the polarity of chain assembly is always the same and which produces several chains at once.

Wooding (1968) supplied sycamore stems with radioactive (^{3}H) glucose for 30 minutes, and then incubated them in non-radioactive glucose for various lengths of time. After a period of 30 minutes of "chasing" it was shown chemically that radioactivity was present in cellulose, but not in pectin or hemicellulose. A parallel study by autoradiographic electron microscopy showed a localization of radioactivity at the periphery of the cell body, and within the inner part of the wall. No radioactivity could be detected within the cytoplasm, nucleus or dictyosomes. This provides strong evidence for the location of the sites of cellulose synthesis at, or very close to, the plasmalemma, and illustrates the centrifugal movement of label within the wall as new polysaccharide is laid down at its inner face.

There is considerable, if less direct, evidence from electron microscopy for the localization of cellulose biosynthesis at the plasmalemma. Northcote (1969c) has shown that in the developing cell plate of sycamore callus cells, there is often an asymmetrical gradation of the state of development of the plate across the cell, so that the sites of vesicular fusion and micro-fibrillar deposition can be separated and studied independently. Observa-tion and comparison of these two sites show that microfibrillar deposition is greatest away from the sites of vesicular fusion, and that the microfibrils in very young primary wall seem to be connected to particles with which they are intermixed. Studies of freeze-etched preparations of cells of higher plants (Northcote, 1968; 1969a and b) have shown the presence of particles on the outer surface of plasmalemmata, which occasionally show a linear pattern of arrangement. Similar particles are present in the microfibrillar parts of the wall, and it appears to be from these that microfibrils radiate (Northcote, 1969; Northcote and Lewis, 1968).

It is, thus, well established that cellulose is synthesized at the plasmalemma and is incorporated into the proximal part of the wall. The structure that produces it lies on the outer face of the plasmalemma and appears as a particle, quite different from the vesicles and tubules of the Golgi apparatus. However, this leaves several questions unanswered. Firstly, it provides no explanation of how GDPGlc (or UDPGlc), which is, presumably, synthesized in the cytoplasm, reaches a synthetic apparatus situated on the exterior of the plasmalemma. Some special mechanism must exist for its transport, either by way of a pump or a facilitated channel, or in the form of some intermediate to which the inner layers of the membrane are freely permeable. Colvin's (1961) glycolipid could be such an intermediate and transfer glucosyl groups from GDPGlc within the cell to cellulose chains at the outer surface of the plasmalemma. A second problem is that some mechanism must exist to terminate the growth of cellulose chains, since they show a limiting size (see Chapter 4) of about $1 \cdot 5 \times 10^3$ residues in primary wall, and $11 \cdot 5 \times 10^3$ residues in secondary wall. These degrees of polymerization are so large that it is difficult to envisage a mechanism which senses the extent of the growth of the chains, unless it can act at a distance from the site of synthesis. In any case it would have to detect the growth of chains deeply embedded in a highly ordered microfibrillar array. It is, perhaps, more likely that timing, rather than spatial sensing, is the limitation upon the size of cellulose molecules. If the units that perform the synthesis of cellulose are grouped together to form the particles that are visible in the electron micrscope, and each unit produces a cellulose chain (so that the whole particle produces one or more microfibrils) it is feasible to introduce or remove these units independently of each other, while perpetuating the whole, particulate array. Each unit might start to produce a cellulose chain immediately it is introduced or activated, and could continue until a specific time interval had elapsed. Then the cellulose chain could be released, perhaps by the removal of some or all of the synthetic unit, or by the action of a specific glycosidase acting at the plasmalemma after the appropriate time interval (and with the removal of the synthetic unit). The appearance of a new unit would enable a new cellulose chain to be started and to interdigitate with its neighbours. Such a model would relate cellulose production to the turnover of a component of the plasmalemma, but not necessarily to gross protein synthesis, or even to total new synthesis of a synthetic unit. The difference in the degree of polymerization observed between celluloses of primary and secondary wall would thus be explicable in terms of a change in the relative rates of chain assembly and turnover of the synthetic units.

The spatial organization of the pattern of microfibrillar deposition takes the form of successive layers of microfibrils lying in planes nearly parallel to the plasmalemma, with the microfibrils of each layer parallel to each other, but not to other layers. This highly ordered arrangement seems to be related to the patterns of microtubules that lie just beneath the plasmalemma at the time of deposition of the cellulose fibrils. Where the deposition of cellulose is in the form of a specific pattern, as in the secondary thickening of xylem vessels, electron microscopic examination reveals patterns of microtubules that lie just over the sites of wall deposition, and then between them (Wooding and Northcote, 1964; Hepler and Newcombe, 1967). No direct contact of microtubules and microfibrils occurs, and it may be that the deposition of cellulose takes place in directions pre-determined by the laying-down of matrix polysaccharides, which originated from the Golgi apparatus in vesicles that passed between, and were aligned by, the microtubules.

2. SYNTHESIS OF PECTINS

The processes by which the molecule of pectin is assembled, are inherently very different from those involved in the synthesis of cellulose. The galacturonan chain is closely analogous to that of cellulose in its stereochemistry, but it is only part of a heteropolysaccharide. Precursors of sugars other than galacturonic acid have to be supplied and their glycosyl residues transferred into the growing pectin molecule at the correct site and in the correct configuration and linkage. Large neutral sugar "blocks" have to be fabricated, in class II pectinic acids, and these themselves are heterogeneous, internally branched and do not have a random distribution of their component sugars. Moreover, the whole structure contains methyl ester, methoxyl and acetyl groups, the insertion of which appears to be somewhat ordered.

On account of the fact that the appreciation of the fundamental structure of pectins is relatively recent, there has been an excessive emphasis upon the importance of methyl ester groups in the structure, physical properties and biological function of these molecules, while components such as the neutral sugars have received very scant attention. Consequently the pectic polysaccharides continue to offer one of the most open fields in the study of plant cell walls, for the correlation of structural, metabolic and functional investigations.

Villemez et al. (1965) obtained a particular preparation from seedlings of the mung bean, which incorporated ^{14}C from UDP-$[^{14}C]$-GalA into a polysaccharide which was identified as a galacturonan, though any content of neutral sugars it might have had was not characterized. Addition of the labelled sugar was presumably onto an endogenous primer. The methyl ester of this sugar nucleotide was not incorporated, and there was, thus, no evidence for a methylation of galacturonic acid prior to its incorporation into a pectin, though UDPGalA is an effective precursor of the macromolecule. The source of methyl ester groups or uronides, in plant polysaccharide biosynthesis generally, is the 1-C pool of the cell. Thus the metabolism of folic acid and its derivatives is intimately associated with pectin biosynthesis. In 1967 Kauss et al. showed that the methyl ester groups of a pectin could arise by a transmethylation from S-adenosyl methionine, and Kauss and Hassid (1967a) reported the partial isolation and characterization of the enzyme activity responsible for this. In the same year, Kauss and Hassid also reported that the methyl ester group of 4-O-methyl-D-glucuronic acid in hemicellulose B can be synthesized by means of a transmethylation from S-adenosyl methionine (Kauss and Hassid, 1967b). Methanol was shown to be a 1-C precursor for pectin biosynthesis by Roberts et al. (1967) and again, either S-adenosyl methionine or 5-N-methyl tetrahydrofolate appeared to be the donor of the methyl group.

In none of these cases is there any evidence for an incorporation of methyl groups before the assembly of the polysaccharide; in every case it occurs afterwards, and no enzyme is yet known to incorporate galac-turonosyl methyl ester groups into a pectin. However, it is likely from the evidence of Stoddart and Northcote (1967, see below) that at least part of this incorporation takes place very soon after, or possibly in synchrony with, polyuronide assembly.

Methyl ester groups can also be removed from pectinic acids after their assembly as as a result of the action of pectin pectyl hydrolases (pectin methyl esterases) which occur in cell walls and elsewhere. Many attempts have been made to correlate change in the activity of these enzymes with alterations in the mechanical properties of cell walls, and these are discussed in detail in Chapter 6, but no clear connection has yet emerged.

Almost no attention has been paid to the question of the incorporation of acetyl groups, but, by analogy with other acetylations in vivo, it is likely that they are derived from 2-C metabolism, and that acetyl-coenzyme A is their immediate precursor. Since acetylated sugar nucleotides have not been detected in plant cells, it is probable that the incorporation of acetyl groups, to form the acetyl esters of sugar hydroxyl groups, takes place after macromolecular synthesis.

The studies of polyuronide biosynthesis described above, while clearly showing the biosynthesis of pectins, did not give definitive evidence of the classes of pectinic acid synthesized and their neutral sugar content. The chemical analysis of the pectins of callus and cambial tissues (Stoddart *et al.*, 1967), discussed in Chapter 4, had shown a change in the type of pectin deposited with the progress of growth and differentiation, and this suggested that the various types of pectic polysaccharide might be metabolically related. Stoddart and Northcote (1967a and b) used [^{14}C]-glucose as a label in "pulse-chase" experiments in which the sugar was supplied to sycamore (*Acer pseudoplatanus*) callus cells in liquid suspension tissue culture. A "pulse" of the radioactive sugar was "chased" through the cells with unlabelled glucose for varying lengths of time, and the pectins that became labelled were extracted with sodium hexametaphosphate, fractionated by electrophoresis upon glass-fibre paper and the distribution of radioactivity analysed and measured. The various components that became labelled were identified by chemical and enzymatic modification. Since the chemistry of the system was already well defined, the chemical nature of each component could be specified with considerable precision (Fig. 7).

The first component that was labelled, after a pulse of one hour and an equal chase, was shown to be a partially esterified galacturonan, on the basis of its becoming more neutral upon treatment with diazomethane (methylation) or ethylene oxide (hydroxylethylation), and more acidic

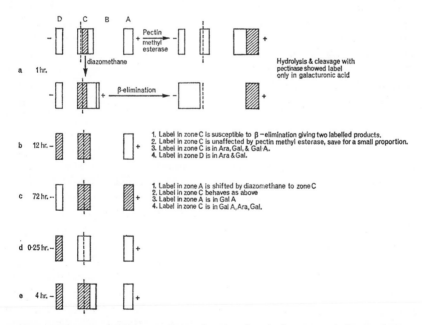

Fig. 7. Summary of the evidence for the chemical nature of the products identified in Stoddart and Northcote's experiments (1967). For details see the text. Glucose is the label in a—c, arabinose in d and e.

upon de-esterification with pectin pectyl hydrolase of orange peel (E.C. 3.1.1.11.). All the radioactivity was liberated by pectinase (E.C. 3.2.1.15.) and was shown to be in galacturonic acid only. Transelimination of the material esterified by diazomethane led to the production of only one acidic labelled component. This proved that the original material could not have contained labelled uronic acid near neutral blocks, since, had it done so, an appreciable amount (20–30%) of the label would have been associated with an almost neutral fragment (see Chapter 4). Thus, either there were no neutral blocks present in the molecule, or the polyuronide chain was assembled in sections, and only those destined to lack neutral blocks were labelled at this stage. There is no evidence to support the latter model, and it is very unlikely to be true on the basis of what is known of the mechanisms of synthesis of other, linear macromolecules.

With longer periods of chasing, radioactivity became associated with the neutral blocks, with the pectic arabinan-galactan and with neutral sugars, especially galactose and arabinose, while the label disappeared from the first labelled component. Quantitative measurements showed that galacturonan was moving from the material first labelled into acidic (type I) pectinic acids by de-esterification, and into the branched (type II) pectinic acids by the addition of blocks of neutral sugar. Similarly, it appeared that radioactivity could be passing from the arabinan-galactan to the neutral blocks of the type II pectinic acids in the form of arabinose and galactose. Pulse-chase experiments with [14C]-L-arabinose as a label confirmed that sugars were passing from the neutral polysaccharide into pectinic acid. The distribution of the label transferred after feeding glucose, and the composition of the neutral blocks of the pectinic acid are different from the composition of the arabinan-galactan in that their arabinose content is higher. This implies that sugars are added to, say, the outside of a branched neutral polymer, in which arabinose is generally more external than galactose, and that parts of this outer label are then transferred. Thus, the interior of the arabinan-galactan labels more slowly than its exterior, and, once labelled, that label turns over more slowly than that on the outside. This is exactly analogous to the entry of [14C]-glucose into the glycogen molecule. It is not known how the neutral sugars are removed from the arabinan-galactan and added to the pectinic acid, but it is tempting to suggest that a transfer of pre-fabricated units is involved.

The particulate preparation of Villemez *et al.* (1965), which incorporates [14C]-GalA from UDPGalA into pectin, is very likely to be derived from the Golgi apparatus, like several other enzymes of sugar nucleotide metabolism, while the subsequent saponification and additions of neutral sugars, described by Stoddart and Northcote (1967a), could be taking place at the plasmalemma or within the wall. Beyond these suggestions, and evidence for the involvement of a lipid-membrane complex in pectin synthesis (Kauss *et al.*, 1969), neither study gives any evidence for the location of the enzymes of pectic biosynthesis within, or upon, the cell. However, direct evidence for the participation of the Golgi apparatus in the synthesis of pectins was obtained in the extremely elegant autoradiographic studies of Northcote and Pickett-Heaps (1966). These authors subjected the young tips of the roots of wheat seedlings to a "pulse" of radioactively labelled (^{3}H) glucose, and then "chased" the activity with

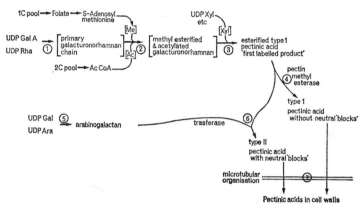

Fig. 8. General scheme of pectin biosynthesis on present evidence (for details see the text).

unlabelled glucose for varying lengths of time. The radioactivity present within the root-cap cells was localized by electron microscopy and autoradiography, and characterized by microchemical means. Within five minutes the cisternae of the dictyosomes showed labelling which could be identified as being a polysaccharide. Upon chasing, this radioactivity showed a centrifugal movement from the dictyosomes into vesicular and tubular structures lying nearer to the plasmalemma. Ultimately, these fused with the plasmalemma, and the radioactivity finally came to be within the thick layer of pectic mucilage which overlies the root-cap cells and acts as a protection and lubricant for the extension of the root through the soil. The isolated polysaccharide had the properties of a pectinic acid (type II) and showed labelling of galacturonosyl, galactosyl and arabinosyl residues. Even though, in this study, the actual assembly of the galacturonan chains is not absolutely proved to take place within the cisternae of the Golgi apparatus, their transfer to the dictyosomes must be extremely rapid and their site of synthesis must lie very close to the dictyosomes, or within them.

The involvement of the Golgi apparatus in the biosynthesis of pectins is wholly in accord with the histochemical and ultrastructural evidence of the cell plate formation (see Chapter 4). There is circumstantial evidence that the cell plate is rich in an incompletely esterified poly-acid which is almost certainly pectic in nature, and the synthesis of the plate is associated with intense activity within the Golgi apparatus and the fusion of vesicles derived from it, which show a characteristic form in tissue culture cells of sycamore, for example (Northcote, 1969c). It should be borne in mind, however, that the endoplasmic reticulum is also very abundant and active at the site of the formation and extension of the cell plate, while Bisalputra (1965) has described the formation of the pectic layer in *Scenedesmus*, as involving vesicles derived from the nuclear membrane.

The details of the regulation of pectin metabolism are not known, but something can be said of its relation to cellular differentiation and growth, and there is a considerable amount of descriptive data about ultrastructural events associated with its deposition. A great deal of effort has been expended in trying to relate changes in the composition of pectic substances to the action of plant growth hormones, and a critique of this will be given in the following chapter.

Pectin synthesis and deposition is restricted to the formation of the cell plate and to primary wall development, which is that phase of growth during which the cell wall shows a degree of plastic extensibility. The evidence for the rôle of the Golgi apparatus in the transport, and, probably the synthesis of pectin, has been considered above, but the pattern of movement and utilization of the vesicles, and tubules that contain pectin is itself subject to control. This control is mediated by, or reflected in, the distribution of microtubules within the cytoplasm of the cell. As the cell plate forms, vesicular and tubular elements derived from the Golgi apparatus are brought up to the site of their fusion, between ranks of aligned microtubules. The plate grows by extension at its edges, brought about by continued fusion, and the direction of growth of these edges again appears to be associated with ordered masses of microtubules. Eventually the cell plate fuses with the old wall of the parent cell, at sites which seem to be associated beforehand with a band of microtubules that lie in a plane parallel to the plasmalemma and a little below it. (Pickett-Heaps and Northcote, 1966a and b; Burgess and Northcote, 1967). Microtubules are associated with wall assembly generally, not just with the deposition of pectin, so their rôle is probably rather remote, in metabolic terms, from the detailed organization of the structure of pectin.

Burgess and Northcote (1967, 1968, 1969) have studied the early phases of microtubular assembly and organization in the meristems of the root tip of wheat, after synchronizing the cells with 5-aminouracil, and have established a close connection between the sites of microtubular formation and activity of the endoplasmic reticulum.

There are still a great many problems associated with the mechanisms by which plant cells assemble the pectic substances, and several of these are now amenable to experimental attack. A particularly important question is that of the rôle of the rhamnosyl residues in the main galacturonan chains. Chemical evidence suggests that they are not arranged in a random fashion, but tend to occur in groups and are often substituted. Transelimination of pectinic acids of type II leads to the isolation of almost all the rhamnose in the fragment which contains the neutral blocks, and this strongly suggests that it could be associated with the affixation of these, either directly as a site of attachment, or indirectly as part of a site of recognition. Certain galacturonosyl residues are also substituted, and it is very possible that the nature of the substituent in this case is different from that with rhamnose. Both these questions could be answered by the application of existing chemical and enzymatic techniques. More difficult are the problems of the arrangement of methyl ester groups in natural pectinic acids, which are certainly somewhat ordered, and the detailed structure of the neutral blocks. It is already clear that both their structure and abundance are variable, and that they can have great effects on the physical properties of pectins (see Chapter 6). It is not yet known how the "primer" for galacturonan synthesis arises, or even certain that such a primer is needed *in vivo*, though, by analogy with other polysaccharides, it is to be expected. A possible approach to this and to other structural and biosynthetic problems that relate to pectin, and to some other wall poly-saccharides, is to study the metabolism of gums. These can be regarded as overproduced or defective components of the wall, which are conse-quently lost from it. Thus, a series of gum-producing varieties within a species are closely analogous to a series of mutant bacterial cells, and, as in the microbiological case, it should be possible to use these mutants for biosynthetic investigations.

3. SYNTHESIS OF OTHER POLYSACCHARIDES OF THE CELL WALL

Feingold *et al.* (1959) showed that a preparation from asparagus would add xylosyl residues to oligosaccharides of D-xylose, using UDPXyl as the xylosyl donor. This observation was of special interest because of the unusually small size of the acceptor molecule, or primer, compared with those normally required for the synthesis of polysaccharide. Bailey and Hassid (1966) used the immature corn-cob to study the synthesis of xylans, and isolated a particulate preparation which would incorporate xylosyl residues into what appeared to be true xylan (with a β-1,4-linked structure) again with UDPXyl as donor. UDPAra was also a glycosyl donor for the same polymer and the L-arabinosyl residues incorporated were in the furanoside configuration, again indicating that this was a true xylan. Kauss and Hassid (1967a and b) were able to show that S-adenosyl methionine acted as the precursor of the methyl group of 4-O-methyl-D-glucuronic acid of hemicellulose B of the grape-vine. Thus this transmethylation is similar to that involved in the formation of pectic esters.

Autoradiography has been applied to the investigation of the site of xylan synthesis in vascular bundles. Pickett-Heaps (1966) studied incorporation of xylans into the wall of the xylem of wheat, and Northcote and Wooding (1966) used sycamore similarly. Northcote and Wooding (1968) describe parallel studies upon phloem. In each case the xylan appears first of all in the dictyosomes and then moves to tubules and vesicles of the Golgi apparatus, before its final deposition in the wall. Thus it appears that, as in the case of the pectic substances, there is an initial assembly of xylan within, or very close to, the dictyosomes. Subsequently it is moved to the wall, where its deposition seems to be under microtubular control, essentially as with pectin. The similarities in the metabolism of xylans and pectins extend also to the class of sugar nucleotides to which their precursors belong.

It is difficult to make general distinctions between xylans of primary and secondary walls, since there is so much species variation in their structures, but it is usually the case that the latter contain rather more glucuronosyl residues. As with pectins, certain classes of gums are very closely similar to the xylans of cell walls and could offer a useful tool for metabolic investigations.

Glucomannans, unlike xylans, seem to belong to the cellulose-like polysaccharides, and are synthesized by the same cell-free, particulate preparations (Barber et al., 1964; Elbein and Hassid, 1966). It is questionable whether the mucilaginous glucomannans, such as that of *Lilium henryi*, belong to the same class as these, and the galactoglucomannans of gymnosperms do not. Passeron et al. (1964) described ADPGal and ADPMan in corn grains, but their function is not known. Mannose is present in traces in pectins, but its precursor and site of addition are unidentified.

Virtually nothing is known of the metabolism of arabino-galactans, save for that of the pectic arabino-galactan described by Stoddart and Northcote (1967a).

Feingold et al. (1958) obtained extracts of mung bean seedlings that would synthesize a β-1,3-glucan, if supplied with UDPGlc, and Flowers et al. (1968) repeated and extended this work, both in the mung bean and in *Lupinus albus*. The polysaccharide produced was characterized as callose (the trace of uronic acid shown to be present in callose by Aspinall and Kessler in 1957 was presumably supplied from an endogenous donor). This enzyme system is remarkable in several respects. Unlike most such particulate preparations, the enzyme can be solubilized by digitonin, with retention of activity, and could be activated by saccharides, such as glucose, cellobiose and laminaribiose, without their incorporation into the product. Goldemberg and Marechal (1963) and Marechal and Goldemberg

(1964) have demonstrated a very similar enzyme in *Euglena gracilis*, which synthesizes the storage polysaccharide paramylon (again a β-1,3-glucan). The enzyme can be solubilized with deoxycholate and shows a similar pattern of activation. It was found that destruction with snail-gut juice of any β-1,3-glucan links present in the preparation left the enzymatic activity quite unaffected. Thus the enzyme can either synthesize a polysaccharide without the need of a primer, which would be the first known case of this in a plant, or it uses something other than a β-1,3-glucan-linked primer, or the primer is protected from the snail-gut enzymes. It would be of great interest if the callose synthetase also showed these strange properties.

Callose is deposited in the sieve-plates of phloem, and on the sieve-tube side of the pores between sieve-tubes and companion cells (Wooding and Northcote, 1964; Northcote and Wooding, 1966; 1968). It is laid down between developing sieve-tubes around a plasmadesma, through which a piece of endoplasmic reticulum passes, in the form of two tapered masses of polysaccharide, the narrow ends of which come to fuse in the region of the middle lamella. The plasmalemmata of the two cells are continuous through the plasmadesma. By elaboration of the endoplasmic reticulum lying within the callose masses, a more complex, folded shape is produced, with simultaneous erosion of the inner part of the fused mass. This eventually breaks through to the two cells to form the pore of a sieve-plate, and the knot of endoplasmic reticulum disappears. During callose deposition there is no cellulose, or hemicellulose, laid down at the same site (Esau *et al.*, 1962), though normal deposition occurs elsewhere in the wall of the young sieve-element. This pattern of deposition of callose is especially interesting, since it shows the action of the endoplasmic reticulum as an agent of erosion of polysaccharide, and shows another rôle of this organelle in the control of wall morphology (see above).

Though the sieve-tube appears to have extensively atrophied cell contents, it retains an ability to synthesize callose as Wooding (1966, 1968) and Northcote and Wooding (1966) have shown in sycamore and *Pinus pinea* by autoradiography. Thus the site of callose synthesis must be located either at the plasmalemma or within the wall.

It is possible that callose synthetase is activated *in vivo* by the release of oligosaccharides from degraded and eroded wall materials. This could explain some of the properties outlined above, and also account for the rapid deposition of callose often observed after trauma.

4. SYNTHESIS OF CELL WALL GLYCOPROTEIN

In the course of a recent series of studies by Chrispeels and his collaborators, a great deal of new information has been gained about the mechanisms of synthesis of the class of glycoprotein peculiar to the cell walls of higher plants.

Following the investigation by Cleland (1968) of the metabolism of proline and hydroxyproline in the oat coleoptile, Chrispeels (1969) made a detailed study of the kinetics of the incorporation and metabolism of ^{14}C-proline in the phloem parenchyma of carrot tubers (*Daucus carota*). Slices of the tissue metabolized some of the proline to hydroxyproline, and label appeared both in the slices and the incubation medium. Chrispeels (1969) distinguished a fraction of hydroxyproline in labelled protein which was soluble in aqueous trichloroacetic acid, from one which was not. This soluble fraction showed characteristics of labelling which led him to regard it as a precursor of cell-wall material. In order to study its metabolism in more detail, Chrispeels (1970a) examined the effects of blocking the hydroxylation of proline with αα-dipyridyl. This blockage was reversible if ferrous iron was applied and the effect persisted after inhibition of protein synthesis with cycloheximide. Measurement of the temperature coefficient of the hydroxylation reaction gave a value of 2·2, which was suggestive of an enzymatic process, and the site of the reaction was cytoplasmic. Clearly the hydroxylation of proline took place after assembly of the polypeptide chain, and the reaction was catalysed by an enzyme located somewhere in the cytoplasm of the cells. The time-course of the reaction led Chrispeels (1970a) to suggest an initial, spatial separation of enzyme and peptidyl substrate.

A later stage in the metabolism of hydroxyproline was studied by Doerschung and Chrispeels (1970) who carried out "pulse-chase" experiments with uniformly labelled ^{14}C-proline and watched the secretion of the labelled hydroxyproline-rich proteins. They found that the synthesis of this type of protein was not simply coupled to general protein synthesis, which again suggests a hydroxylation of proline after assembly of the polypeptide, and the process of secretion of the glycoproteins into the cell wall was rigidly dependent upon the activity of the respiratory chain. This implies a strong energy-dependence of the transport of the glycoproteins to, and through, the plasmalemma.

Chrispeels (1970b) considers that the major stages in the biosynthesis of the glycoproteins of the carrot cell wall are first, the assembly of the polypeptide chains; second, the enzymatic hydroxylation of some of the prolyl residues of these chains; third, the glycosylation of most of the hydroxylprolyl residues with arabinosyl groups; and, fourth, the transport of the glycoproteins to the cell wall.

Autoradiography with the optical microscope has been applied to the system by Sadava and Chrispeels (1971). They localized radioactivity supplied as ^{3}H-proline in both proline and hydroxyproline of the cell wall and cytoplasm. Hydroxyproline accounted only for 15% of the label in the cytoplasm, but for 52% of that in the wall. Autoradiography was performed on plasmolysed cells, to separate the walls and cytoplasms, and a wide distribution of the label was seen within the walls. All the radioactivity in walls and cytoplasm was removed by pronase. Sadava and Chrispeels (1971) point out that they could not exclude the possibility that regions of the plasmalemma remained adherent to the wall. Such adhesions are to be expected, especially at sites of cellulose synthesis.

C. Cell Membranes

On account of the difficulty in isolating and characterizing glycoproteins and glycolipids of the plasma membranes of plant cells, independently of components of the cell wall (see Chapter 4), nothing definite can yet be said of their metabolism. A certain amount is known, however, of the occurrence and metabolism of what are probably the low-molecular weight precursors of such molecules.

Most of the sugars (Salton, 1965) and sugar nucleotides (Ginsburg, 1964) that are likely to be involved in their synthesis have been identified in extracts of plant tissues, and several have been considered above. Among others are UDP-N-acetylglucosamine (Solms and Hassid, 1957) and UDP-N-acetylgalactosamine (Gonsalez and Pontis, 1963), and Roberts (1970) has shown the conversion of D-glucosamine to UDP-N-acetyl-glucosamine in seedlings. One notable absence is that of sialic acid, either as free sugar or sugar nucleotide.

Roberts et al. (1971) have recently studied the incorporation of ^{14}C-D-glucosamine into high molecular weight materials of corn (*Zea mays*), sycamore callus (*Acer pseudoplatanus*), duckweed (*Lemna minor*), broad bean (*Vicia falsa*) and barley (*Hordeum vulgare*). In duckweed some label appeared in glucose, but in the other tissues it appeared largely in glucosamine, released upon acid hydrolysis. The compounds labelled in corn root and sycamore cells were examined in some detail. They were susceptible to degradation by pronase and trypsin (especially after performate oxidation), were hydrolysed in 0·5 M potassium hydroxide, were soluble in water (but not ethanol, 5% trichloroacetic acid or concentrated aqueous ammonium sulphate) and were mostly negatively charged at pH 8·2. Radioactive peptides were isolated from pronase-digests and were proved to contain glucosamine and amino acids. Thus, it was unequivocally demonstrated that a synthesis of glycoprotein was taking place and that D-glucosamine is a good precursor as in animal cells. However, the solubility of the glycoproteins suggests that they might not be of membraneous origin. Many soluble glycoproteins of plants are known and Roberts et al. (1971) cite several of these.

In summary, the synthesis of glycoproteins in plants is likely to be very similar to that of animal cells and probably involves the same synthetic sequence and the same organelles. Likewise, the metabolism of glycolipids is broadly like that of animals. What is lacking, is direct evidence that any specifically membraneous glycoprotein or glycolipid is assembled in the same ways as those of animal plasmalemmata.

References

ALBERSHEIM, P. (1963). *J. biol. Chem.* **238**, 1608–1610.

ASPINALL, G. O. and KESSLER, G. (1957). *Chemy. Ind.* 1296.

BARBER, G. A. and HASSID, W. Z. (1964). *Biochim. biophys. Acta* **86**, 397–399.

BARBER, G. A. and HASSID, W. Z. (1965). *Nature, Lond.* **207**, 295–296.

BARBER, G. A., ELBEIN, A. D. and HASSID, W. Z. (1964). *J. biol. Chem.* **239**, 4056–4061.

BAILEY, R. W. and HASSID, W. Z. (1966). *Proc. natn. Acad. Sci. U.S.A.* **56**, 1586–1593.

BEAMS, H. W. and KESSEL, R. G. (1968). *Int. Rev. Cytol.* **23**, 209–276.

BENEDETTI, E. L. and EMMELOT, P. (1967). *J. Cell Sci.* **2**, 492–512.

BENNETT, G. (1970). *J. Cell Biol.* **45**, 668–673.

BISALPUTRA, T. (1965). *Can. J. Bot.* **43**, 1549–1552.

BOSMANN, H. B. and WINSTON, R. A. (1970). *J. Cell Biol.* **45**, 23–33.

BOSMANN, H. B., HAGOPIAN, A. and EYLAR, E. H. (1968). *Archs. Biochem. Biophys.* **128**, 51–69.

BOSMANN, H. B., HAGOPIAN, A. and EYLAR, E. H. (1969). *Archs. Biochem. Biophys.* **130**, 573–583.

BROWN, J. C. (1972). *J. Supramolecular Structure* **1**, 1–7.

BRUMMOND, D. O. and GIBBONS, A. P. (1964). *Biochem. biophys. Res. Commun.* **17**, 156–159.

BRUMMOND, D. O. and GIBBONS, A. P. (1965). *Biochem. Z.* **342**, 308–318.

BURGESS, J. and NORTHCOTE, D. H. (1967). *Planta* **75**, 319–326.

BURGESS, J. and NORTHCOTE, D. H. (1968). *Planta* **80**, 1–14.

BURGESS, J. and NORTHCOTE, D. H. (1969). *J. Cell Sci.* **5**, 433–451.

CACCAM, J. F. and EYLAR, E. H. (1970). *Archs. Biochem. Biophys.* **137**, 315–324.

CACCAM, J. F., JACKSON, J. J. and EYLAR, E. H. (1969). *Biochem. biophys. Res. Commun.* **35**, 505–511.

CASTANERA, E. G. and HASSID, W. Z. (1965). *Archs. Biochem. Biophys.* **110**, 462–474.

CHEETHAM, R. D., MORRÉ, D. J. and YUNGHANS, W. N. (1970). *J. Cell Biol.* **44**, 492–500.

CHRISPEELS, M. J. (1969). *Pl. Physiol., Lancaster* **44**, 1187–1193.

CHRISPEELS, M. J. (1970a). *Pl. Physiol., Lancaster* **45**, 223–227.

CHRISPEELS, M. J. (1970b). *Biochem. biophys. Res. Commun.* **39**, 732–737.

CLELAND, R. (1968). *Pl. Physiol., Lancaster* **43**, 865–870.

COLVIN, J. R. (1961). *Can. J. Biochem. Physiol.* **39**, 1921–1926.

COOK, G. M. W. (1972). *In* "Lysosomes in Biology and Pathology", (J. T. Dingle, ed.), pp. 237–277, North Holland, Amsterdam.

COOK, G. M. W., LAICO, M. T. and EYLAR, E. H. (1965). *Proc. natn. Acad. Sci. U.S.A.* **54**, 247–252.

DALLNER, G., SIEKEVITZ, P. and PALADE, G. E. (1966a). *J. Cell Biol.* **30**, 73–96.
DALLNER, G., SIEKEVITZ, P. and PALADE, G. E. (1966b). *J. Cell Biol.* **30**, 97–117.
DALTON, A. J. (1961). *In* "The Cell", (J. Brachet and A. E. Mirsky, eds), pp. 603–619, Academic Press, New York.
DIETZ, M. and ALBERSHEIM, P. (1966). *Biochem. biophys. Res. Commun.* **19**, 598–603.
DOERSCHUNG, M. R. and CHRISPEELS, M. J. (1970). *Pl. Physiol., Lancaster* **46**, 363–366.
ELBEIN, A. D. and HASSID, W. Z. (1966). *Biochem. biophys. Res. Commun.* **23**, 311–318.
ELBEIN, A. D., BARBER, G. A. and HASSID, W. Z. (1964). *J. Am. Chem. Soc.* **86**, 309–310.
ENGLISH, P. D., DIETZ, M. and ALBERSHEIM, P. (1966). *Science, N.Y.* **151**, 198–199.
ESAU, K., CHEADLE, V. I. and RISELEY, E. B. (1962). *Bot. Gaz.* **123**, 233–243.
EVANS, W. H. and GURD, J. W. (1971). *Biochem. J.* **125**, 615–624.
EYLAR, E. H. (1965). *J. theor. Biol.* **10**, 89–113.
EYLAR, E. H. and COOK, G. M. W. (1965). *Proc. natn. Acad. Sci. U.S.A.* **54**, 1678–1685.
EYLAR, E. H. and MATIOLI, G. T. (1965). *Science, N.Y.* **147**, 869–870.
FEINGOLD, D. S., NEUFELD, E. F. and HASSID, W. Z. (1958). *J. biol. Chem.* **233**, 783–788.
FEINGOLD, D. S., NEUFELD, E. F. and HASSID, W. Z. (1959). *J. biol. Chem.* **234**, 488–489.
FISHMAN, P. H., McFARLAND, V. W., MORA, P. T. and BRADY, R. O. (1972). *Biochem. biophys. Res. Commun.* **48**, 48–57.
FLEISCHER, B., FLEISCHER, S. and OZAWA, H. (1969). *J. Cell Biol.* **43**, 59–79.
FLOWERS, H. M., BATRA, K. K., KEMP, J. and HASSID, W. Z. (1968). *Pl. Physiol., Lancaster* **43**, 1703–1709.
GINSBURG, V. (1964). *Adv. Enzymol.* **26**, 35–88.
GOLDEMBURG, S. H. and MARECHAL, L. R. (1963). *Biochim. biophys. Acta* **71**, 743–744.
GOLGI, C. (1898a). *Archs. ital. Biol.* **30**, 60–71.
GOLGI, C. (1898b). *Archs. ital. Biol.* **30**, 278–286.
GONSALEZ, N. S. and PONTIS, H. G. (1963). *Biochim. biophys. Acta* **69**, 179–181.
GRIMES, W. J. (1970). *Biochemistry, N.Y.* **9**, 5083–5092.
GRUBNER, K. A. and HOFMANN-OSTENHOF, O. (1966). *Hoppe-Seyl. Z. Physiol. Chem.* **347**, 278–279.
HADDAD, A., SMITH, M. D., HERSCOVICS, A., NADLER, N. J. and LEBLOND, C. P. (1971). *J. Cell Biol.* **49**, 856–882.
HAGOPIAN, A., BOSMANN, H. B. and EYLAR, E. H. (1968). *Archs. Biochem. Biophys.* **128**, 387–396.
HAGOPIAN, A. and EYLAR, E. H. (1968). *Archs. Biochem. Biophys.* **128**, 422–433.
HALLINAN, T., MURTY, C. N. and GRANT, J. H. (1968). *Archs. Biochem. Biophys.* **125**, 715–720.
HARRIS, P. J. and NORTHCOTE, D. H. (1970). *Biochem. J.* **120**, 479–491.
HASSID, W. Z. (1967). *A. Rev. Pl. Physiol.* **18**, 253–280.

HAYDEN, G. A., CROWLEY, G. M. and JAMIESON, G. A. (1970). *J. biol. Chem.* **245**, 5827–5832.

HELGELAND, L. (1965). *Biochim. biophys. Acta* **101**, 106–112.

HEPLER, P. K. and NEWCOMBE, E. H. (1967). *J. Ultrastruct. Res.* **19**, 498–513.

HIRANO, H., PARKHOUSE, B., NICOLSON, G. L., LENNOX, E. S. and SINGER, S. J. (1972). *Proc. natn. Acad. Sci. U.S.A.* **69**, 2945–2949.

HOLTZMAN, J. I., GRAM, T. E. and GILLETTE, J. R. (1970). *Archs. Biochem. Biophys.* **138**, 199–207.

HUGHES, R. C., SANFORD, B. and JEANLOZ, R. W. (1972). *Proc. natn. Acad. Sci. U.S.A.* **69**, 942–945.

HUNT, L. T. and DAYHOFF, M. O. (1970). *Biochem. biophys. Res. Commun.* **39**, 757–765.

IKEHARA, Y., MOLNAR, J. and CHAO, H. (1971). *Biochim. biophys. Acta* **247**, 486–495.

ITO, S. (1969). *Fedn. Proc. Fedn. Am. Socs. exp. Biol.* **28**, 12–25.

KAUSS, H. and HASSID, W. Z. (1967a). *J. biol. Chem.* **242**, 3449–3453.

KAUSS, H. and HASSID, W. Z. (1967b). *J. biol. Chem.* **242**, 1680–1684.

KAUSS, H., SWANSON, A. L. and HASSID W. Z. (1967). *Biochem. biophys. Res. Commun.* **26**, 234–240.

KAUSS, H., SWANSON, A. L., ARNOLD, R. and ODZUCK, W. (1969). *Biochim. biophys. Acta* **192**, 55–61.

KINDIG, D. A. and KIRSTEN, W. H. (1967). *Science, N.Y.* **155**, 1543–1545.

KOBATA, A., GROLLMAN, E. F., TORAIN, B. F. and GINSBURG, V. (1970). *In* "Blood and Tissue Antigens", (D. Aminoff ed.), pp. 497–504, Academic Press, New York.

KORNFELD, S. and GINSBURG, V. (1966). *Expl. Cell Res.* **41**, 592–600.

KORNFELD, S. and GREGORY, W. (1968). *Biochim. biophys. Acta* **158**, 468–470.

KORNFELD, R. and NOLL, C. (1968). *Biochim. biophys. Acta* **170**, 77–87.

KRAEMER, P. M. (1967a). *J. cell. Physiol.* **69**, 199–207.

KRAEMER, P. M. (1967b). *J. Cell Biol.* **33**, 197–200.

LAICO, M. T. and EYLAR, E. H. (1966). *Fedn. Proc. Fedn. Am. Socs. exp. Biol.* **25**, 587.

LANGLEY, O. K. and AMBROSE, E. J. (1967). *Biochem. J.* **102**, 367–372.

LAWFORD, G. R. and SCHACTER, H. (1966). *J. biol. Chem.* **241**, 5408–5418.

LEELAVATHI, D. E., ESTES, L. W., FEINGOLD, D. S. and LOMBARDI, B. (1970). *Biochim. biophys. Acta* **211**, 124–138.

LELOIR, L. F. (1964). *In* "Proceedings of the Plenary Sessions 6th Intern. Congs. Biochem." pp. 15–29, Federation of American Societies for Experimental Biology, New York.

LOEWUS, F. A. (1965). *Fedn Proc. Fedn Am. Socs. exp. Biol.* **24**, 855–862.

LOEWUS, F. A. and KELLY, S. (1962). *Biochem. biophys. Res. Commun.* **7**, 204–208.

LOEWUS, F. A., KELLY, S. and NEUFELD, E. F. (1962). *Proc. natn. Acad. Sci. U.S.A.* **48**, 421–425.

McGUIRE, E. J. and ROSEMAN, S. (1967). *J. biol. Chem.* **242**, 3745–3747.

MARECHAL, L. R. and GOLDEMBURG, S. H. (1964). *J. biol. Chem.* **239**, 3163–3167.

MARSHALL, R. D. (1972). *A. Rev. Biochem.* **41**, 673–702.

MAYHEW, E. (1966). *J. gen. Physiol.* **49**, 717–725.

MILNER, Y. and AVIGAD, G. (1965). *Nature, Lond.* **206**, 825.

MOLLENHAUER, H. H. and MORRÉ, D. J. (1966). *A. Rev. Pl. Physiol.* **17**, 27–46.

MOLNAR, J. (1967). *Biochemistry, N.Y.* **6**, 3064–3076.

MOLNAR, J., ROBINSON, G. B. and WINZLER, R. J. (1964). *J. biol. Chem.* **239**, 3157–3162.

MOLNAR, J., LUTES, R. A. and WINZLER, R. J. (1965a). *Cancer Res.* **25**, 1438–1445.

MOLNAR, J., TEEGARDEN, D. W. and WINZLER, R. J. (1965b). *Cancer Res.* **25**, 1860–1866.

MORRÉ, D. J., HAMILTON, R. L., MOLLENHAUER, H. H., MAHLEY, R. W., CUNNINGHAM, W. P., CHEETHAM, R. D. and LEQUIRE, V. S. (1970). *J. Cell Biol.* **44**, 484–491.

MORRÉ, D. J., MERLIN, L. M. and KEENAN, T. W. (1969). *Biochem. biophys. Res. Commun.* **37**, 813–819.

MORRÉ, D. J. and MOLLENHAUER, H. H. (1964). *J. Cell Biol.* **23**, 295–305.

MORRÉ, D. J., MOLLENHAUER, H. H. and CHAMBERS, J. E. (1965). *Expl. Cell Res.* **38**, 672–675.

NEUFELD, E. F. and HALL, C. W. (1965). *Biochem. biophys. Res. Commun.* **19**, 456–461.

NEUTRA, M. and LEBLOND, C. P. (1966a). *J. Cell Biol.* **30**, 119–136.

NEUTRA, M. and LEBLOND, C. P. (1966b). *J. Cell Biol.* **30**, 137–150.

NICOLSON, G. L. and SINGER, S. J. (1971). *Proc. natn. Acad. Sci. U.S.A.* **68**, 942–945.

NORTHCOTE, D. H. (1963). *Int. Rev. Cytol.* **14**, 223–265.

NORTHCOTE, D. H. (1968). *In* "Plant Cell Organelles", (J. B. Andham ed.), pp. 179–197, Academic Press, London and New York.

NORTHCOTE, D. H. (1969a). *Symp. Soc. gen. Microbiol.* **19**, 333–349.

NORTHCOTE, D. H. (1969b). *Proc. R. Soc.* B. **173**, 21–30.

NORTHCOTE, D. H. (1969c). *In* "Essays in Biochemistry", Vol. 5, (P. N. Campbell and G. D. Greville, eds), pp. 90–137, Academic Press, London.

NORTHCOTE, D. H. (1971). *Endeavour* **30**, 26–33.

NORTHCOTE, D. H. and LEWIS, D. R. (1968). *J. Cell Sci.* **3**, 199–206.

NORTHCOTE, D. H. and PICKETT-HEAPS, J. D. (1966). *Biochem. J.* **98**, 159–167.

NORTHCOTE, D. H. and WOODING, F. B. P. (1966). *Proc. R. Soc.* B. **163**, 524–535.

NORTHCOTE, D. H. and WOODING, F. B. P. (1968). *Sci. Prog. Oxford* **56**, 35–58.

OMURA, T., SIEKEVITZ, P. and PALADE, G. E. (1967). *J. biol. Chem.* **242**, 2389–2396.

ORDIN, L. and HALL, M. A. (1967). *Pl. Physiol., Lancaster* **42**, 205–212.

PASSERON, S., RECONDO, E. and DANKERT, M. (1964). *Biochim. biophys. Acta* **89**, 372–374.

PICKETT-HEAPS, J. D. (1966). *Planta* **71**, 1–14.

PICKETT-HEAPS, J. D. and NORTHCOTE, D. H. (1966a). *J. Cell Sci.* **1**, 109–120.

PICKETT-HEAPS, J. D. and NORTHCOTE, D. H. (1966b). *J. Cell Sci.* **1**, 121–128.

RAMBOURG, A., HERNANDEZ, W. and LEBLOND, C. P. (1969). *J. Cell Biol.* **40**, 395–414.

REITH, A., OFTEBRO, R. and SELJELID, R. (1970). *Expl. Cell Res.* **59**, 167–170.

ROBERTS, R. M. (1970). *Pl. Physiol., Lancaster* **45**, 263–267.

ROBERTS, R. M. and LOEWUS, F. A. (1966). *Pl. Physiol., Lancaster* **41**, 1489–1498.
ROBERTS, R. M., SHAH, R. H., GOLEBIEWSKI, A. and LOEWUS, F. A. (1967). *Pl. Physiol., Lancaster* **42**, 1737–1742.
ROBERTS, R. M., CONNOR, A. B. and CETORELLI, J. J. (1971). *Biochem. J.* **125**, 999–1008.
ROSEMAN, S. (1962). *Fedn Proc. Fedn. Am. Soc. exp Biol.* **21**, 1075–1083.
ROSEMAN, S. (1970). *Chem. Phys. Lipids* **5**, 270–297.
ROSENBERG, S. A. and EINSTEIN, A. B. (1972). *J. Cell Biol.* **53**, 466–473.
SADAVA, D. and CHRISPEELS, M. J. (1971). *Science, N.Y.* **165**, 299–300.
SALTON, M. R. J. (1965). *A. Rev. Biochem.* **34**, 143–174.
SARCIONE, E. J. and CARMODY, P. J. (1966). *Biochem. biophys. Res. Commun.* **22**, 689–694.
SCHACTER, H., JABBAL, I., HUDGIN, R. L., PINTERIC, L., McGUIRE, E. J. and ROSEMAN, S. (1970). *J. biol. Chem.* **245**, 1090–1100.
SINOHARA, H. and SKY-PECK, H. H. (1965). *Biochim. biophys. Acta* **101**, 90–96.
SOLMS, J. and HASSID, W. Z. (1957). *J. biol. Chem.* **228**, 357–364.
STODDART, R. W. and NORTHCOTE, D. H. (1967a). *Biochem. J.* **105**, 45–59.
STODDART, R. W. and NORTHCOTE, D. H. (1967b). *Biochem. J.* **105**, 61–63.
STODDART, R. W., BARRETT, A. J. and NORTHCOTE, D. H. (1967). *Biochem. J.* **102**, 194–204.
STROMINGER, J. L. and MAPSON, L. W. (1957). *Biochem. J.* **66**, 567–572.
VILLEMEZ, C. L., LIN, T. Y. and HASSID, W. Z. (1965). *Proc. natn. Acad. Sci. U.S.A.* **54**, 1626–1632.
WAGNER, R. R. and CYNKIN, M. A. (1969a). *Archs. Biochem. Biophys.* **129**, 242–247.
WAGNER, R. R. and CYNKIN, M. A. (1969b). *Biochem. biophys. Res. Commun.* **35**, 139–143.
WARREN, L. (1966). *In* "Glycoproteins, their Composition, Structure and Function", (A. Gottschalk, ed.), pp. 570–593, B.B.A. Library series vol. 5, Elsevier, Amsterdam.
WARREN, L. (1969). *In* "Current Topics in Developmental Biology", (A. A. Moscona and A. Monroy, eds), pp. 197–222, Academic Press, New York.
WARREN, L. and GLICK, M. C. (1968). *J. Cell Biol.* **37**, 729–746.
WATKINS, W. M. (1970). *In* "Blood and Tissue Antigens", (D. Aminoff, ed.), pp. 441–453, Academic Press, New York.
WHALEY, W. G., DAUWALDER, M. and KEPHART, J. E. (1972). *Science, N.Y.* **175**, 596–599.
WHUR, P., HERSCOVICS, A. and LEBLOND, C. P. (1969). *J. Cell Biol.* **43**, 289–311.
WINZLER, R. J. (1970). *Int. Rev. Cytol.* **29**, 77–125.
WOODING, F. B. P. (1966). *Planta* **69**, 230–243.
WOODING, F. B. P. (1968). *J. Cell Sci.* **3**, 71–80.
WOODING, F. B. P. and NORTHCOTE, D. H. (1964). *J. Cell Biol.* **23**, 327–337.

6

Functional Importance of Surface Heterosaccharides in Cellular Behaviour

I. Functions of Surface Carbohydrates of Animal Cells

One of the most challenging questions in the study of cell surface heterosaccharides must be "what is the biological rôle that these materials play in the cell membrane"? In many respects this is the aspect of the subject on which there is the least amount of definitive knowledge, and consequently much of this chapter must be speculative. As will be evident from the earlier chapters, the evidence for their presence at the periphery of the plasma membrane is overwhelming and they are, therefore, likely to be of importance in a wide range of phenomena specifically associated with the cell surface. The range of cellular interactions involving the cellular periphery is great, comprising for example: the morphogenetic movement of cells during development of the embryo, viral interaction with host cells, the interaction of bacteria and particles with phagocytes; a wide range of immunological reactions as well as the problem of malignant transformation. An examination of such phenomena indicates that a high degree of molecular specificity in the cell surface is necessary to explain many of these processes. The possibility that surface heterosaccharides may be responsible for providing this degree of molecular specificity has been argued elsewhere (Cook, 1968), and it is the rôle of membrane glycosubstances in the processes of cellular interaction, especially cell recognition, which will be further examined here.

A. Cell–Cell Interactions: Some General Considerations

The glycoproteins and glycolipids would appear to be ideally suited for providing the cell with recognition properties, because of the wide range of molecular structure which it is possible to achieve even in small oligosaccharides. In addition to the differences in sequential arrangement of the constituent monosaccharides, additional diversity of structure can be obtained by branching within the oligosaccharide portion of the macromolecule as well as by differences in the anomeric configurations of the constituent glycosyl units. The diversity of structure possible within the oligosaccharide groups of a glycosylated molecule is considerably greater than that which could be obtained with an equivalent number of amino acid residues present in an oligopeptide. This is not to argue that, in a glycoprotein present at the cell surface, the polypeptide portion does not contribute to the particular property bestowed on the cell by such a macromolecule. Indeed, in the case of the M and N blood-group antigens, Lisowska and Morawiecki (1967) showed that the blocking of free amino groups abolished, but guanidation enhanced, M blood-group activity, suggesting that the influence of both polypeptide and carbohydrate structures in the immunodominant groups are recognized by the appropriate antisera. Further, Pardoe et al. (1971) have taken this into account when suggesting a possible biosynthetic pathway for the synthesis of the M and N blood-group antigens.

If glycosubstances provide the cell with a recognition surface, one might expect to be able to demonstrate that a large number of different glycoproteins are associated with various plasma membranes, and Brown (1972) has made the point that no biochemical study has yet established an upper limit to the relative diversity of glycoprotein species present at the surface of a dividing animal cell. Brown (1972) suggests that when one considers the large number of different antigenic species characteristic of cell surfaces, one might suspect that the number of different glycoproteins present may be correspondingly large. The same author goes on to say that the human erythrocyte with "only one major glycoprotein" (but see Chapter 3) indicates that a simpler pattern may be the general rule. Assuming that the red cell is a valid model on which to make such a generalization, a limited number of glycoproteins may not necessarily exclude the diversity of structure necessary to explain the differing biological properties possessed by a cellular surface. Any one glycoprotein may possess various carbohydrate groups with differing specificity. In the red cell glycoprotein, for example, the structures largely responsible for phytohaemagglutinin binding and M and N blood-group activity are associated with different oligosaccharide residues, present in the same macromolecule. Indeed the experiments devised by Brown (1972) for examining membrane glycoproteins, by labelling these macromolecules with radioactive glucosamine and classifying their labelled tryptic glycopeptides into four main classes by chromatography on columns of DEAE cellulose, supports this idea, if it is assumed that labelling is even. He has studied a number of cell types including mouse L-cells, primary chick embryo myoblast cultures, as well as primary cultures of chick spinal cord and sympathetic ganglion, and shown that glycopeptide "spectra" differed among all of the various cell types examined. Each type had quite significantly different relative amounts of the four glycopeptides. Naturally, as Brown (1972) acknowledges, these four classes may conceal other levels of heterogeneity that can only be revealed by detailed chemical studies. These results, which show heterogeneity among cell types, are not incompatible with the involvement of glycoproteins of the cell surface in specific cell-cell recognition processes, and may indicate either that a range of different glycoproteins are present, or that if only a few types are present there is at least considerable variation within their structures from cell to cell. Brown (1971) has applied this technique to a study of neuroblastoma C-1300 cells. The differentiation of these cells is manifested in culture by changes in the cell surface. Undifferentiated cells are roughly spherical in shape and attach to the substrate, flatten and extend long processes which resemble neurons. This process of differentiation is promoted by the addition of 8×10^{-6} M 5-bromodeoxyuridine to the cultures, while an excess of thymidine reverses this effect. Undifferentiated neuroblastoma cells were shown to contain glycopeptides 3 and 4, and to lack classes 1 and 2, while cells differentiated under the effect of 5-bromodeoxyuridine show (in addition to classes 3 and 4) glycopeptide 1, indicating that this glycopeptide, or mixture of glycopeptides, is characteristic of differentiated neuroblastoma cells. The addition of excess thymidine to cultures containing 5-bromodeoxyuridine showed less than 2% of differentiated cells, with a corresponding decrease in the yield of glycopeptide 1. That the results could stem from a unique dependence on the use of 5-bromodeoxyuridine seems unlikely, since cells which differentiate in the absence of this compound, albeit few (approximately 5%), contain small amounts of glycopeptide 1. It is possible that this glycopeptide fragment or mixture may represent a part of the cell surface which is involved in the morphological changes, such as adherence to the substrate, which accompany differentiation. This is a very interesting system and, as Brown (1971) points out, his technique lends itself to others where cellular differentiation can be manipulated in culture.

An alternative approach to the examination of whether cells possess a unique pattern of glycoproteins, without resorting to the use of proteases, has been published by Glossmann and Neville (1971).

A comparative study was made of three different cell surfaces of the rat, namely liver, kidney brush border and erythrocyte, by the use of acrylamide-gel electrophoresis in the presence of sodium dodecyl sulphate and a discontinuous buffer system. Bearing in mind the difficulties associated with the determination of the molecular weight of glycoproteins by the use of gels and sodium dodecyl sulphate (see Chapter 3) Glossmann and Neville (1971) were able to make some very useful observations. They point out that each cell surface contains between six and eleven different glycoprotein subunits, though most of the carbohydrate is present in only one to three of these. Further, they were able to show that each cell surface has a unique glycoprotein subunit composition, though some subunits of identical mobility (i.e. identical apparent size) were found in all three membranes examined. These authors pose the questions of whether different sized glycoprotein subunits have different biological functions, and whether the subunits of identical size have the same function. However, Glossmann and Neville (1971), aware that the functional specificity of glycoproteins probably resides in the carbohydrate sequences, point out that "carbohydrate sequence" could well be substituted for "function" in propounding the relationships between the various glycoprotein subunits. If carbohydrate sequence and subunit size can be correlated, it might reasonably be assumed that those glycoproteins of common size of subunit share a function common to all the membranes. Glossmann and Neville (1971) cite histocompatibility antigens and lectin binding sites, whose exposure at the cell surface is related to mitosis in normal cells, as possible candidates for this. Those molecules which are unique to the membrane type may well be responsible for properties peculiar to the cell that possesses them. With regard to this, the authors indicate that carbohydrates have been implicated in cell adhesion, and of the cell types they investigated only the liver cell membrane normally displays three-dimensional adhesion *in vivo*. Hence there is much presumptive evidence to indicate a diversity of glycoprotein structures within plasma membranes, which would be in accord with the hypothesis that they provide the cell with a recognizable surface. However, to substantiate this suggestion, a demonstration of a direct involvement of these macromolecules in a recognition process is essential.

B. Cell Recognition

Gesner and Ginsburg (1964) presented evidence which would support the idea that the integrity of oligosaccharide structures on the surface of the lymphocyte is necessary for their normal circulation. By treating ^{32}P labelled thoracic duct lymphocytes *in vitro* with a mixture of glycosidases, obtained from *Clostridium perfringens*, and following their fate on reimplantation in recipient animals, they (Gesner and Ginsburg, 1964) were able to demonstrate that such treatment profoundly affects their circulation. With increasing enzyme treatment less radioactivity appeared in the spleen; assay of this organ was used as a convenient method for determining whether the treatment with the glycosidases affected the fate of lymphocytes in the bodies of the recipients. Of course, it can be argued that removal of sugars from the surface of the lymphocyte may not just be a question of altering those molecules which complementarily interact with those on the surface of endothelial cells in the post-capillary venules of lymphoid tissue, which is the event which controls the selective emigration of lymphocytes from the blood into lymphoid tissue. Such modification of surface heterosaccharides on the lymphocyte might make the cells more susceptible to removal by reticuloendothelial cells. Nevertheless, the suggestions of Gesner and Ginsburg (1964) are important, as is their observation that D-glucose is normally absent from the heterosaccharides of mammalian cell surfaces. This is consonant with a recognition-site hypothesis. Citing the way in which haptens interfere with antibody-antigen interactions, Gesner and Ginsburg (1964) argue that a surface composed of D-glucosyl residues is unlikely to provide an efficient surface for recognition, since it would be impaired by free D-glucose in the body fluid. Evolutionary selection would tend to eliminate D-glucose as a component of these surfaces.

Following from this work, Woodruff and Gesner (1968) were able to demonstrate that treatment of lymphocytes from the thoracic duct of the rat with trypsin also prevents their normal circulation, even though they are viable. Cells were labelled on this occasion with ^{51}Cr. At early intervals (four hours) following transfusion of the cells back into the animal, accumulation of radioactivity in the lymph nodes was abolished whilst uptake in the spleen was not decreased, and recovery of radioactivity in the lungs and liver was not greatly altered. Later the trypsin-treated cells appeared to "home" to the lymph nodes. This suggests that the cells require certain membrane constituents sensitive to tryptic cleavage to be present at the cell surface, but that after a time interval the cell surface has (presumably as a result of membrane turnover: Chapter 5) been renewed. The finding that the cells were able to regain their ability to accumulate selectively in the lymph nodes, and could recirculate to the lymph nodes, argues against trypsin's killing the cells. The demonstration that trypsin does remove or modify material necessary for the selective emigration of lymphocytes is very important. In particular, the ability of these cells to emigrate into the white pulp of the spleen, by passing between endothelial cells of the marginal sinus, is not impaired by such treatment. Further to this, Woodruff and Gesner (1968) also studied the action of neuraminidase on the homing properties of rat lymphocytes. Treatment with this enzyme causes the cells to accumulate in the liver with a decrease in uptake in lymph nodes and spleen. Subsequently, many of the enzyme-treated cells leave the liver and concentrate in the lymph nodes. It is suggested that sialic acid residues may play an important rôle in the normal distribution pattern of lymphocytes in the body. Certainly the combined evidence for effects of cleavage of peptide and glycoside links makes it likely that glycoproteins are involved in the homing properties of lymphocytes and, as such, further reinforces the view that these macromolecules have an important rôle to play in the phenomena of cellular interaction.

As well as the studies above, Cox and Gesner (1965) demonstrated that the addition of various sugars to the culture medium may alter the cell surface in a number of ways in several different mammalian cell lines. As in the case of the homing of lymphocytes, a number of different explanations of these results are feasible. The authors suggest that sugars may bind to specific, complementary sites on the cell surface, and that this leads to subsequent morphological and metabolic changes. Further studies by these investigators (Cox and Gesner, 1967; 1968), on the effects of L-fucose on BHK cells transformed by oncogenic virus, adds weight to this view. The effects of this sugar on the cells do not seem to be directly associated with viral infection, as a parallel appears to exist between a cell's susceptibility to alteration by the sugar and its property of being inhibited by contact with a normal cell. Here again it is suggested that the effects of the added sugar are mediated by means of a complementary binding mechanism. For example, fucose might be acting like a constituent of the normal cell surface, by combining with a complementary site on susceptible cells to produce the observed morphological and metabolic changes.

One of the most attractive pieces of evidence in favour of the view that specific cell associations are caused by the interaction of complementary glycosubstances at the cell surfaces comes from the work of Crandall and Brock (1968). They studied the sexual fusion of the yeast, *Hansenula wingei*, two opposite mating types of which (strain 5 and 21) agglutinate upon mixing. Crandall and Brock (1968) were able to isolate from strain 5 cells an agglutination factor derived from the cell surface which they characterized as a glycoprotein, whose only constituent sugar is mannose. Strain 21 cells do not contain an agglutinin. However, they do contain a specific cell-surface component, called the 21-factor, also a mannan-protein, which neutralizes the 5-factor. Apparently these specific cell-surface molecules neutralize each other in a manner analogous to antibodies and antigens. Whether similar mechanisms can be demonstrated in other systems remains to be seen, especially since *Hansenula* is not an animal. Even so, the work of Crandall and Brock (1968) is valuable, because not only does it emphasize the importance of carbohydrate-containing materials in the phenomena of cell recognition, but it serves to underline the formal resemblance of cell to cell and immunological interactions.

C. Intercellular Adhesion

The importance of surface heterosaccharides in the interaction of cell and cell has been discussed above in terms of those macromolecules that provide the cell with considerable variation in surface structures, with great economy of means, commensurate with the large number of specific interactions which take place at the cell periphery. A particularly important surface property of certain tissue cells is that of adhesion. Adhesion, as an example of cell interaction, is involved in several fundamental biological processes and its rôle in morphogenesis has received particular attention. Cellular adhesiveness may change as the cell differentiates, and hence an understanding of the cell surface is directly relevant to embryological studies (Moscona, 1963). Changes in adhesion are also likely to be important in the malignant process, and Coman (1953) has demonstrated, by microdissection techniques, that malignant cells are less adhesive than normal cells.

Evidence that surface heterosaccharides may be involved in the adhesion of at least one type of animal cell has come from the work of Oppenheimer *et al.* (1969). These investigators found that single cells obtained by tryptic treatment of "embryoid bodies", the ascites-grown form (subline 402AX) of a mouse teratoma (strain 129 teratoma), aggregate in a complex tissue culture medium (Medium 199) but not in Hanks' balanced salts solution. The Medium 199 contains, in addition to salts and glucose, 51 different components, and Oppenheimer *et al.* (1969) were able to demonstrate, by adding these components to Hanks' solution, that only one of these— L-glutamine—was the active component in the complex culture medium necessary for aggregation to proceed. Medium 199 in which L-glutamine was omitted prevented aggregation. It was found that of a number of compounds tested, only D-glucosamine or D-mannosamine could effectively replace L-glutamine. Addition of the glutamine antagonists, 6-diazo-5-oxo-norleucine and O-diazo-acetylserine, blocked the action of L-glutamine, but were ineffective against D-glucosamine or D-mannosamine. D-Glucose is the precursor of all amino sugars (see Chapter 5), the essential intermediate being D-glucosamine-6-phosphate. This intermediate is synthesized from D-fructose-6-phosphate and L-glutamine by a transamidase which is inhibited by 6-diazo-5-oxo-norleucine and azaserine. Oppenheimer *et al.* (1969) suggest that the response of the teratoma cells to L-glutamine may be explained by the fact that such cells require this amino acid for the synthesis of amino sugars, which in turn are utilized for the synthesis of those macromolecules of the cell surface which are involved in intercellular adhesion. The ability of D-glucosamine and D-mannosamine to replace L-glutamine strengthens this interpretation, since D-glucosamine and D-mannosamine can be converted to D-glucosamine-6-phosphate and thence to other amino sugars, presumably by phosphorylation with hexokinase. In the absence of N-acetyl hexosamine the cells will be unable to complete the synthesis of the oligosaccharide component of an essential surface glycoprotein or glycolipid. However, it should be remembered that aggregation in buffered saline solutions deficient in L-glutamine has been observed by many workers (see for example Edwards and Campbell, 1971), and as such the requirement for this amino acid in the above system might well represent a special case.

The mechanism by which complex carbohydrates at the surfaces of opposing cells could mediate the formation of intercellular adhesions is of fundamental importance. Such molecules could interact in a complementary manner as described by Crandall and Brock (1968) for mating types of the yeast *Hansenula wingei,* and Moscona (1968) has long held the view that factors promoting aggregation of sponges and vertebrate cells may well be glycoproteins. These derive from the cell surface and cross-link opposing cells in a manner analogous to that by which they complementarily bind cells in the native state. More recently, however, Roseman (1970) has extended this concept by suggesting that the complementary molecules

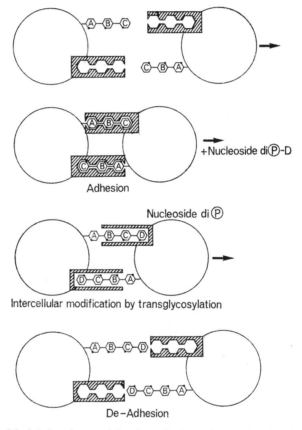

Fig. 1. Model for the participation of glycosyl transferases in cellular adhesion (after Roseman, 1970). Cells bearing both glycosyl transferases and the appropriate heterosaccharide acceptors are suggested to adhere upon formation of enzyme/substrate complexes between opposing surfaces. It is proposed that modification of the carbohydrate acceptor site by transglycosylation would weaken the enzyme substrate to afford de-adhesion (from Kemp *et al.,* 1973).

are enzymes and substrates on the opposing cell surfaces: in particular glycosyl transferases and the complex carbohydrates of the cell surface. In this mechanism, a glycosyl transferase on the surface of one cell may not only be involved in the synthesis of surface heterosaccharides, but may also be able to bind to the appropriate saccharide acceptor on an apposing cell. This process would result in the adherence of the two cells. The specificity of the initial adhesive recognition would depend on the high degree of specificity of the enzyme for the particular acceptor, as well as for the donor molecule. According to the suggested mechanism, on transfer of the appropriate saccharide residue from the sugar nucleotide donor to the acceptor on the apposing cell surface, the enzyme substrate complex dissociates and the cells separate. The transferase acceptor model is particularly attractive because by completing a glycosyl transferase reaction, and thus causing a small change in surface chemistry, the cells become detached. Such a mechanism could explain such surface phenomena as contact inhibition (Abercrombie and Heaysman, 1954).

An essential feature of the Roseman (1970) hypothesis is the suggestion that glycosyl transferases are localized at the cell surface. The work of Roth *et al.* (1971) with chicken-embryo neural retinal cells indicates that glycosyl transferases are located in part on the outer surface of the plasma membrane. They demonstrated that intact embryonic chick neural retinal cells are able to catalyze transfer of radioactively labelled galactose from uridine diphosphate galactose to endogenous acceptors of high molecular weight, as well as to various exogenous acceptors. There was no evidence of appreciable uptake of sugar nucleotide. Only appropriate galactosyl acceptors interfered with the adhesive specificity of the cells studied—an important observation if galactosyl transferase mediates intercellular adhesion. In a subsequent study Roth and White (1972) showed that non-adhesive Balb/c 3T12 (mouse) cells were able to transfer galactose from the donor molecule to acceptors situated on the same cell as the enzyme (*cis*-glycosylation). On the other hand, the adhesive counterparts, Balb/c 3T3 cells, were able to transfer galactose from the donor sugar nucleotide to the appropriate acceptors on adjacent cells (*trans*-glycosylation). However, *trans*-glycosylation does not allow of a detectable distinction between its initiating adhesion, or being a consequence of it.

The phenomenon of platelet: collagen adhesion (Jamieson *et al.*, 1971) may represent a special case of the more general mechanism of Roseman (1970) for intercellular adhesion. Collagen contains carbohydrate as the disaccharide glucosyl-galactose linked via the hydroxyl group of δ-hydroxy-lysine (see Chapter 3). Jamieson *et al.* (1971) point out that several of these saccharides have been shown (Spiro, 1969) to be incomplete, and a wide range (3–60%) of single galactosyl residues are present in mammalian collagens. This group has found that collagen: glucosyl (Barber and Jamieson,

1971a) and galactosyl (Barber and Jamieson, 1971b) transferases reside in the plasma membrane of the platelet and the former enzyme appears to be located on the outer surface of this membrane. The known inhibitors of platelet:collagen adhesion also inhibit glucosyltransferase activity. These results strongly indicate that the formation of an enzyme-acceptor complex is the primary step in haemostasis. This suggestion is strengthened by the fact that the platelets themselves have no glycose acceptor activity for these highly specific enzymes. On present evidence the collagen:galactosyl transferase is probably not implicated in platelet:collagen adhesion, although the collagen:glucosyl transferase is involved (Barber and Jamieson, 1971b).

More recently Bosmann (1972) has suggested that a sialyl transferase located on the surface of the platelet may be important in mediation of cell adhesion. He suggests, in addition, that the low but measurable neuraminidase activity associated with the platelet may be responsible for regenerating substrate for the sialyl transferase. By the combined action of glycosyl transferase and sublethal autolysis by glycosidase, one may arrive at an understanding of how a true intercellular adhesion may be achieved by a mechanism of the type that Roseman proposes. However, secondary mechanisms may have to be worked out to account for long-term stable adhesion.

In addition to the studies above, which implicate the complex carbohydrates of the cell surface in intercellular adhesion, other investigators have been impressed by the likely importance of sialosyl residues in this process. This is to be expected when one considers the contribution that this molecule makes to the surface charge of cells. The removal of sialic acid from the cell periphery might increase cell adhesiveness by lowering the electrostatic repulsion between cells, though the local dielectric constant will also influence this. The contribution of sialic acid to intercellular adhesion may be tested by the enzymatic removal of this sugar with neuraminidase. However, the use of this enzyme in such studies is not entirely without difficulties of interpretation. Berwick and Coman (1962) showed that neuraminidase was without effect on the mutual contact of tumour cells. Kemp (1968, 1970), however, found that trypsin-dissociated chick muscle cells allowed to adhere in the presence of neuraminidase showed a reduction in aggregation. On the other hand, McQuiddy and Lilien (1971) found that the aggregation of embryonic chick retinal cells, as well as muscle cells, was not affected by the presence of neuraminidase, and failed to confirm that sialic acids were important for the aggregation of embryonic muscle cells. It is difficult to account for these various differences especially within the same cell type; Kemp et al. (1973) have, however, pointed out that the discrepancies may lie in the differing experimental techniques used.

Vicker and Edwards (1972) have studied the effect of neuraminidase on the aggregation of BHK21 (Clone 13) cells and the polyoma virus-transformed variant. Bacterial neuraminidase increased the adhesiveness of normal BHK21 cells, while the polyoma transformed cells were almost unaffected. The influence of the enzyme on the normal cells can be most clearly demonstrated when neuraminidase is added to pre-aggregated suspensions of normal cells. Neuraminidase treated monolayer cultures, however, still show a more extensive response than untreated cells on being suspended and tested for aggregation in the absence of the enzyme. The extent of the aggregation of normal cells was found to increase with the growth density of the cultures, neuraminidase increases aggregation at all densities and though the effect is proportionately greater at lower densities, the enzyme does not eliminate the dependence of the extent of aggregation on growth density. The polyoma-transformed cells are less adhesive than the normal, and respond only to neuraminidase treatment to a very small degree. Though these results could be interpreted in terms of electrostatic phenomena, the finding (Forrester et al., 1964) that neuraminidase eliminates the differences in the electrophoretic mobility between normal and polyoma-transformed BHK21 cells argues against such an interpretation and, in any case, strong repulsion between sialosyl groups would not arise in an aqueous environment of high dielectric constant. Vicker and Edwards (1972) prefer to interpret their results in terms of the Roseman (1970) enzyme-substrate theory of intercellular adhesion. The transformed cells, they consider, are less adhesive because of a lack of sialyl transferase. In this respect it is interesting to note that Grimes (1970) has demonstrated that SV40 transformed 3T3 cells have a reduced level of sialyl transferase when compared with normal cells. Spontaneously transformed 3T3 cells have levels of activity intermediate between normal and transformed cells. A similar result was obtained with Balb/c cells and those transformed with SV40 virus. Alternatively, Vicker and Edwards (1972) suggest that a loss of adhesive function of an oligosaccharide component could be as the result of its being present in the polyoma-transformed cells as a shorter chain than is the case with the normal cells. They cite the results of Wu et al. (1969) which indicates that shorter oligosaccharide chains may exist in the membrane glycoproteins of SV40 transformed 3T3 cells. Viral transformation apparently alters the composition of the carbohydrate moieties of glycolipids to shorter chains (Hakomori and Murakami, 1968; Mora et al., 1969). Vicker and Edwards (1972) also point out that chain-length changes could also account for the density-dependence of aggregation of the normal BHK21 cells. It has been demonstrated (Hakomori, 1970) that the level of ceramide trihexoside increases in these cells with increase in culture density, while this same glycolipid is absent in the polyoma transformed cells. Similar changes for

different glycolipids have been demonstrated with NIL 2 hamster fibroblasts by Robbins and Macpherson (1971).

D. Rôle of Glycoproteins in Growth Regulation

The characteristic of the malignant cell is its ability to invade and to
destroy normal, healthy tissue. Malignant cells do not necessarily show as
high a rate of division and growth as their normal counterparts, but they
do show a lack of response to those constraints that limit the growth of
normal cells. In particular they show a lack of contact inhibition of growth,
that is, they continue to divide beyond that cell density at which their
normal counterparts stop proliferation. It is likely that the cellular surface
plays a fundamental rôle, both in invasion and in the pattern of growth of
such cells. These features of the malignant cell surface have been discussed
in the previous section. Earlier (Chapter 2) the ability of normal cells to
bind lectins during mitosis (Fox *et al.*, 1971) or following transformation
has been described. Burger and Noonan (1970) have used a lectin (concanavalin A) to investigate the phenomenon of contact inhibition of growth
(or density-dependent inhibition of growth); which is directly related to the
formation of both benign and malignant tumours. These workers (Burger
and Noonan, 1970) demonstrated that the growth pattern of polyoma virus
transformed 3T3 fibroblasts, which are insensitive to contact inhibition of
growth, could be restored to that of normal cells, which are sensitive to
contact inhibition of growth, by treatment with monovalent concanavalin
A. The monovalent concanavalin A was prepared by treating the lectin
with proteolytic enzymes, either chymotrypsin or more preferably trypsin.
Though Burger and Noonan (1970) showed that trypsinized concanavalin A,
which radically increases the amount of lectin necessary to bring about
agglutination of the virally transformed cells, behaved in the analytical
ultracentrifuge as a slightly slower moving peak than the intact molecule,
they did not give any further details of the characterization of their fragment. In view of the complexity of this molecule (see Chapter 2) it is unfortunate that greater detail was not provided by these authors. The
finding that a fragment of plant protein, when bound to the surface
of transformed cells, restores growth control is a remarkable result. That
a plant protein, presumably quite different in composition from the protein
components of the normal mouse fibroblast surface, should produce this
effect led Burger and Noonan (1970) to place emphasis on the agglutinin
receptor as being important in contact inhibition of growth. Burger and
Noonan (1970) suggest that the presence or absence of a cover layer over
the agglutinin receptor site may possibly be involved in such surface properties as adhesiveness or membrane flexibility, which are important for
cell mobility and division. Surface components may be important with
regard to the uptake of general nutrients or the attachment of specific

growth factors. It has been suggested (Sheppard, 1971) that the binding of the monovalent plant lectin to the cell surface could alter the activity of adenylate cyclase of the cell. Certainly, changes at the cell surface brought about by proteolysis (which stimulate growth as well as increasing the agglutinability by lectins) may be correlated with a decrease in cellular cyclic AMP (see Burger *et al.*, 1972); though how far a drop in the level of cyclic AMP may act as a message between the cell surface and the nucleus is not known. Warren and his colleagues have also considered the chemistry of carbohydrates of the cell surface with respect to changes which are likely to be related to mechanisms of growth control. In comparative studies using gel filtration of glycopeptides derived from the surface of BHK 21/C13 cells, before and after transformation by Rous sarcoma virus, Buck *et al.* (1970, 1971) were able to demonstrate an enrichment in fucose-containing glycopeptides of apparently higher molecular weight in the transformed cells. In addition Warren *et al.* (1972a) were able to show that such early eluting material was also present in relatively large quantities in the surface of T5[1] transformed chick-embryo fibroblasts grown at the permissive temperature (35°), but was virtually absent in cells grown at the non-permissive temperature (41°). Cells transformed by this tempera-ture-sensitive virus, when grown at 35°, manifest malignancy, but when grown at the non-permissive temperature resemble normal cells. Thus, the sensitivity to temperature of the presence of this fucose-containing glycoprotein correlates with the social behaviour of the cells. In this system no consistent differences were found between the glycolipids of normal and transformed cells.

The presence of this fucose-containing glycoprotein, described by Warren *et al.* (1972a), may well be important in arriving at an understand-ing of the malignant process. Subsequently Warren *et al.* (1972b) describe what they believe to be the reasons for the difference observed in their experiments between normal and malignant cells. They (Warren *et al.*, 1972b) find that this early eluting material can be altered to resemble material from untransformed cells by treatment with neuraminidase. Further, they found, using desialylated early eluting material as an acceptor, a specific sialyl transferase which is greatly increased in the transformed cells; this enzyme is present in greater amounts in cells grown at permissive, as opposed to non-permissive, temperatures. This result might at first sight appear to be at variance with those of other workers discussed above who have shown a decreased sialyl transferase activity, however, Warren *et al.* (1972) suggest this may be due to sialyl transferases other than the one studied by themselves. The biological function of the carbohydrate

---

[1] A temperature-sensitive mutant of the Schmidt-Ruppin (sub–group A) strain of Rous sarcoma virus.

moiety studied by Warren and his colleagues is unknown, though they speculate that it is important in malignancy.

E. Conclusions

The carbohydrate-rich macromolecules at the cell surface present the cell biologist with many interesting problems, and the elucidation of their molecular structure is likely to provide a detailed insight into a wide range of biological surface phenomenon. Their function in blood-group antigens has been discussed (see Chapter 3) and the demonstration of the rôle of sialic acids in the myxovirus receptor site of cells (see Chapter 2) did much to draw early attention to heterosaccharides as a feature of biological membranes. In this book an attempt has been made to show how over the last twelve years the importance of carbohydrates in cell membranes has become apparent, and in this final chapter an explanation as to their wider biological function is offered. No doubt much of what has been written here will have to be modified as more information is obtained. Nevertheless the contention made initially in this book that models of cell membranes and cell surfaces which ignore the rôle of carbohydrates in their structure are incomplete will undoubtedly remain valid.

II. Functions of Surface Carbohydrates of Plant Cells

A. General Considerations

So little is yet known of the structure or functions of those carbohydrates that are associated with the cell membrane, rather than the cell wall, that only the latter will be considered in this chapter, though properties of the membrane will be discussed where relevant to the synthesis and behaviour of the cell wall.

Three general sorts of approach have been used in attempts to define the rôles of the various types of carbohydrates present in plant cell walls. The first, and simplest, of these is to compare the composition, and morphology, of the cell wall, at various stages of its growth, and to correlate the changes that are seen in its structure with known changes in the properties of the periphery of the cell. As explained in Chapter 4, considerable modifications of cell-wall structure and composition attend the differentiation of the higher plant cell, and by increasingly refined techniques of chemical analysis, subtle differences in molecular architecture are being revealed as well as more gross changes in relative quantities of polysaccharides. One note of caution must be sounded, there is a danger inherent in the uncritical application of this type of study. While it is possible to define chemical and morphological changes, and many physical properties of the cell wall, in fairly precise terms, it remains unsafe to correlate them too closely upon a purely observational basis, for any seeming parallels may be quite fortuitous. For this reason it is necessary to seek supporting evidence from studies of other kinds.

The second general class of approach is to define the properties of the cell wall in exact physical terms, and then to examine the changes in these properties that follow defined and selective modification of the wall. Such a modification can be brought about by chemical extraction of polysaccharides, their enzymatic degradation or by interference with their synthesis. Experimental systems of this type offer a particularly useful means of studying the relation between the plant growth hormones and cell-wall metabolism.

The study of the physical and physical-chemical properties of isolated and purified polysaccharides constitutes the third class of approach to the problems of the function of cell walls. While it suffers from the severe criticism that isolated polysaccharides are not in the conformations that they assume within the whole wall, and may be degraded, this line of approach does give information which can be used to refine the conclusions of other studies. They can in particular be used to exclude some models of wall structure which are incompatible with physical-chemical data, and so focus attention on those models which are feasible.

A first step in making any study of function in a cell wall (or other structure) is to find a means of defining and measuring the function as a property of the structure. Thus, the function of cell walls in supporting plant stems has to be defined in terms of the mechanical properties, such as elasticity and tensile strength, of the whole stem and of individual cell walls. Only then does the system become amenable to other than the most crude, qualitative study. Likewise, any investigation of an isolated or modified component of a cell wall should provide chemical data, as exact as possible, upon the nature of the substance under study; while mechanical explorations of cells and stems should always be accompanied by a precise definition of the orientation of the tissue or cell, lest its behaviour be anisotropic. All too few studies have fulfilled all, or even any, of these requirements.

Since there is insufficient space here to give a comprehensive account of all functions of cells walls, only their mechanical, ion-exchange and hydration properties will be considered in any detail.

The bulk properties of the plant cell wall are fairly easily described in general terms. The structure confers a degree of rigidity upon the plant cell, can expand if required to accommodate cellular growth, affords a measure of protection against lysis of the plasmalemma in hypotonic solutions, constitutes an ion-exchanger and has a variety of other properties. Most of these can be described in quite precise physical terms, especially the mechanical properties, but their correlation with the properties of polysaccharides in the wall is very difficult. In general, precise properties and functions cannot with certainty be ascribed to particular carbohydrates, though indications of their rôles can be obtained.

Most discussions of the function of the plant cell wall and the carbohydrates within it have laid emphasis on the importance of the mechanical properties of the wall, and have given little or no consideration to any rôle it may have as a permeability barrier and to any enzymatic activities that it may contain. Moreover, there has been a tendency to attribute specific mechanical properties of cell walls to single types of carbohydrates within them, usually upon rather slight evidence, and many accounts of such studies are made difficult to interpret by ambiguity of terminology. This is not to say that the mechanics of cell walls are unimportant, but that they do not constitute a complete description of the functions of such structures, and are not necessarily the consequence of the physical properties of single polysaccharides, but of associations of polysaccharides.

Plant cell walls show both elastic and plastic properties when subjected to externally applied forces, or to internal forces resulting from cellular swelling. In general, they show greater deformations in response to extensive, rather than compressive, forces and these responses may be anisotropic. The same is true of aggregates of cells, such as plant stems, which have often been used in physiological studies of this type, but the interpretation of properties such as tensile strength is complicated in this situation. A multicellular structure has a resistance to permanent extension, when pulled, which is a result both of cohesion within cell walls and of adhesion between them. The cell walls within a plant stem cannot be regarded as a uniform matrix of amorphous and fibrous polysaccharides: as explained in Chapter 4, they are layered structures in which each major layer is structurally distinct in morphology and chemistry, and in any cell wall the layers tend to be concentric about the cell which produced them. Contact between adjacent cell walls is via the middle lamella (which lies in a plane defined by the original cell plate), except where plasmodesmata break through or where callose is deposited. Thus it seems probable that cellular adhesion in a multicellular plant involves processes which may be rather different from those involved in the cohesion of individual walls. This distinction has, unfortunately, often been neglected. Plant stems are, in any case, highly heterogeneous structures in which several discrete cell-types are found at various stages of growth. The attribution of the mechanical properties of stems to general properties of cell walls and their interactions is, therefore, hazardous, since the vascular bundles of the whole stem undoubtedly contribute a large part of the elastic properties and tensile strength of the stem.

As plant cells differentiate, their walls show changes in their mechanical properties, which reflect the specialization of their functions. These changes throw considerable light upon the possible functions of the various polysaccharides within the wall, since the patterns of their deposition reflect the changing morphology and properties of the walls. The formation of the cell plate represents a stage in wall deposition at which the topology and topography of the future wall is largely defined, though this very juvenile cell wall has no clearly definable mechanical properties. During primary wall development the cell grows in size, often in certain preferred directions, and with it, the cell wall grows both in area and thickness. The meshworks of cellulose microfibrils within primary walls often show signs of having been re-aligned by stretching movements, and primary walls can clearly undergo plastic, as well as elastic deformation. Secondary thickening marks the end of the growth of the cell in volume and of any linear extension of it. The area of a plane near the inner face of the wall does not alter, but the wall itself increases in thickness. There is a loss of plasticity, and of much elasticity, and the cell wall becomes rigid. Lignification accompanies these changes, and leads to a much lessened permeability of the cell wall to water and to aqueous solutes, leading ultimately to the death of the cell within (though other factors may also be involved in this). Not only mechanical factors change during differentiation. Primary wall deposition corresponds to a greatly increased content of negatively charged groups in the wall, unlike secondary growth, while the polysaccharides laid down in the secondary wall generally have a lesser water-regain upon hydration than do those of primary wall, and show a different pattern of hysteresis upon reversible sorption and desorption of water (see below). Both of these properties are likely to be of great physiological, and of some mechanical, significance. The majority of plant cell walls, and all of those which are growing, surround the plasmalemma of a living cell, which may be adding material to the wall, or abstracting components from it all the time. Thus, under normal conditions of growth, one is not dealing with a fixed amount of wall, but an increasing amount. This makes it extremely difficult to separate the contributions of, and the relationships between, the stretching of the wall by the cell and new wall synthesis. A variety of approaches have tried to circumvent this problem, but have often left open the possibility of criticism on the grounds of uncertain physiological significance. These studies have particularly sought a connection between the action of auxins in promoting wall growth, and the necessary increase in the plasticity of the cell wall to permit it. So far, all attempts to show a direct action of auxins in promoting loosening of the cell wall have proved inconclusive, though indirect effects are known. Nevertheless, the ideas underlying this approach still have much currency and cannot simply be dismissed, despite the lack of evidence for them.

Selective extraction of plant cell walls, of the type described in Chapter 4, gives good indication of which components are required to maintain the overall shape of the wall. The removal of lipid and phenols by organic solvents, such as ethanol-benzene mixture, does not disrupt the gross form of the wall, but leads to some shrinkage and corrugation. Removals of pectins, by extraction with chelating agents, likewise does not cause a gross loss of shape, and is usually accompanied by reswelling of the wall. However, the extraction of hemicellulose often does cause some fragmentation of the walls, though their fundamental form remains unless cellulose is degraded. Selective enzymatic degradation gives rather similar results, and is discussed later in this chapter, suggesting that the cohesion of the mature wall, and probably its rigidity and elasticity, is associated largely with cellulose.

B. Growth, Plasticity and Carbohydrate Metabolism in Cell Walls

1. Origins of the Classical Model

Heyn (1931, 1934) and Söding (1931, 1932) considered that the initial stages of cellular growth involved an increase in the plastic extensibility of the wall. That is to say, that the wall undergoes a loosening process as the first step in its growth, and this occurs before any deposition of new wall material. Using temperatures low enough to prevent any new wall synthesis, Heyn and van Overbeek (1931) and Bonner (1934) showed a net wall extension without net wall synthesis, when plant cells were allowed to swell under an osmotic gradient. These experiments are not strictly physiological and, thus, their validity is uncertain, since the extension of cell walls *in vivo* is attended by the deposition of new polysaccharide.

Ray and Reusink (1962) have given evidence that the so-called "plasticity" of plant cell walls really involves the making and breaking of chemical bonds and they term it a "chemorheological" process, in which true plasticity may well play no part. However, they do not specify at what sort of bond strength a "plastic" process becomes "chemorheological", and whether there is any clear distinction to be made.

The classical view of the rôle of polysaccharides in these processes is that formulated by van Overbeek (1939, 1952) who suggested that the binding of calcium ions by the carboxyl groups of pectins could form bridges between adjacent polyuronide chains. It was supposed that these bridges would confer a degree of rigidity upon the cell wall, which could be reversed by the methyl esterification of the carboxyl groups and the consequent prevention of the formation of such ionic bridges. The plastic extensibility of the wall would, thus, result from increased methyl esterification of pectins, while the saponification of these esters would make the wall rigid and locked in its particular configuration. This view is still widely held (Bonner, 1961), though there is now a large body of biochemical evidence against such a model.

2. Experimental Analysis of the Model

Two types of study have been made in attempts to investigate both this model and the general problem of the growth and extension of cell walls. One body of workers has studied the mechanical properties of cell walls, the effects of inhibitors and growth hormones on these, and their relation to general metabolism in cell wall synthesis. Others have chosen to concentrate attention upon the metabolism of pectins in extending walls, and the ability of pectins to bind calcium ions in the appropriate way in *in vitro* systems. The former approach will be considered first.

a. *Use of Metabolic Inhibitors and Hormones*

Several studies have been made of the effects of inhibitors of the synthesis of protein and nucleic acid on the mechanical properties of the cell wall, usually in the *Avena* (oat) coleoptile. A general caveat must be made about such studies, in that many workers do not sufficiently distinguish between extensibility and extension. Hence, it is not always clear that a particular inhibitor has made the wall rigid and incapable of extension under an applied force, or whether it has merely blocked some other process involved in extension, so that the plant cell applies no force to the wall which, therefore, does not extend. In the latter case the wall is extensible, but is not extended, while in the former it does not extend because it is inextensible.

(i) *Inhibitors of Protein Metabolism:* A wide range of inhibitors of protein metabolism have been shown to inhibit the extension of the *Avena* coleoptile. Among them are canavanine (Bonner, 1949), actinomycin D,8-azaguanine (Key, 1964), puromycin (Key, 1964; Noodén and Thimann, 1963), chloramphenicol (Noodén and Thimann, 1963, 1965) and parafluorophenylalanine (Noodén and Thimann, 1963). These act in very different ways upon the enzymes of protein and nucleic acid metabolism, and all are effective only at levels where their effects on protein synthesis are large. There is no reason to suppose that they exert any special action upon the wall, independent of their more general effects. Cleland (1965a) claimed that actinomycin inhibited the extension of the *Avena* coleoptile, but not the auxin-induced extensibility of the tissue, as revealed by stress-strain analysis. Coartney *et al.* (1967) have contested this, and report that in their hands both actinomycin D and cycloheximide inhibit the expansion of the wall and its extensibility. These latter authors cite unpublished results of Masuda and Wade in support of their claim, and it should be noted that they used levels of actinomycin D appreciably lower than those used by Cleland. They suggest that enzymes are required both for wall plasticity and for extension, and that actinomycin D inhibits the synthesis of the specific (messenger-like) RNA that Key and Ingle (1964) claimed to detect.

(ii) *Growth Hormones:* Auxins (indolylacetic acid and its derivatives) are plant growth hormones which promote cellular growth and elongation, and appear to be needed for the growth-promoting effects of gibberellins. They increase wall plasticity by a process which appears to depend upon respiration (Olson and Cleland, 1964) and which is inhibited by blockade of the respiratory chain with cyanide (Cleland, 1965a and b), by uncoupling of respiratory phosphorylation with 2,4-dinitrophenol (Bonner, 1949) and by anoxia. Auxins also increase the general synthesis of new cell wall polysaccharides (Baker and Ray, 1965a, b: Ray and Baker, 1965).

The effects of auxin upon the metabolism of methyl groups of pectin has been studied by Cleland (1963a), who has distinguished auxin-sensitive and auxin-insensitive incorporations of methyl groups from methionine into the hot and cold water soluble pectins of *Avena* coleoptiles. These pectins were not well characterized. Both incorporations were abolished at levels of ethionine at which the extension of the coleoptiles was only partially inhibited. Schrank (1956) has shown that at considerably higher levels of ethionine there is a cessation of cellular extension, which apparently results from competitive inhibition of the utilization of methionine. Most probably this is an effect upon protein synthesis. The results of Cleland, however, clearly establish that the metabolism of methyl groups is separable from changes in the plasticity of cell walls. Norris (1964) found that ATP, supplied externally to the *Avena* coleoptile, would counteract the effects of ethionine upon the elongation of the cell walls. If ethionine was absent, ATP promoted extension.

(iii) *Hydroxyproline:* Lamport (1965) has discussed the possible rôle of the hydroxyproline-rich glycoprotein of plant cell walls (see Chapter 4) in cellular growth, and argues an important function for it in the extension of the cell wall. Because of the association of hydroxyproline with this glycoprotein, some attention has been directed towards the possible effects of this imino acid upon plant growth. Steward *et al.* (1958) reported that it inhibited the growth of carrot callus cells in culture, and Cleland (1965b) has described effects of it upon auxin-induced wall loosening and upon extension in *Avena* coleoptiles (Cleland, 1965b, 1967). In the latter paper, Cleland reports an interesting difference in the action of this imino acid, compared with amino acid antagonists, in that some extension is needed for its effects to become evident and these are enormously potentiated by sucrose. Norris (1967) described the same effects and further claimed some reversal of the inhibition by externally applied ATP, as with ethionine (above). He has suggested that these results imply an action of ethionine and of hydroxyproline upon ATP metabolism, and hence, indirectly, upon protein and RNA synthesis (Norris, 1964, 1967), but the evidence is slim.

(iv) *Stress–Strain Analysis:* In his studies of the effects of indolylacetic acid upon the cell walls of *Avena* coleoptiles, Cleland (1967) made use of the stress-strain analyser to separate elastic (reversible) and plastic (irreversible) properties. The plastic extensibility was dependent upon the rate of extension and the applied force. It also required the presence of the protoplast and was influenced by added auxin, but not sucrose. Indolylacetic acid at 5×10^{-5} M promoted growth and increased the plastic extensibility, but at higher concentration it had no further effect upon the plastic properties and inhibited growth. The elastic extensibility of the wall was inversely proportional to the applied force, was independent of the presence of the protoplast and showed hysteresis. Like the plastic extensibility, the elastic extensibility was affected by auxin, but to a lesser extent. This could be a reflection of the generality of auxin action upon wall carbohydrates, or might imply that the molecules which are influenced by auxin contribute to both types of mechanical properties. If, as Ray and Reusink (1962) contend, plastic deformation of cell walls involves making and breaking bonds, the need for the presence of the protoplast for such extensions is to be expected, and the plasmalemma is, presumably, involved. Cleland (1971) has reviewed the field.

(v) *Metal Ions:* Bivalent metal ions might be inhibitors of cellular extension, if they are involved in bridging between carboxyl groups in the wall. Tagawa and Bonner (1957) and Cooil and Bonner (1957) showed that calcium, strontium and magnesium are inhibitors of the growth of *Avena* coleoptiles, though manganese promotes growth in the presence of auxin (Bonner, 1949). The inhibition by calcium salts is slowly reversible by auxin in the presence of potassium ions, and the inhibition affects both the aerobic and anaerobic influences of auxin upon growth. There is no direct evidence for the involvement of the pectin or calcium-bridging in any of this, though Brewbaker and Kwack (1963), in similar studies of the effect of calcium upon the growth of pollen tubes, imply, but do not prove, a rôle for pectin.

From all these mechanical and biochemical studies of cell walls, no firm evidence has been produced to support the views of van Overbeek (1939, 1952) and Bonner (1961). Though elastic and plastic properties are separable physically, their biochemical separation may not be complete. The effects of metal ions need not be directly upon pectins, while auxins probably act primarily at sites which are metabolically remote from the cell wall. The inhibition of protein synthesis may well contain direct effects on the production of the glycoprotein of the cell wall, but will also contain components from the inhibition of the synthesis of the enzymes required for wall metabolism. At the same time, apart from the separation of the metabolism of methyl groups from wall plasticity, there is no direct evidence against the classical model.

b. *Metabolism of Pectin*

Studies of the metabolism of pectin have shown clearer results. Since indolylacetic acid leads to an increased plastic extensibility of the wall, it might be expected that pectin methylesterase activity (pectin pectyl hydrolase E.C.3.1.1.11) would be affected by treatment of cells with the hormone, and several workers claimed such an effect (Glasziou, 1957; Glasziou and Inglis, 1958; Sacher and Glasziou, 1959; Adamson and Adamson, 1958; Yoda, 1958). However, in a critical study by Jansen *et al.* (1960b) it was found that all such claims could be explained as the result of a non-specific adsorption of protein upon the cell wall, and there remains no evidence for any special effect of the hormone upon pectin methyl-esterase. An investigation of the incorporation of radioactively labelled methyl groups into pectin led Jansen *et al.* (1960a) to the conclusion that indolylacetic acid promotes the incorporation of methyl ester groups into pectin, but does not cause any overall change in the total content of such groups. This need not be contrary to the classical model of cell wall plasticity, for that requires only a making and breaking of salt bridges that could be quite local, so that any changes in methyl ester content would possibly be only transient, and also local. Cleland (1960) found a very similar result in maize, where indolylacetic acid promoted an incorporation of methyl groups, but he concluded that in this case the incorporation was unrelated to growth. Thus, the case for an action of auxins directly upon the incorporation of methyl ester groups into pectin is unproven, and the evidence is, perhaps, rather against it. Auxins do have well defined effects upon total pectin synthesis, in that they promote the synthesis and deposition of the whole molecule (Carlier and Buffel, 1955; Wilson and Skoog, 1954; Wilson, 1961; Ordin *et al.*, 1955—and see also Sato *et al.*, 1957). Albersheim and Bonner (1959) studied the effect in some detail in *Avena* coleoptiles and found that the incorporation of glucose into a water soluble pectin was promoted, while that into a "residual pectin" was not. Neither pectin was well characterized and no evidence was obtained as to their possible metabolic relationships. Auxin showed a similar action upon the incorporation of methyl ester groups from methionine into pectin (Sato *et al.*, 1958) and a close study showed that the incorporation was into a cold water soluble pectin, which was possibly a precursor of cell wall material (Jansen *et al.*, 1960a). Myoinositol incorporation is, likewise, promoted (Albersheim, 1963). The important papers of Baker and Ray (1965a and b) and Ray and Baker (1965) throw considerable light upon these results, and strongly suggest that they are simply one aspect of the general, direct promotion of the synthesis of non-cellulosic polysaccharides that results from the administration of auxins. They also distinguish a separate, indirect effect of auxin in stimulating cellulose synthesis. The mechanism of these effects is not known, but gibberellic acid shows a closely similar effect when administered to pea epicotyl tissue in the presence of indolyl-acetic acid (Maciejewska-Potapczykowa et al., 1961), and it particularly stimulates pectin synthesis. Like auxins, the gibberellins greatly promote cellular growth and elongation.

These are all effects upon a net synthesis of polysaccharide; there is little effect of auxin upon the turnover of polysaccharides in the wall (Matchett and Nance, 1962), though these authors do report that it leads to a marked loss of cellular calcium.[2] Auxin promotes cellulose synthesis (Roy, 1973a, b).

c. Pectins in Adhesion

There does appear to be one fairly clearly defined function of pectins, which is that they are involved in the adhesion of plant cells. The treatment of plant tissues with cellulases generally leads to fragmentation of the cell walls and the release of a protoplast, unless the osmotic potential of the environment causes its lysis. Treatment with polygalacturanases or pectin transeliminases, however, does relatively much less obvious harm to individual cells, but makes aggregates of them fall apart (Doesburg, 1965). This so-called "macerating" effect is well known and of considerable importance in food technology and in plant pathology, and is entirely in accord with the supposed concentration of pectins in the middle lamella of the cell wall. Exactly how pectins are involved in cellular adhesion is not known, but it is tempting to suggest that neutral blocks may be involved, since enzymatic transelimination is sufficient to achieve maceration.

One important conclusion from this is that, whereas cellulose is closely involved in maintaining cohesion within the cell wall, it may be less directly involved with adhesion between cells which is more likely to be associated primarily with pectin. If so, the elastic properties of any multicellular plant structure do, indeed, involve two different types of bonding, one within the walls, the other between them.

3. IMPLICATIONS OF STRUCTURAL STUDIES ON PECTINS

The new information gained about the structure of pectins during the last decade (see Chapter 4) makes necessary a total re-appraisal of the questions of plasticity and extension of cell walls. The classical model of van Overbeek treats pectins simply as linear polyuronides of variable esterification. It makes no allowances for the presence of sugars other than galacturonic acid, for branching or for acetylation. All of these could, certainly, be accommodated within an elaborated model, but they could add nothing to it without further postulation of functions for pectin. It would be particularly difficult to accommodate the observation of Stoddart

---

[2] Rubery and Northcote (1970) report that cultures of sycamore callus tissue deprived of auxin incorporate less arabinose into their pectins than normal. The effect of this are difficult to predict in exact molecular terms.

et al. (1967), that there are two distinct types of pectinic acid, the ratio of amounts of which reflect the state of growth of the cell, unless further functions are postulated for the neutral "blocks" of pectins.

The studies of Schweiger (1962, 1963, 1964, 1966) upon the binding of metal ions to polyuronides throw further doubt upon the classical model of cell wall plasticity. He has produced clear evidence that the binding of calcium ions to polygalacturonic acids is intramolecular rather than intermolecular in solution,[3] and involves co-ordination of the metal ions with hydroxyl groups within the polysaccharide. In alginic acid (a non-random, linear copolymer of mannuronic and guluronic acids) interchain bonding is preferred, but there is no indication that these molecules are involved in the plastic properties of those algal walls in which they are found.

An investigation of the physical properties of a polygalacturonic acid (of molecular weight about 31,600) in solution by Stoddart *et al.* (1969) showed that a variety of physical properties of this polymer were very sensitive to the titration of its carboxyl groups. The binding of ruthenium red was examined as a function of pH, and both the position of the absorption maximum and the value of the extinction were shown to be titrated in parallel with the carboxyl groups. Interaction of ruthenium red with the polymer was shown to continue after titration of the carboxyl groups, which suggests that it stains by a mechanism more complex than is usually supposed. The sedimentation coefficient and Drude constant were also pH dependent, and differed slightly from the acid-base titration curves in their point of inflection. Clearly the conformation of the molecule is strongly, but not solely, dependent upon the carboxyl groups it contains, and some of its properties alter after the discharge of more ionised groups than are required for the alteration of others. It is noteworthy that the physical properties of degraded pectins used commercially are well known to be dependent upon the degree and pattern of esterification and upon acetylation. Acetylation tends to reduce gel formation, presumably by preventing hydrogen bonding, and the effects of esterification are to be found on a variety of properties related to gelling. Doesburg (1965) has reviewed the field. If a molecule as small and simple as that studied by Stoddart *et al.* is affected in such a complex way, it is very likely that a native pectin molecule will behave in a way even more intricate and difficult to interpret. However, it is clear that even if calcium ions are bound in the manner Schweiger has shown, they will probably cause large changes in the conformation of pectins. Calcium ions could, thus, influence the mechanical properties of the cell wall, without being involved in direct bridging, so long as the pectins interact with other wall components, or themselves, by different means of bonding.

---

[3] Though Kohn (1971) contests this very cogently.

A major contribution to the cohesion of plant cell walls, and probably to their adhesion also, arises from the hydrogen bonding of the polysaccharides that compose them. In the case of the cellulose microfibril, the importance of hydrogen bonding is very obvious and well known— and widely exploited in, for example, paper and cotton thread. In the other polysaccharides the significance of this type of link has received less attention, especially with regard to its importance in the cell wall. It is tempting to suggest that the neutral blocks of pectins, polyuronide chains themselves and all the hemicelluloses are involved in specific hydrogen bonding, from which much of the detailed architecture of cell walls arises. If this is so, then the importance of those pectins laid down at the very start of wall deposition may lie in their forming hydrogen-bonded templates, upon which the cellulose microfibrils come to lie in specified patterns, and the ability of pectins to undergo large changes of conformation may well be important in this. Also, if much of the cohesion of cell walls lies in their being hydrogen-bonded, and cell wall plasticity involves the rearrangement of such cross-links, lignification will have two major effects. First, it will make the environment of hydrogen-bonded structures much more hydrophobic, and so make the cell wall much more rigid by increasing massively the effective bond-strengths of these hydrogen bonds. Second, it will make these bonds much less easy to cleave and rearrange, by virtue of their increased strength, and will also tend to increase steric hindrance to their approach by enzymes. Thus, wall plasticity will be dramatically reduced. Moreover, such a model makes no postulate as to the size of lignin molecules, it merely requires that they are hydrophobic, while it explains the known changes in the properties of cell walls associated with secondary thickening. It should be noted, that on this basis the seeming lack of pectin metabolism in secondary walls need not be correlated with the loss of the plasticity of these walls, but might have quite a different significance.

C. The Permeability of Cell Walls

Because the plant cell wall is rich both in charged groups (especially carboxyl groups) and in hydroxyl groups, it is capable of interacting with a wide range of aqueous solutes, especially metal ions. Where a true and effectively ideal thermodynamic equilibrium exists between the cell membrane, the cell wall and the external environment (which may be considered to be approximately infinite), the concentration of a given solute at the plasmalemma will not be affected by the presence of the cell wall. If the external environment is finite, or if the system is not at equilibrium, then the presence of the cell wall may have a marked effect—as it will also do if the system departs far from ideality. Thus, in any real situation, the cell wall will influence the concentration of nutrients, hormones, toxins and other substances (to which it is permeable) at the plasmalemma. Substances of high molecular weight cannot, in general, penetrate far into the cell wall, and for these it is the wall, rather than the plasmalemma, which is the major permeability barrier of the cell.

The interaction of metal ions with the carbohydrate of the cell wall is of two major types: ionic interaction with uronide carboxyl groups, and co-ordination with hydroxyl groups in a wide variety of saccharide residues. The former type of interaction leads to a loss of net charge in the wall, and so is regarded as an ion-exchange phenomenon and has to be allowed for in some types of ion-flux study. The latter interaction need not be associated with any alteration of charge and will show a generally lesser relation to pH. Moreover, it will tend to be selective for those metal ions which have the appropriate shape and size of ligand-field to fit into the lattice of hydroxyl groups in the cell wall. Thus, from what is known of the interaction of metal ions with free polysaccharides, one might, for example, expect cupric ions to interact preferentially with certain xylans.

Little study has been made of the relative contributions of the various cell wall polysaccharides to ion-binding. Ion-exchange is largely associated with galacturonosyl groups in pectins and, to some extent, with glucuronosyl groups in xylans. Cellulose can contribute nothing. Co-ordinate binding is likely to be far more general, and cellulose could certainly contribute to that. The protein of the cell wall is likely to be important in both processes, and will contribute most of the positive charges of the wall. Lignification will affect the strengths of any bonds formed which have any polar character, by altering the effective dielectric constant of their environment.

Considerations of this type are of importance in any discussion of the uptake of minerals by roots, their transport and distribution within the plant and their deposition in specialized sites (e.g. the "windows" of calcium oxalate in *Fenestraria* spp.). They are also important in any analysis of the mechanisms of homœostasis in aquatic, and especially marine, plants. It is particularly worth bearing in mind that the type of cell wall predominantly dealt with in this book is universal among higher, terrestrial plants, but not among algae. Though some (such as the *Characeae*) do possess walls like those of terrestrial plants, many algal walls are quite different and can, for example, contain sulphated polysaccharides. Indeed, some marine algae have walls that are osmotically unstable in fresh water, where they swell to excess, become disrupted and fragment.

Plant cell walls will also affect the approach of nutrients, such as amino acids and sugars, and hormones, such as auxins, gibberellins and kinins, to the cell membrane. In particular, they are likely to interact with sugars and cyclitols by way of hydrogen bonding. This may lead to an important regulating rôle of the cell wall in differentiation, for, as Northcote (1963) has pointed out, the cell wall not only reflects the state of differentiation of the cell, but may regulate its further differentiation. If the development of the cell is regulated by its supply of nutrients, growth factors and hormones, then anything that influences the levels of some, or all, of these at the plasmalemma will tend to regulate differentiation, especially if there is a differential influence upon concentrations.

The studies of D. H. Northcote and the late R. A. Jeffs lend support to this view (Jeffs and Northcote, 1966, 1967). They supplied nutrients to cuboidal blocks of callus tissues of sycamore and bean, in the form of a substrate of medium solidified with agar. By placing wedges of agar

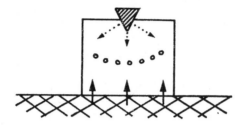

Fig. 2. Schematic diagram of the experiments of Jeffs and Northcote (1966). A block of undifferentiated sycamore or bean callus tissue was maintained upon a solid, nutrient medium with added sucrose, the concentration of which was effectively constant. A wedge of agar containing IAA or 2,4-D (both growth hormones) was applied to the top of the block. Differentiation of vascular nodules was followed chemically and histologically. They formed on a locus which was at a distance from the base and wedge such that the ratio of concentrations of growth hormone to sucrose were constant, as measured radiochemically. Absolute concentrations were of less importance.

containing growth hormones on the top of the blocks (Fig. 2) they were able to induce differentiation at specific sites within the blocks of undifferentiated tissue. It was shown by histological and chemical analysis that this differentiation was into essentially normal vascular tissue. The crucial determinant seemed to be the ratio of the concentrations of nutrient (sucrose, itself also a growth factor) and hormones (indolylacetic acid, or the artificial hormone 2,4-dichlorophenoxyacetic acid), rather than their absolute concentrations. In a heterogeneous system, such as a normal stem, the differential permeability of cell walls to such factors, itself a function of the states of development of the various cells, would play an important part in determining which cells would undergo further specializations. Factors other than the permeability of cell walls are also involved, such as the geometry of the arrangement and the separation of sources of growth factor, but wall permeability could still be a very important influence.

There is a second way in which the cell wall may influence differentiation. The capacity of a cell wall to undergo further modification in its form and properties is dependent upon its existing mechanical properties, and the possibilities that remain for their further specialization. If differentiation is already too far advanced along one route for the properties of the wall to be sufficiently capable of alteration, it may be that the wall itself plays a direct part in limiting further development. This implies that information relating to the state of the wall can reach the cellular interior perhaps by way of its permeability or mechanical properties. Alternatively, it could be that the cell cannot specialize anyway, whether or not the wall is present, but it is very difficult to distinguish between these possibilities, experimentally, at present.

D. The Hydration of Cell Walls

Save where a plant cell is plasmolysed, the cell wall, the water and the solutes associated with it constitute the external, local environment, or microenvironment, of the plasmalemma. For this reason, the rôle of the cell wall in retaining and organizing water is of great importance in considerations of general function of the plasmalemma, as well as specific problems of water balance.

Terrestrial plants have to withstand two distinctive types of constraint. They have to support themselves in a medium of low specific gravity, while still being able to resist sizeable lateral forces from wind, and they have to be able to exist in conditions where the partial pressure of water vapour may be low, or almost zero in some cases. Though species vary greatly in the extent of their adaptation to these constraints, all terrestrial plants have to have some degree of mechanical strength and drought resistance, however slight. For this reason it is particularly striking that all terrestrial vascular plants seem to have a common general type of cell-wall structure, and it is likely that a part of the special functions of this structure is to retain water near the plasmalemma. Some polysaccharides and glycoproteins are highly hydroxylated, and may also bear charged groups, and they are likely to be involved in any specialized retention of water by the cell wall.

1. THEORY OF HYDRATION

The uptake of water by polymers is conventionally described in terms of the relative weight of water sorbed, expressed as a "percentage water regain", and the relative humidity, also expressed as a percentage. at a fixed temperature. The results are usually plotted as sorption isotherms and desorption isotherms (Fig. 3), which generally show a hysteresis. For almost all biological polymers the sorption isotherms are sigmoidal, as they are for many synthetic polymers. A variety of models have been put forward to explain the shape of these isotherms and to enable useful quantities to be calculated from them. Unfortunately some models, such as that of Hailwood and Horrobin (1946), give rise to general equations to sigmoid curves, so that any set of data will fit the curve, given the assumption of suitable constants. Though these constants are related to interesting thermodynamic quantities by the model, they are not capable of independent measurement and so the values assumed for the constants cannot be checked. For this reason no attempt will be made here to extract precise figures from sorption isotherms, but some generally useful conclusions will be drawn.

Most models proposed for the mechanism of sorption treat the first stages of the process as the formation of a monomolecular layer of strongly bound water upon the polymer at low relative humidity, followed by the condensation of further layers of water upon this first layer. This is exactly similar to the model of Brunauer et al. (1938) for the sorption of gases upon metals, and is susceptible to the same, mathematical treatment. At high relative humidities more complex interactions occur, such as swelling of the lattice of the polymer, which expose more binding sites for water, so that the initial smooth curve becomes upturned as the sorption rises steeply. Various terms have been introduced to allow for this effect in calculations, but they generally give rise to uncheckable equations, or equations which do not give good fits to experimental data. When applied to desorption data, most models do not fit very well. A possible explanation is that weakly bound "water of hydration" is lost more rapidly than "structural water" (i.e. heavily hydrogen-bonded water) so that the relative concentration of the latter tends to increase. Hence, as the relative humidity was lowered, the ratio of the fractions of the two types of bound water would alter. The structure would gradually collapse as structural water was lost, so altering the activities, in the thermodynamic sense, of the two types of water as the humidity fell. Thus the causes of the anomalies in sorption would persist in desorption down to very low relative humidities and water contents, so that even at the lowest regains the simple model might not be applicable. It is notable that the sorption properties of polymers are somewhat dependent upon the past history of the samples of polymer chosen, and so care must be taken in drawing any conclusions from small differences.

2. Hydration of Cell Wall Polysaccharides

Cellulose takes up water in a way typical of polysaccharides, and in no way especially remarkable. Across most of the range of relative humidity cellulose has a regain of about 4–8 %, which rises to the order of 25 % at high relative humidity. These figures are quite similar to those of several other polysaccharides, including yeast mannan (Northcote, 1953). Pectins are distinctly different and show some variation between types of pectin. In Fig. 3 sorption and desorption data are shown for polygalacturonic acid, apple fruit pectinic acid and the two types of pectinic acid isolated from sycamore callus tissue (Stoddart, unpublished). The behaviour of polygalacturonic acid is not unlike that of neutral polysaccharides, save that the sorption isotherm is initially nearly linear.

The upswing of the curve at high relative humidity is not pronounced, but is perceptible above 50% relative humidity. There is a fair amount of hysteresis between sorption and desorption. Apple fruit pectinic acid has a higher water regain for sorption at all relative humidities, and especially at higher ones. The sorption isotherm is more clearly sigmoidal than for polygalacturonic acid and is swinging upward below 40% relative humidity. Hysteresis is less marked than for polygalacturonic acid, but the desorption regains are less than the values for this polysaccharide at relative humidities below about 65%. Hence there may be more weakly bound water in the apple fruit pectin than in polygalacturonic acid, but the latter has more "structural water".

The sycamore pectins are especially interesting. Both show exceptionally large regains at higher relative humidities, and a sharp upswing of the sorption curves from about 30% relative humidity. The weakly acidic (type II) pectinic acid has a very low initial water regain until the upswing begins, but its regain at a relative humidity of 90% is 51%, almost double that of polygalacturonic acid, while the strongly acidic (type I) pectinic acid takes up water to a regain of 76% at this relative humidity, but has high values for its regain under all conditions. Such regains are very large indeed, and in the case of the acidic pectinic acid an exceptionally large proportion of the water sorped is resistant to desorption, even at a relative humidity of 6·5%. Thus a great deal of the water taken up has become "structural water". The weakly acidic (type II) pectinic acid resembles the apple fruit pectinic acid, in that its desorption curve is quite close to that for its sorption of water.

Two general features emerge: the presence of neutral oliosaccharide blocks in the pectins leads to a much reduced hysteresis, relative to polygalacturonides that lack these blocks, and the pectins from an actively growing tissue have very large water regains at high relative humidities. The inadequacy of polygalacturonic acid as a model for pectin *in vivo* is well illustrated by these data.

If the sorption and retention of water by these polysaccharides in the intact cell wall is even approximately like their behaviour in isolation, these results have considerable implications. Even though pectins are quantitatively minor components of primary walls, their ability to take up water so greatly exceeds that of cellulose, that at high humidities they will bind at least a third of the water in the wall, since hemicelluloses do not

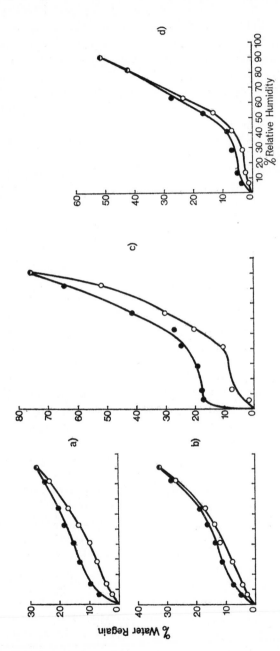

Fig. 3. Hydration of Polyuronides. (a) Polygalacturonic acid (as studied by Stoddart *et al.*, 1969). (b) Apple fruit pectinic acid (as prepared by Barrett and Northcote, 1965). (c) Sycamore pectinic acid of type I. (d) Sycamore pectinic acid of type II (both as prepared by Stoddart *et al.*, (1967).

generally show such huge water-regains. At low humidities (below 50%) they would still account for about one quarter of the water of the wall. Secondary thickening would change the picture entirely, so that in the whole wall, after this process, the pectins could hardly hold more than 5% of the total water content of the wall, at high humidity (above 75% relative humidity). At lower humidities their contribution would be minute. In any case, at secondary thickening there is no further deposition of pectin, and that in the primary wall would be removed from the environment of the plasmalemma by the new secondary wall below it. Hence, secondary thickening does represent a real and extensive change in the microenvironment of the plasmalemma.

The greater part of the body of a terrestrial plant consists of cells, the walls of which must be saturated, or nearly saturated, with water under normal conditions. At surfaces, in roots (especially root hairs) and under conditions of stress the cell walls may not be saturated with water, or in hydration-equilibrium with their environment. We know little of this state in terms of the location of water in the cell wall and its effects on the plasmalemma and wall structure, but data of this type should be borne in mind, for it implies that the partition of water within the cell wall may be a function of the external environment of the wall.

3. STRUCTURAL ASPECTS OF HYDRATION

The effect of neutral sugars upon the hydration properties of pectins is especially important, since it is probably the first physical property of the molecule, shown to be influenced by these residues, which may be of physiological significance. It is known that the methyl esterification of polyuronides does not affect their sorption of water (Palmer et al., 1948), and the most probable explanation of this is that ionization of these groups is irrelevant to any interaction that they may have with water. The binding of water to hydroxyl groups in polysaccharides is by way of hydrogen bonds, and similar bonds are involved in the interaction of sugars with each other. Solms (1960) has postulated hydrogen bonding as the major cause of gel formation in pectin, and it could also affect the precipitation of pectins with long-chain aliphatic ammonium salts, which is not simple micellar (Scott, 1961). Palmer et al. (1947b) have used X-ray crystallography of fibres of a degraded pectin to show the rôle of structural water. In their preparation of "sodium pectate" the spacing of the chains of carbohydrate was constant from 61% (weight for weight) to 20% (weight for weight) of water, but at lower water contents the chains became closer to each other. Crystallinity was lost at 10% (weight for weight) water and regained at 14%. They considered that 24% of the water entered crystallites, and that the remainder was associated with non-crystalline regions. The hysteresis of sorption and desorption was caused by that water bound in the crystallites (Palmer *et al.*, 1948). In a more recent study Rentjes *et al.* (1962) have shown the presence of an infra-red absorption band in pectin which they consider to arise from bound water, and which is resistant to heating at 105° for two hours. It thus seems clear that pectins can bind water very strongly and in at least two ways, and that some of this water is itself intimately associated with the structure and conformation of the polyuronide chains. Neutral sugars affect this, but esterification of carboxyl groups does not have a substantial effect upon hydration, though it does lead to a loss of order in crystallites, presumably by a direct effect. (Palmer *et al.*, 1947a.)

E. Future Approaches

Where plant cell walls show a morphological specialization which reflects some unusual physiological function, it is likely that the carbohydrate composition (and, often, structure) is also modified. For example, the polysaccharide secreted by the Golgi apparatus of the root-cap cells is a type of pectin, which probably has a lubricatory and protective function (though it could also be involved in the sequestration of metal ions). In cases such as this the polysaccharides and glycoproteins of the wall have acquired new functions by a secondary modification of their structure or deposition, and these functions can easily be determined by comparison with other cell types. It is far more difficult to define the older and more fundamental rôles of these molecules, because the more basic the function of a component of a cell wall, the more cells will possess it, and possibly no cell will be found which lacks it. Any interference with the synthesis of such a molecule may be lethal, so rendering useless a major approach to the problem of its function. Only by the most careful study of plant cells cultured in artificial media, where they can be protected against normal physiological limitations, can these difficulties be overcome. Most probably it will be from the study of wall resynthesis by isolated plant protoplasts in culture, that the first definitive description of the rôle of carbohydrates in cell wall, and in plant cell membrane, function will come.

Albersheim's group (Talmadge *et al.*, 1973; Bauer *et al.*, 1973; Keegstra *et al.*, 1973) have advanced a new hypothesis for the structure of the primary walls of sycamore callus cells, on the basis of fragments released from such walls and their polysaccharides by endopolygalacturonase, alkali and pronase. Covalent linkages of wall glycoproteins and pectins are proposed, and of pectins and xyloglucans. Wall plasticity is supposed to arise from the slipping of xyloglucan chains past cellulose. Though the model is by no means proved, and the nature of the linkage of pectin and glycoprotein via neutral sugars is obscure, it is important as a first attempt to incorporate new chemical data into a mechanical model of the wall.

References

ABERCROMBIE, M. and HEAYSMAN, J. E. M. (1954). *Expl. Cell Res.* **6**, 293–306.

ADAMSON, D. and ADAMSON, H. (1958). *Science, N.Y.* **128**, 532–533.

ALBERSHEIM, P. (1963). *J. biol. Chem.* **238**, 1608–1610.

ALBERSHEIM, P. and BONNER, J. (1959). *J. biol. Chem.* **234**, 3105–3108.

BAKER, D. B. and RAY, P. M. (1965a). *Pl. Physiol., Lancaster* **40**, 345–352.

BAKER, D. B. and RAY, P. M. (1965b). *Pl. Physiol., Lancaster* **40**, 360–368.

BARBER, A. J. and JAMIESON, G. A. (1971a). *Biochim. biophys. Acta* **252**, 533–545.

BARBER, A. J. and JAMIESON, G. A. (1971b). *Biochim. biophys. Acta* **252**, 546–552.

BAUER, W. D., TALMADGE, K. W., KEEGSTRA, K. and ALBERSHEIM, P. (1973). *Pl. Physiol., Lancaster* **51**, 174–187.

BERWICK, L. and COMAN, D. R. (1962). *Cancer Res.* **22**, 982–986.

BONNER, J. (1934). *Proc. natn. Acad. Sci. U.S.A.* **20**, 393–397.

BONNER, J. (1949). *Am. J. Bot.* **36**, 323–332.

BONNER, J. (1961). *In* "Proceedings of the 4th International Conference Plant Growth Regulation", p. 307, Ames, Iowa, The Iowa University Press.

BOSMANN, H. B. (1972). *Biochim. biophys. Acta* **279**, 456–474.

BREWBAKER, J. L. and KWACK, B. H. (1963). "Pollen Physiology Fertilization", symposium, Nijmegen, Neth. (pub. 1964), p. 143. (*Chem. Abs.* 1965, **62**, 10836b.)

BROWN, J. C. (1971). *Expl. Cell Res.* **69**, 440–442.

BROWN, J. C. (1972). *J. Supramolecular Structure* **1**, 1–7.

BRUNAUER, S., EMMETT, P. H. and TELLER, E. (1938). *J. Am. chem. Soc.* **60**, 309–319.

BUCK, C. A., GLICK, M. C. and WARREN, L. (1970). *Biochemistry, N.Y.* **9**, 4567–4576.

BUCK, C. A., GLICK, M. C. and WARREN, L. (1971). *Science, N.Y.* **172**, 169–171.

BURGER, M. M., BOMBIK, B. M., BRECKENRIDGE, B. McL. and SHEPPARD, J. R. (1972). *Nature New Biology* **239**, 161–163.

BURGER, M. M. and NOONAN, K. D. (1970). *Nature, Lond.* **228**, 512–515.

CARLIER, A. and BUFFEL, K. (1955). *Acta bot. néerl.* **4**, 551–564.

CLELAND, R. (1960). *Nature, Lond.* **185**, 44.

CLELAND, R. (1963). *Pl. Physiol., Lancaster* **38**, 12–18.

CLELAND, R. (1965a). *Pl. Physiol., Lancaster* **40**, 595–600.

CLELAND, R. (1965b). *Pl. Physiol., Lancaster* **40**, lxi.

CLELAND, R. (1967). *Pl. Physiol., Lancaster* **42**, 271–274.

CLELAND, R. (1971). *A. Rev. Pl. Physiol.* **22**, 197–222.

COARTNEY, J. S., MORRÉ, D. J. and KEY, J. L. (1967). *Pl. Physiol., Lancaster* **42**, 434–439.

COOIL, B. J. and BONNER, J. (1957). *Planta* **48**, 696–723.

COOK, G. M. W. (1968). *Biol. Rev.* **43**, 363–391.

COMAN, D. R. (1953). *Cancer Res.* **13**, 397–404.

COX, R. P. and GESNER, B. M. (1965). *Proc. natn. Acad. Sci. U.S.A.* **54**, 1571–1579.

COX, R. P. and GESNER, B. M. (1967). *Cancer Res.* **27**, 974–979.

COX, R. P. and GESNER, B. M. (1968). *Expl. Cell Res.* **49**, 682–686.

CRANDALL, M. A. and BROCK, T. D. (1968). *Science, N.Y.* **161**, 473–475.

DOESBURG, J. J. (1965). "Pectic Substances in Fresh and Preserved Fruits and Vegetables", Institute for Res. on Storage and Processing of Horticultural Produce, Wageningen, Netherlands.

EDWARDS, J. G. and CAMPBELL, J. A. (1971). *J. Cell Sci.* **8**, 53–72.

FORRESTER, J. A., AMBROSE, E. J. and STOKER, M. G. P. (1964). *Nature, Lond.* **201**, 945–946.

FOX, T. O., SHEPPARD, J. R. and BURGER, M. M. (1971). *Proc. natn. Acad. Sci. U.S.A.* **68**, 244–247.

GESNER, B. M. and GINSBURG, V. (1964). *Proc. natn. Acad. Sci. U.S.A.* **52**, 750–755.

GLASZIOU, K. T. (1957). *Aust. J. biol. Sci.* **10**, 426–434.

GLASZIOU. K. T. and INGLIS, S. D. (1958). *Aust. J. biol. Sci.* **11**, 127–141.

GLOSSMANN, H. and NEVILLE, D. M. JNR. (1971). *J. biol. Chem.* **246**, 6339–6346.

GRIMES, W. J. (1970). *Biochemistry, N.Y.* **9**, 5083–5092.

HAILWOOD, A. J. and HORROBIN, S. (1946). *Trans. Faraday Soc.* **42B**, 84–102.

HAKOMORI, S. (1970). *Proc. natn. Acad. Sci. U.S.A.* **67**, 1741–1747.

HAKOMORI, S. and MURAKAMI, W. T. (1968). *Proc. natn. Acad. Sci. U.S.A.* **59**, 254–262.

HEYN, A. N. J. (1931). *Rec. Trav. bot. néerl.* **28**, 113–241.

HEYN, A. N. J. (1934). *Jb. wiss. Bot.* **79**, 753–789.

HEYN, A. N. J. and VAN OVERBEEK, J. (1931). *Proc. K. Akad. Wet. Amsterdam* **34**, 1190–1195.

JAMIESON, G. A., URBAN, C. L. and BARBER, A. J. (1971). *Nature New Biology* **234**, 5–7.

JANSEN, E. F., JANG, R., ALBERSHEIM, P. and BONNER, J. (1960a). *Pl. Physiol., Lancaster* **35**, 87–97.

JANSEN, E. F., JANG, R. and BONNER, J. (1960b). *Pl. Physiol., Lancaster* **35**, 567–574.

JEFFS, R. A. and NORTHCOTE, D. H. (1966). *Biochem. J.* **101**, 146–152.

JEFFS, R. A. and NORTHCOTE, D. H. (1967). *J. Cell Sci.* **2**, 77–88.

KEEGSTRA, K., TALMADGE, K. W., BAUER, W. D. and ALBERSHEIM, P. (1973). *Pl. Physiol., Lancaster* **51**, 188–196.

KEMP, R. B. (1968). *Nature, Lond.* **218**, 1255–1256.

KEMP, R. B. (1970). *J. Cell Sci.* **6**, 751–766.

KEMP, R. B., LLOYD, C. W. and COOK, G. M. W. (1973). "Progress in Surface and Membrane Science" (J. F. Danielli, M. D. Rosenberg and D. A. Cadenhead, eds), **7**, 271–318, Academic Press, London.

KEY, J. L. (1964). *Pl. Physiol., Lancaster* **39**, 365–370.

KEY, J. L. and INGLE, J. (1964). *Proc. natn. Acad. Sci. U.S.A.* **52**, 1382–1388.

KOHN, R. (1971). *Carbohydrate Res.* **20**, 351–356.

LAMPORT, D. T. A. (1965). *Adv. bot. Res.* **2**, 151–218.

LISOWSKA, E. and MORAWIECKI, A. (1967). *European J. Biochem.* **3**, 237–241.

MATCHETT, W. H. and NANCE, J. F. (1962). *Am. J. Bot.* **49**, 311–319.

MCQUIDDY, P. and LILIEN, J. (1971). *J. Cell Sci.* **9**, 823–833.

MACIEJEWSKA-POTAPCZYKOWA, W., WILUSZ, T. and LUKASIAK, H. (1961). *Acta Soc. Bot. Pol.* **30**, 43–51.

MORA, P. T., BRADY, R. O., BRADLEY, R. M. and MCFARLAND, V. W. (1969). *Proc. natn. Acad. Sci. U.S.A.* **63**, 1290–1296.

MOSCONA, A. A. (1963). *Proc. natn. Acad. Sci. U.S.A.* **49**, 742–747.

MOSCONA, A. A. (1968). *Devl. Biol.* **18**, 250–277.

NOODÉN, L. D. and THIMANN, K. V. (1963). *Proc. natn. Acad. Sci. U.S.A.* **50**, 194–200.

NOODÉN, L. D. and THIMANN, K. V. (1965). *Pl. Physiol., Lancaster* **40**, 193–201.

NORRIS, W. E. (1964). *Archs. Biochem. Biophys.* **108**, 352–355.

NORRIS, W. E. (1967). *Pl. Physiol., Lancaster* **42**, 481–486.

NORTHCOTE, D. H. (1953). *Biochim. biophys. Acta* **11**, 471–479.

NORTHCOTE, D. H. (1963). *Int. Rev. Cytol.* **14**, 223–265.

OLSON, A. C. and CLELAND, R. (1964). *Pl. Physiol., Lancaster* **39**, v.

OPPENHEIMER, S. B., EDIDIN, M., ORR, C. W. and ROSEMAN, S. (1969). *Proc. natn. Acad. Sci. U.S.A.* **63**, 1395–1402.

ORDIN, L., CLELAND, R. and BONNER, J. (1955). *Proc. natn. Acad. Sci. U.S.A.* **41**, 1023–1029.

PALMER, K. J., MERRILL, R. C. and BALLANTYNE, M. (1948). *J. Am. chem. Soc.* **70**, 570–577.

PALMER, K. J., MERRILL, R. C., OWENS, H. S. and BALLANTYNE, M. (1947a). *J. Phys. and Colloid Chem.* **51**, 710–720.

PALMER, K. J., SHAW, T. M. and BALLANTYNE, M. (1947b). *J. Polymer Sci.* **2**, 318–328.

PARDOE, G. I., UHLENBRUCK, G. and REIFENBURG, U. (1971). *Med. Lab. Tech.* **28**, 255–283.

RAY, P. M. (1973a). *Pl. Physiol., Lancaster* **51**, 601–608.

RAY, P. M. (1973b). *Pl. Physiol., Lancaster* **51**, 609–614.

RAY, P. M. and BAKER, D. B. (1965). *Pl. Physiol., Lancaster* **40**, 353–360.

RAY, P. M. and REUSINK, A. W. (1962). *Devl. Biol.* **4**, 377–397.

RENTJES, M., MUSCO, D. D. and JOSEPH, G. H. (1962). *J. Fd. Sci.* **27**, 441–445.

ROBBINS, P. W. and MACPHERSON, I. A. (1971). *Nature, Lond.* **229**, 569–570.

ROSEMAN, S. (1970). *Chem. Phys. Lipids* **5**, 270–297.

ROTH, S., MCGUIRE, E. J. and ROSEMAN, S. (1971). *J. Cell Biol.* **51**, 536–547.

ROTH, S. and WHITE, D. (1972). *Proc. natn. Acad. Sci. U.S.A.* **69**, 485–489.

RUBERY, P. H. and NORTHCOTE, D. H. (1970). *Biochim. biophys. Acta* **222**, 95–108.

SACHER, J. A. and GLASZIOU, K. T. (1959). *Nature, Lond.* **183**, 757–758.

SATO, C. S., BYERRUM, R. U., ALBERSHEIM, P. and BONNER, J. (1958). *J. biol. Chem.* **233**, 128–131.

SATO, C. S., BYERRUM, R. U. and BALL, C. D. (1957). *J. biol. Chem.* **224**, 717–723.

SCHRANK, A. R. (1956). *Archs. Biochem. Biophys.* **61**, 348–355.

SCHWEIGER, R. G. (1962). *J. org. Chem.* **27**, 1789–1791.

SCHWEIGER, R. G. (1963). *Kolloid Z.* **196**, 47–53.

SCHWEIGER, R. G. (1964). *J. org. Chem.* **29**, 2973–2975.

SCHWEIGER, R. G. (1966). *Kolloid Z.* **208**, 28–31.

SCOTT, J. E. (1961). *Biochem. J.* **81**, 418–424.

SHEPPARD, J. R. (1971). *Proc. natn. Acad. Sci. U.S.A.* **68**, 1316–1320.

SÖDING, H. (1931). *Jb. wiss. Bot.* **74**, 127–151.

Söding, H. (1932). *Ber. dt. bot. Ges.* **50**, 117–123.

Solms, J. (1960). *Advances chem. Series* **25**, 37. (*Chem. Abs.* 1960, **54**, 19097c.)

Spiro, R. G. (1969). *J. biol. Chem.* **244**, 602–612.

Steward, F. C., Pollard, J. K., Patchett, A. A. and Witkop, B. (1958). *Biochim. biophys. Acta* **28**, 309–317.

Stoddart, R. W., Barrett, A. J. and Northcote, D. H. (1967). *Biochem. J.* **102**, 194–204.

Stoddart, R. W., Spires, I. P. C. and Tipton, K. F. (1969). *Biochem. J.* **114**, 863–870.

Tagawa, T. and Bonner, J. (1957). *Pl. Physiol., Lanacaster* **32**, 207–212.

Talmadge, K. W., Keegstra, K., Bauer, W. D. and Albersheim, P. (1973). *Pl. Physiol., Lancaster* **51**, 158–173.

van Overbeek, J. (1939). *Bot. Rev.* **5**, 655–681.

van Overbeek, J. (1952). *A. Rev. Pl. Physiol.* **3**, 87–108.

Vicker, M. G. and Edwards, J. G. (1972). *J. Cell Sci.* **10**, 759–768.

Warren, L., Critchley, D. and Macpherson, I. (1972a). *Nature, Lond.* **235**, 275–278.

Warren, L., Fuhrer, J. P. and Buck, C. A. (1972b). *Proc. natn. Acad. Sci. U.S.A.* **69**, 1838–1842.

Wilson, C. M. (1961). *Pl. Physiol., Lancaster* **36**, 336–341.

Wilson, C. M. and Skoog, F. (1954). *Physiologia Pl.* **7**, 204–211

Woodruff, J. and Gesner, B. M. (1968). *Science, N.Y.* **161**, 176.–178.

Wu, H. C., Meezan, E., Black, P. H. and Robbins, P. W. (1969). *Biochemistry, N.Y.* **8**, 2509–2517.

Yoda, S. (1958). *Bot. Mag., Tokyo* **71**, 1–6.

Author Index

Numbers in italics indicate those pages where references are given in full.

Subject Index

A

Acanthamoeba spp., 51
Acer pseudoplatanus, *see* sycamore
Acetic acid
  as solvent of membranes, 134
Acetobacter xylinum, 234
N-Acetylgalactosamine
  in "A" glycoprotein, 148
  in "A" glycolipid, 126, 128
  in alkali-labile saccharide of erythrocyte, 149
  in blood platelets, 154
  in cerebrosides, 125, 126, 128
  in chondroitin, 104
  in erythocyte glycoproteins, 144, 146, 147, 149
  in gangliosides, 119, 125, 130
  interaction with ricin II, 86
  interaction with soybean agglutinin, 86
  in "M" and "N" substances, 146, 147
  periodate-Schiff positive, 67
  structure, 101
  in viral-binding saccharide, 149
N-Acetylgalactosamine-4-sulphate
  in chondroitin sulphate A, 104, 105
  B, 104, 105
N-Acetylgalactosamine-6-sulphate
  in chondroitin sulphate C, 104, 105
N-Acetylglucosamine
  in "A", "B", and "H" glycoproteins, 148
  in "A", "B", "H", "Le^a" and "Le^b" glycolipids, 126, 128
  in *Agaricus bisporus* agglutinin receptor site, 152
  in alkali-stable saccharide of erythrocyte, 150, 151
  in blood platelets, 154

N-Acetylglucosamine (*cont.*)
  in cerebrosides, 126, 127, 128
  in chitan, 102
  in chitin, 99, 101, 102
  in erythrocyte glycoproteins, 144, 146, 147
  in hyaluronic acid, 106, 107
  interaction with *Dolichos biflorus* lectin, 87
  interaction with wheat germ agglutinin, 86
  in "M" and "N" substances, 146, 147
  structure, 101
N-Acetylglucosamine-6-sulphate
  in keratan sulphate, 106
N-Acetylglucosamine: galactosyl transferase, 216, 218
  techniques for estimating activity of, 218
β-D-N-Acetylhexosaminidase
  in Tay-Sachs disease, 129
N-Acetyl-lactosamine synthetase
  see N-Acetylglucosamine: galactosyl transferase
Acid phosphatase
  action on tumour cells, 85
Actinomycin D, 275
Adeno-12 virus
  transformation of embryo hamster cells, 72
Adenosine diphosphate glucose (ADPGlc)
  metabolism, 231, 232, 233
  function, 232
  synthesis, 231, 233
Adenosine triphosphatase, *see* ATPases
S-Adenosyl methionine
  in methylation of hemicellulose B, 242, 247